The People's Peking Man

The People's
Peking Man

*Popular Science and Human Identity
in Twentieth-Century China*

SIGRID SCHMALZER

The University of Chicago Press   Chicago and London

SIGRID SCHMALZER is assistant professor of history at the
University of Massachusetts, Amherst.

The University of Chicago Press, Chicago 60637
The University of Chicago Press, Ltd., London
© 2008 by The University of Chicago
All rights reserved. Published 2008
Printed in the United States of America

17 16 15 14 13 12 11 10 09 08   1 2 3 4 5

ISBN-13: 978-0-226-73859-8 (cloth)
ISBN-13: 978-0-226-73860-4 (paper)
ISBN-10: 0-226-73859-0 (cloth)
ISBN-10: 0-226-73860-4 (paper)

Library of Congress Cataloging-in-Publication Data

Schmalzer, Sigrid.
        The people's Peking man : popular science and human identity
    in twentieth-century China / Sigrid Schmalzer.
            p. cm.
        Includes bibliographical references and index.
        ISBN-13: 978-0-226-73859-8 (cloth : alk. paper)
        ISBN-10: 0-226-73859-0 (cloth : alk. paper)
        ISBN-13: 978-0-226-73860-4 (pbk : alk. paper)
        ISBN-10: 0-226-73860-4 (pbk : alk. paper)
        1. Peking man. 2. Paleoanthropology—China. 3. Communism
    and science—China. I. Title.
    GN284.7.S36 2008
    569.90951—dc22   2007046296

惟人萬物之靈

Contents

Acknowledgments

Over the course of researching and writing this book, I have acquired an enormous number of debts. I am afraid I will never be able to repay them beyond these heartfelt acknowledgments.

At the University of California, San Diego

First I thank my Ph.D. advisers, Joseph Esherick and Paul Pickowicz. Whether in the stacks with Joe or reviewing a paper "line by line" with Paul, I was always certain I was receiving the best education imaginable. They spared no effort in creating the remarkably vibrant and supportive intellectual community that is the graduate program in modern Chinese history at the University of California, San Diego.

My other three committee members each deserve a special thanks. Marta Hanson was a vital link between Chinese studies and science studies; her energy and breadth of knowledge made her a constant source of new ideas and inspiration. Martha Lampland provided much important feedback on critical aspects of my work, including anthropology, science in socialist contexts, and fieldwork practices. Naomi Oreskes always put her finger on the crucial problems, but never made me feel foolish for not having seen them myself. For their contibutions I also thank Suzanne Cahill, Tak Fujitani, Adrian Johns, Susan Leigh Star, Stefan Tanaka, Robert Westman, and Bin Wong (University of California, Irvine).

Among my many helpful student colleagues, I would

especially like to acknowledge Jeremy Brown, Susan Fernsebner, Christian Hess, Matthew Johnson, Liu Lu, Sarah Malena, Cecily McCaffrey, Elena Songster, Donald Wallace, and Adam Warren. They read my papers, offered much-needed advice, and were wonderful friends and colleagues. For research assistance throughout graduate school, I am indebted to Ye Wa. Her helpfulness with Chinese-language documents and her knowledge of everything from agriculture to archaeology provided advantages I wish I could have taken with me to Massachusetts.

I am also grateful for the help of UCSD librarians Jim Cheng and Richard Wang, and the interlibrary loan department; they put a world of resources at my fingertips. Mary Allen, Ivonne Avila, Betty Gunderson, Carol Larkin, and Julie Scales helped me out of administrative scrapes.

In China

I am tremendously fortunate to have spent a year as the guest of that truly cosmopolitan place, the Institute of Vertebrate Paleontology and Paleoanthropology. At IVPP, Liú Wǔ patiently wrote three letters of invitation in order to satisfy visa requirements before he had even met me. Once I arrived in Běijīng, he made me feel completely welcome at the institute and facilitated my research. Also at IVPP, Gāo Xīng, Wú Xīnzhì, and Zhāng Mímàn took time out of their busy schedules for several interviews in addition to helping with logistics. The following scientists, technicians, artists, and museum workers at IVPP offered interviews and sometimes materials as well: Cài Bíngxī, Chén Zǔyín, Duàn Shūqín, Guó Jiànwēi, Hóu Yāméi, Fù Huálíng, Huáng Wànbō, Huáng Wèiwén, Lǐ Chāoróng, Lǐ Róngshān, Lín Shènglóng, Liú Shífān, Lù Qìngwǔ, Qí Guóqín, Shěn Wénlóng, Tóng Hàowén, Wáng Shùqín, Wèi Qí, Wú Màolín, Xú Qīnqí, Xǔ Xiǎopíng, Yàn Défā, Yuán Zhènxīn, Yóu Yùzhù, Zhāng Lìfēn, Zhāng Sēnshuǐ, and Zhào Zhōngyì. Jiǎ Yùzhāng, the son of Jiǎ Lánpō, was very generous in sharing his father's files. Péi Shēn, son of Péi Wénzhōng, offered helpful reminiscences. Fèng Xiǎobō, now of Húběi Provincial Institute of Archaeology, dropped everything to travel to the heart of Shēnnóngjià with me. I am also grateful to the library and archive staff at IVPP, including Cáo Yíng and Wǔ Lǐliū, and finally to the graduate students Shàng Hóng and Wú Xiùjiě for their friendship and helpfulness.

Zhōu Guóxīng of the Natural History Museum in Běijīng deserves special thanks for his consistent willingness to help. I am also grateful to the following people in Běijīng for granting interviews and providing materials: Zhāng Fēng at the Agricultural Science and Technology Press,

Jiǎ Zǐwén and Wáng Huìméi at the Chinese Association for Science and Technology, Yán Shí at the Popular Science Press, Léi Qǐhóng and Shí Shùnkē at the Chinese Research Institute for Science Popularization, and Wáng Fāngchén of the Strange and Rare Animals Exploration and Investigation Committee. Zhāng Jiǔchén and Zhāng Lí of the Institute for the History of Natural Science took time to introduce me to the institute and share insights and resources.

Also in Běijīng, Zhào Chāo and Wáng Yànxiá's family helped with logistics. Yun-chiahn Chen, Kate Lingley, Elena Songster, Suzanne Thomas, Lisa Tran, and Yáng Yáng and her family provided much friendship and assistance of all kinds. Zhāng Xiàoyán transcribed many interviews beautifully at a special rate.

Many others outside Běijīng also provided interviews and guided tours of important sites. I thank Jīn Lì, Chén Chún, and Lǐ Huī of Fùdàn University; Gāo Qiáng of the Bànpō Neolithic Museum in Xī'ān; Lán Jiàn and Liú Ruìqiáng in Lántián; Jiāng Yǒng, Lí Guóhuá, Rǎn Chāo, Shàng Chángchūn, Yuán Yùháo, and Zhāng Jīnxīng in Shénnóngjià; Lǐ Àipíng, Mín Zé, and Yè Miáo in Wǔhàn; Liú Wén and Luó Ānhú of the Liǔzhōu Museum; Gāo Fēng and Jí Xuépíng of the Archaeological Institute in Kūnmíng; Mǎ Wéndǒu of the Yúnnán Provincial Museum; Jiāng Chū, Yáng Qīng, and Yáng Shàoxiáng of the Yuánmóu Man Exhibition Hall; Lǐ Zìxiù in Yuánmóu; Lǐ Xùwén, Lǐ Hóngjiǔ, Tián Xiǎowén, and Yáng Xī in Lìjiāng; Chén Wēngliáng and Xú Yǒngqìng at the Shànghǎi Natural History Museum; and Táo Yǔnhàn at the Shànghǎi Science and Technology Museum.

Finally, I thank the fifty-four people who completed surveys and wrote letters detailing their experiences as readers of *Fossils* magazine and as hobbyists. Of those, the following gave permission to mention them by name: Huáng Déxiāng, Jiāng Chénguāng, Léi Xiǎo, Lǐ Chūnyuán, Lǐ Hóuwén, Lǐ Xiùcǎi, Liú Bìfēn, Liú Yǒng, Shàn Yuánshēng, Shěn Ēnhuá, Sūn Qīngmín, Yuán Zhìyì, and Zhào Shān.

At the University of Massachusetts, Amherst

I could not have landed better than I have at the University of Massachusetts, where the history faculty has been immensely supportive—from the moment they appealed to the dean to add a special position for me to this semester, when they granted a leave from teaching and service so I could finish the book. For their advice, encouragement, and support, I am especially grateful to Audrey Altstadt, Chris Appy, Anne Broadbridge, Richard Chu, Alvin Cohen (Asian Languages and Literatures),

Dick Minear, Brian Ogilvie, Larry Owens, Steve Platt, and Heather Cox Richardson. I thank the Five-College Science Studies group—especially Jeffrey Ramsey and Mike Dietrich—for taking time to read and discuss chapter 8. In the library, Sharon Domier, Jim Kelly, and the interlibrary loan staff have all leapt to help at a moment's notice. A Faculty Research Grant supported a month of follow-up interviewing in 2005.

At Wesleyan University

Returning to the Connecticut River Valley has allowed me to reconnect with my roots at Wesleyan, where I spent four (noncontiguous) years of excitement and delight that instilled in me a desire to remain on college campuses for the rest of my life. My thesis advisers, Vera Schwarcz and Sue Fisher, devoted much time to guiding my studies and offered me a great deal of encouragement along the way. With their help, I began to participate in the long process of bridging the fields of Chinese studies and science studies. Bill Johnston was another important source of inspiration, and Steve Angle invited me back in 2005 to talk at the beloved Freeman Center. Finally, Joe Rouse recently offered a gentle criticism that made a subtle but great difference in the finished book.

At the University of Chicago Press

Since Pete Beatty first found my proposal on his desk and promptly sent me a warm and professional response, Chicago has treated me like royalty. I am deeply indebted to Pete, Catherine Rice, Christie Henry, and, most important, Karen Darling for their tag-team support. Erik Carlson patiently answered many questions and corrected many errors of grammar and style, a task made more difficult by the cumbersome fonts. Chuck Hayford, Fan-ti Fan, and one anonymous reader offered very generous and helpfully critical comments on the submitted draft. David Goodrich of Birdtrack Press is responsible for the beautiful typesetting job.

And Beyond . . .

Richard Kutner gave me early lessons in time management. For helping kindle my interest in China, I thank Morris Rossabi. David Branner introduced me to classical Chinese and insisted on the use of tone marks in *pīnyīn*. Ann Waltner included me in her renowned classical Chinese reading group. Charlotte Furth offered much useful advice and suggested I apply to UCSD.

Dennis Etler introduced me to IVPP and shared seven hours of taped interviews with Jiǎ Lánpō. Alice Conklin, Fa-ti Fan, Amy Hwei-shuan Feng, Laiguo Long, Barry Sautman, Wen-ching Sung, and Nadine Weidman generously shared unpublished work. Michael Schoenhals swallowed his disgust at the egregious imperialist behavior of my homeland and shared his extraordinary knowledge of Cultural Revolution documents to help me improve chapter 5. Laurence Schneider thoughtfully read the entire manuscript and provided many useful comments on content and style. For their helpful suggestions, I thank David DeGusta, Ingrid Fryklund, Mark Lewis, Perry Link, Nakayama Shigeru, Jesse Richmond, Jessica Riskin, Grace Shen, Steve Smith, Matthew Sommer, Sharon Traweek, Fred Wakeman, and Kären Wigen.

I greatly appreciate the assistance of librarians at Stanford, Yale, and the University of California, Berkeley. The librarians at Harvard-Yenching, especially James Cheng, Nobuhiko Abe, Matthew Bilder, and Eiji Kuge, deserve a special thanks for their tolerance of my voluminous requests.

The National Science Foundation's Graduate Student Fellowship, Institute of International Education's Fulbright Fellowship, and Social Science Research Council's International Dissertation Research Fellowship provided generous financial support.

Portions of chapters 2 and 3 first appeared in "'The Very First Lesson': Teaching about Human Evolution in 1950s China," reprinted by permission of the publisher from Dilemmas of Victory: The Early Years of The People's Republic of China, edited by Jeremy Brown and Paul G. Pickowicz, pp. 232–255, Cambridge, Mass., Harvard University Press, Copyright © 2007 by the President and Fellows of Harvard College. An earlier version of chapter 5 first appeared as "Labor Created Humanity: Cultural Revolution Science on Its Own Terms," in The Chinese Cultural Revolution as History, edited by Joseph Esherick, Paul Pickowicz, and Andrew Walder, © 2006 by the Board of Trustees of the Leland Stanford Jr. University, all rights reserved.

Taraivina Costello and Maria Zepeda gave me insight into late-night popular science. For her friendship and advice on what goes on in the mind of an editor, I thank Michelle Frey. Page Bridgens helped me proofread. Mikola De Roo, Cindy Heller, Colleen Scott, Deepali Panjabi, and Táng Fèngqíng are there when I need them most. The San Diego Coalition for Peace and Justice and Northampton Committee to Stop the War in Iraq have given me faith in humanity.

My parents, Emily and Victor, have supported me and my education these thirty-five years and counting; I cannot begin to thank them. Linda and Roy Close welcomed me into their home in 1991 and have been

Conventions

Rendering Chinese

In Chinese, the surname precedes the given name. I have followed this rule except in cases where people of Chinese ancestry, in the course of living or publishing in the West, have themselves reversed their names. Where Romanization of Chinese is necessary, I use the *pīnyīn* system, complete with tone marks. Tones are essential to the Chinese language, and readers who hope to discuss this subject in Chinese will benefit from knowledge of the correct pronunciation.

In most cases where both the English and Chinese is required in the text (as in titles and terms), I have chosen to use Chinese characters alone rather than *pīnyīn*. This is because adding both *pīnyīn* and characters would be cumbersome in the text, and while the pronunciation can almost always be derived from the characters, the reverse is by no stretch of the imagination true. Moreover, because they represent meaning, Chinese characters are useful for all readers of Chinese regardless of dialect, and in many cases for readers of Japanese and Korean as well. I have found no way to adopt the same principles in the notes and bibliography, where subsequent references and alphabetic order make Romanization necessary. However, the index does provide characters.

Terms

To prevent confusion, I offer a brief explanation of some of the terms used in this book. The term "popular science" has multiple meanings, and so I have endeavored to be specific in most cases. I use "science dissemination" to describe top-down efforts to spread scientific knowledge to the general population. For the production of scientific knowledge wholly or in part by nonscientists, I employ "popular participation in science" or "mass science," depending on the context. By "popular culture," I mean ideas, beliefs, and practices generated and maintained through social networks other than those of the state and state-supported cultural apparatus.

I have chosen for various reasons to retain a number of actors' categories used by the Chinese socialist state. ("Actors' categories" are those employed by historical actors under study; they may be contrasted with "analytical categories" used by the researcher in ways the historical actors would not necessarily have recognized.) I use "peasants" instead of "farmers" because of the degree to which these people are still tied to land they cannot own, often with little legal opportunity for migration or change of employment. Embarrassingly for the socialist state, "peasants" remains the most accurate term.

I also employ the term "class" despite the confusions caused by socialist reorganizing and the Chinese state's unorthodox uses of the term.[1] While, in Max Weber's terms, status or power might be more appropriate than class to analyze these relations, the exclusive use of such language would render the history utterly divorced from the experiences of the historical actors themselves.

I use "the masses" exclusively as an actors' category. Just as Máo denied the existence of a "universal human nature," I deny the existence of "the masses." Nonetheless, it is impossible to provide an understanding of Máo-era popular science without using this term. The same is true of "superstition."

One of the chief protagonists in this story has had a great many names, but I will stick to "Peking Man" except in special circumstances. Amadeus Grabau gave the fossils the informal name "the Peking Man" in 1926 (since frequently misspelled as "Pekin Man"), and Davidson Black provided them with the Latin moniker *Sinanthropus pekinensis*, literally "China-human of Peking." Scientists at the time also sometimes referred to the fossils fondly as "Nellie." The modern synthesis in biology (circa 1950; see chapter 8) greatly simplified the human family tree and categorized Peking Man as a subspecies of *Homo erectus*.

Chinese names for the fossils are also numerous, and many forms continue to be used interchangeably:

---

1. On this subject, see Richard Curt Kraus, *Class Conflict in Chinese Socialism* (New York: Columbia University Press, 1981).

- Běijīng Human (北京人, identical to the term for living human natives of Běijīng)
- Běijīng Ape-Human (北京猿人; "ape-human" is the literal translation of the Latin *Pithecanthropus*)
- China Ape-Human (中國猿人, less common since the discovery of many other *Homo erectus* sites in China)
- China Human (震旦人; at least one republican-era author used this term to translate *Sinanthropus*—震旦 was a term in Buddhist texts meaning "China")[2]
- *Homo erectus* (直立人, rarely used in dissemination materials, and typically only since the 1980s)

Since "Peking Man" is a well established name, I have decided not to confuse matters further by adopting the term "Běijīng Man."

For the same reason, I have resisted the great temptation to change "Peking Man" to "Peking Human," despite the lack of gender specificity in the Chinese names. To placate my feminist sensibilities, I have adopted the pronoun "she" to refer to the fossils. This is especially appropriate since the first skullcap, unearthed in 1929, was identified as a female, and, following tradition, reconstructed busts representing Peking Man are typically also female. Chinese authors occasionally refer to Peking Man as "she."

I have also consistently eschewed "man" and "mankind" for "human" and "humanity," since the Chinese terms are not gender specific. The German word "Mensch" is also gender neutral, so I have rendered Engels's famous treatise as "The Part Played by Labor in the Transition from Ape to Human."

2. Huá Wéidān, *Zhèndàn rén yǔ Zhōukǒudiàn wénhuà yīcè* [Peking Man and Zhōukǒudiàn Culture] (Shànghǎi: Shāngwù yìnshūguǎn, 1937 [1936]).

Introduction

Beginning in 1918, scientists exploring a mining town near Běijīng followed local legends through a lair of fox spirits and a trove of dragon bones to emerge in the late 1920s with the richest evidence of human evolution the world had ever seen: Peking Man. They began in an excavated crater of a limestone quarry where one pillar of clay filled with bird bones remained standing. According to local informants, the area had once been the home of foxes with an insatiable appetite for neighborhood chickens. Over time, the foxes became evil spirits, and one poor soul who interfered with them succeeded only in losing his own mind.[1] Foreign scientists were not afraid of fox spirits but recognized the power such legends held as clues in the investigation of natural history.

"Chicken Bone Hill" did not turn up anything very interesting, but while they were excavating there one day, a local man brought news of a more promising sort. "There's no use staying here any longer," he told them. "Not far from here there is a place where you can collect much larger and better dragons' bones." The scientists had heard of dragon bones before—in their lexicon, these were "fossils." Chinese herbalists had long recognized their medicinal value, and apothecaries paid good money to the farmers and miners who knew where to find them.[2] Adapting this

1. J. Gunnar Andersson, *Children of the Yellow Earth* (London: Kegan Paul, 1934), 96. Fox spirits are very common in Chinese folklore. They are typically pranksters and shape-shifters and often seek sex, food, or other favors from humans. Robert Ford Campany, *Strange Writing: Anomaly Accounts in Early Medieval China* (Albany: State University of New York Press, 1996), 254, 361, 389.

2. Students of Chinese history may be more familiar with dragon bones as

long-standing economic practice to their own purposes, the scientists followed the man to a limestone fissure containing many dragon bones.[3] The place came to be called Dragon Bone Hill (龍骨山) by the Chinese and foreign scientists who gathered to participate in the excavation of this scientific treasure-house.[4] In 1926, a foreign scientist found what appeared to be a human tooth, which scientists boldly dubbed Peking Man. Justifying this optimism, in 1929 a Chinese scientist and his team raised from the earth the first complete skullcap of a 500,000-year-old human female.

———

In 1956 a visitor to the Peking Man exhibition hall at Zhōukǒudiàn wrote in the guest book, "The ironclad evidence of our ancestors' fossils has taught me that 'labor created humanity.' I have no doubts whatsoever anymore. From now on I will participate even more positively in every kind of labor and dedicate myself entirely to constructing our communist society and making the most of the precious property passed down to us from our ancestors." Another penned, "Our great ancestral country is so lovely. It was created for and passed down to us by our ancestors. I should study the hard labor through which human ancestors participated in the work of building socialism. I feel deeply the greatness and loveliness of the ancestral country."[5] It would seem that the new socialist state had succeeded in its massive efforts to convince people of the veracity of the Marxist scientific perspective on human evolution and of China's special place in that story. From picture books to thought-reform sessions, scientific and political elites had used every form of persuasion to drive out the superstitious notion that a deity had created humans out of dust or clay. Instead, they encouraged the masses to embrace the world-famous Peking Man fossils as ancestors of the Chinese people and as embodiments of Frederick Engels's celebrated theory that labor created humanity.

---

they appear in the story of the earliest discovery of oracle bones—tortoise shells and animal scapula inscribed with Chinese characters and used for divination in the Shāng dynasty (eighteenth to eleventh centuries B.C.). The discovery occurred at the turn of the twentieth century in an apothecary's inventory of dragon bones.

3. Andersson, *Children*, 96–97.

4. Péi Wénzhōng, *Zhōukǒudiàn dòngxué céng cǎijué jì* [Record of Excavations of the Zhōukǒudiàn Cave Strata] (Běijīng: Dìzhàn chūbǎnshè, 2001 [1934]), 7.

5. Zhōukǒudiàn Peking Man Site Exhibition Hall Guest Book, [22?] June 1956 and 2 December 1956.

In 2002 I received a letter from a Mr. Fāng Lì of rural Shǎnxī Province. He had responded to an ad I had placed in *Fossils* (化石, a popular science magazine published by the Institute of Vertebrate Paleontology and Paleoanthropology [IVPP] in Běijīng) and was now returning the survey I had mailed him. He was sixty-five years old, an "unsuccessful farmer," an avid reader of *Fossils*, and a devotee of an illegal *qìgōng* sect whose leader had been arrested in 1999, the same year the government cracked down on the more famous sect known as Fǎlún Gōng.[6] In writing to me care of IVPP and enclosing photographs of his fossil collection, he was at once fulfilling our "predestined relationship" (緣) and hoping to make a valuable contribution to China's scientific research. *Qìgōng* cosmology, folk religion, evolutionary theory, communist ideology, Chinese classical texts, and science fiction all fit together seamlessly in Fāng's understanding of where humans came from and where we are headed.

"I believe," he wrote, "that humans originated on Yáo Mountain in my province of Shǎnxī. On Yáo Mountain, the Queen Mother[7] gave birth to the ten thousand things. On Yáo Mountain there are two springs of pure water, which are the two breasts of the Queen Mother. Use them and they will not be exhausted.[8] The people of Shǎnxī originated on Yáo Mountain. The people of China originated in Shǎnxī. Over the course of several ecological transformations, they went out across the globe in search of environments suitable for survival. And so I say that Indians, Soviets, Native Americans, and British people are all the descendants of Chinese people. We all come originally from one mother. . . . We humans will in the end realize communism—international communism. From each according to his abilities; to each according to his needs. One hundred families will unite to become one family. All strife, invasion, and war will be extinguished. We will improve the environment . . . and use technology to conquer one planet after another. . . . Placing value on harmony[9] is a precious heirloom of the Chinese people. Uniting two to make one; uniting humans with heaven; uniting humans, earth, and

---

6. *Qìgōng* is a set of practices—including physical exercises, breath control, and meditation—designed to maintain and improve health. Its principles are based in Chinese medical and martial arts traditions; many *qìgōng* schools also emphasize spiritual or religious aspects of the practices.

7. A Daoist deity.

8. An allusion to the Daoist classic *Dào dé jīng*.

9. A quotation from the Confucian *Analects*.

heaven; uniting human and human. . . . This is the motion of the universe and the law of development. It is the law of historical progress." He closed with an invitation to visit his home and view his fossils; together we could climb Yáo Mountain.

How has evolutionary science shaped what it means to be human? Who is privileged to speak on the subject of human identity, using what forms of knowledge? These important and contentious questions are debated throughout the modern world. For readers relatively unfamiliar with Chinese history, this book will be an opportunity to explore common concerns in a refreshingly different context. China scholars, I hope, will find that Chinese history looks different when such issues are placed at the center of analysis.

Many people have claimed the authority to interpret human origins. Who "we" are and where "we" came from are questions that matter deeply because they are fundamental in shaping core identities and loyalties. In the United States today, the most conspicuous struggle in this arena has been between evolutionists and creationists—recently demonstrated when residents of Dover, Pennsylvania, put intelligent design on trial. Science popularizers have eagerly sought to define human origins in indisputably scientific terms, since people who understand themselves thus are wedded to science in a very powerful way. For similar reasons, religious leaders have often been equally adamant in their desire to introduce or preserve their own interpretations of human origins.

Beyond this familiar conflict, other identities and social movements have contributed to the contested field of human origins, producing a diverse array of meanings associated with it. Nationalist movements have tapped into theories of human evolution to make claims about a group's solidarity, superiority, or ties to a specific geographic area. Humanists, on the other hand, have used scientific evidence to trumpet the common heritage and essential sameness of all peoples. Local people have often maintained their own origin stories, sometimes buttressed by fossils found near them. Feminist scholars have challenged theories that place males at the center of evolutionary change while sidelining females.

Most if not all of these themes are well represented in twentieth-century Chinese history, but they have played out in distinctive ways. Darwinism, and with it knowledge of human evolution, came in the final years of the Qīng dynasty (1644–1911) to a China in the midst of imperialist invasion. Evolution stood for progress and national salvation, the potential for China to leave behind a past of "superstition" and weakness and to become scientific and strong. The opposition between evolutionary science and older, "unscientific" ways of thinking was thus present from the outset in China as in the West, and it became only more central after the founding of the socialist state in 1949. A key difference, however, lay in the relative weakness of Christianity in China even among Westernized circles. China had no indigenous belief system that suffered quite so devastating a challenge from evolutionism as Christianity did in the West, or that wielded such political influence. One consequence of this difference was that scientists and the state could be—and were—far more forceful in their efforts to replace religious interpretations of human origins with scientific ones.

Another important difference was the way in which the perceived lack of a scientific culture was intertwined with feelings of national inferiority. This contributed to the already highly charged significance that social Darwinist ideas had for national identity. Early twentieth-century Chinese intellectuals struggled to use such ideas to explain Chinese weakness and chart a course for China's future. At the same time, however, China had an international reputation as a once-great civilization with ancient roots. The result was a profound confusion about the value of the Chinese past and China's place in the modern world that continues to consume Chinese people today and thus continues to shape the ways they think about human evolution.

Added to these distinctions is the position China has held in the history of paleoanthropology—the study of human origins and evolution. When an international group of scientists discovered the Peking Man fossils in the late 1920s, Běijīng quickly became the center of attention for scientists interested in human evolution from all over the world. Eighty years later, African fossils have eclipsed those of China, but Peking Man still holds the allegiance of Chinese scientists and laypeople alike.

There are, moreover, a few central themes in the Chinese narrative that are unlikely to have occurred to the majority of Western readers. The influence of socialism generally, and Máo's unique social and ideological vision in particular, framed familiar issues in startling ways. These elements of difference and even surprise serve as telling reminders of the specificity and contingency of our own experiences.

Although few raised under capitalism will recognize the slogan "Labor created humanity," it is very familiar to virtually everyone educated in socialist countries. The source is Frederick Engels's elegant but unfinished

1876 essay "The Part Played by Labor in the Transition from Ape to Human." School textbooks, museum exhibits, and other materials produced in socialist countries have used this thesis to emphasize the role of manual labor in human evolution and thus to underscore labor's primacy to the human condition and the centrality of laboring people in human development. In early socialist China, human evolution was the "very first lesson" party officials imparted in political classes designed to bring the population up to speed on socialism and materialism. So reported Cuthbert O'Gara, a Catholic bishop who was in China in 1949 and found it horrifying that communists were indoctrinating the Chinese population with Darwinism.[10]

Scholarship on the People's Republic of China has not noted the significance of this "first lesson," but there are several striking points to make even at the outset. First is the state's direct, sustained involvement in promoting a specific scientific understanding of human origins and evolution. Science dissemination (*kēxué pǔjí*, 科學普及, or *kēpǔ*, 科普) was central to the ideological mission of Chinese socialists beginning before the 1949 revolution, and it remains so today. Like recent movements in the United Kingdom and United States to overcome knowledge "deficits" and improve the "public understanding of science," *kēpǔ* in China descended from the nineteenth-century professionalization of science: this was when the English language gained the term "scientist," and with it a sense that professionals with intellectual expertise and institutional authority should produce scientific knowledge and should then "diffuse" it to create an enlightened citizenry. "Popular science" was thus a category that helped distinguish and elevate the newly emerging "professional science."[11] While similar in some respects to its Western cousins, *kēpǔ* in China has been far more centrally and

10. Cuthbert M. O'Gara, *The Surrender to Secularism* (St. Louis: Cardinal Mindszenty Foundation, 1989 [1967]), 11.
11. On the origins of "professional" and "popular" science in the Victorian era, see Bernard Lightman, "The Visual Theology of Victorian Popularizers of Science: From Reverent Eye to Chemical Retina," *Isis* 91, no. 4 (2000): 651–680; and James A. Secord, *Victorian Sensation: The Extraordinary Publication, Reception, and Secret Authorship of "Vestiges of the Natural History of Creation"* (Chicago: University of Chicago Press, 2000). The literature on the "public understanding of science" movement in the United Kingdom and United States is vast, and the subfield has its own journal: *Public Understanding of Science*. Useful critical perspectives include Alan Irwin and Brian Wynne, eds., *Misunderstanding Science? The Public Reconstruction of Science and Technology* (New York: Cambridge University Press, 2003 [1996]); Jane L. Lehr, "Social Justice Pedagogies and Scientific Knowledge: Remaking Citizenship in the Non-science Classroom" (Ph.D. diss.: Virginia Tech University, 2006); Stephen Hilgartner, "The Dominant View of Popularization: Conceptual Problems, Political Uses," *Social Studies of Science* 20 (1990): 519–539; and Bruce V. Lewenstein, "The Meaning of 'Public Understanding of Science' in the United States after World War II," *Public Understanding of Science* 1 (1992): 45–68.

explicitly a political mission of the state. The goal in China has been to use evidence and reasoning to convince Chinese people to discard their previous superstitious beliefs and embrace a scientific worldview consistent with the philosophical and political teachings of Marxism. For this purpose, Engels's twist on the story of human evolution was of particular utility and inspiration.

The second point is the explicit attention given to the political implications of theories of human evolution in media created for the general public—even children's picture books. Such propaganda is inevitably one sided, but it offers exposure to a way of thinking about science that in the United States is rare outside scholarly literature. Rather than treating science as a pure realm that is (or should be) isolated from external influences, Máo-era *kēpǔ* presented science as inseparable from its social and political contexts.

Finally, historians of science have acknowledged race, nation, and gender as instrumental in shaping Western understandings of human origins and evolution.[12] Class, on the other hand, has received far less attention, either because it has been less influential in Western paleoanthropology or because it is less interesting to scholarly critics (and probably both). In contrast, the principle that labor created humanity placed class firmly at the center of the discourse in socialist countries like China.

Among socialist countries, China was extraordinary in the vehemence with which official ideology rejected the notion of a "universal human nature"; it focused instead on what Donald Munro has called "the malleability of man"—the notion that different social classes have different natures and that human nature can change.[13] This commitment was present already in Máo's writings from the early 1940s and gained strength in later decades with attacks on literature and philosophy that purported to identify characteristics (for example, love) that defined humanity and were shared across all classes of people. The alleged problem with such writings was the way they took a bourgeois understanding of human identity and claimed it was not bourgeois but universal. In this regard, Máo's position was similar to more recent, Western scholarship

12. See, for example, Donna Haraway, *Primate Visions: Gender, Race, and Nature in the World of Modern Science* (New York: Routledge, 1989); Stephen Jay Gould, *Ontogeny and Phylogeny* (Cambridge, Mass.: Harvard University Press, Belknap Press, 1977); Londa Schiebinger, *Nature's Body: Gender in the Making of Modern Science* (Boston: Beacon Press, 1994); and Robert N. Proctor, "Three Roots of Human Recency: Molecular Anthropology, the Refigured Acheulean, and the UNESCO Response to Auschwitz," *Current Anthropology* 44, no. 2 (2003): 213–239.

13. Donald J. Munro, *The Concept of Man in Contemporary China* (Ann Arbor: University of Michigan Press, 1977).

on science and human identity. Donna Haraway, for example, has demonstrated in her study of primatology that what purports to be discourse on the human in fact works to reify more specific constructions of race and gender. Máo and Haraway thus both criticized liberal humanists who failed to recognize the significance of specific social identities and their entanglement with ideas about what it means to be human. A key difference between them, of course, is that Haraway does not seek to establish the existence of separate natures for different kinds of people; she merely shows that such concepts are embedded within ostensibly humanist discourses.[14]

Socialist-era China offers a fascinating example from the other side—from a political-historical context in which the tendency was rather to emphasize social identities to such an extent that humanism itself was denied. This case thus warrants a different corrective, one that demonstrates the existence, rhetoric not withstanding, of a distinct human identity. While scholars in the humanities were unable to explore questions of humanism, their colleagues in the natural sciences were busy creating a human identity based primarily—though not exclusively—on Engels's theory that labor created humanity. The Máo-era state commissioned this human identity, which in turn served some of the state's highest political and ideological priorities. Recognition of this precedent provides a richer historical context for understanding post-Máo struggles over humanism (人道主義), from defining human rights (人權) to reviving a "humanistic spirit" (人文精神) in the face of growing commercialism to interpreting the meaning behind Hú Jǐntāo's adopted principle of "human-centered" (以人爲本) development.[15] Indeed, the notion that

14. On the entanglement of humanism and racism, see also Etienne Balibar and Immanuel Wallerstein, *Race, Nation, Class: Ambiguous Identities* (London: Verso, 1991 [1988]).
15. The literatures on humanism and human rights in China are extensive and wide ranging. The following examples represent a few of the many approaches scholars have taken to the issues: Zhāng Zīzhāng, *Rénxìng yǔ "kàngyì wénxué"* [Human Nature and "Protest Literature"] (Táiběi: Yòushī wénhuà shìyè gōngsī, 1984); Stephen Angle, *Human Rights and Chinese Thought: A Cross-Cultural Inquiry* (Cambridge: Cambridge University Press, 2002); Geremie Barmé, *In the Red: On Contemporary Chinese Culture* (New York: Columbia University Press, 1999); Shiping Hua, *Scientism and Humanism: Two Cultures in Post-Mao China, 1978–1989* (Albany, N.Y.: State University of New York Press, 1995). For translated primary sources, see Stephen Angle and Marina Svensson, *The Chinese Human Rights Reader: Documents and Commentary, 1900–2000* (Armonk, N.Y.: M. E. Sharpe, 2001); and Geremie Barmé and John Minford, eds., *Seeds of Fire: Chinese Voices of Conscience* (New York: Hill and Wang, 1988). For a collection of early post-Máo articles on humanism, see Zhōngguó shèhuì kēxuéyuàn zhéxué yánjiūsuǒ, *Rénxìng, réndàozhǔyì wèntí táolùn jí* [Collected Discussions on Questions of Human Nature and Humanism] (Běijīng: Rénmín chūbǎnshè, 1983). For a bibliography of Chinese articles (1977–1981) on Marxism and humanism, see "Mǎkèsīzhǔyì wényì lǐlùn yánjiū" biānjí bù, ed., *Lùn rénxìng hé réndàozhǔyì* [On Human Nature and Humanism] (Běijīng: Guāngmíng rìbào chūbǎnshè, 1982), 202–216.

labor created humanity has been alternately a useful point of departure, subject for critique, and answer to the widely asked question that underlies these debates: What is human nature (人性 or 人的本質)?[16]

Also characteristic of Chinese socialism was the seriousness with which Máo and some of his followers pursued what I call the class politics of knowledge, specifically by attacking intellectual elitism and seeking to overthrow the division between mental and manual labor. Particularly during the Great Leap Forward (1958–1960) and the Cultural Revolution (1966–1976), political leaders called upon scientists to "follow the masses" and implement "mass science." This was one especially radical example of a historical phenomenon common at least since the professionalization of science in the nineteenth century. If we think of the top-down diffusion of knowledge represented by "science dissemination" as one aspect of popular science, bottom-up challenges to professional elitism constitute its historical twin. From Victorian-era assertions that science is based on "common sense" and thus belongs to the "Everyman," to recent scholarly efforts to reconstruct a "people's history of science," to late-night radio shows where callers relate their paranormal experiences in the face of "expert" disbelief, critics have called into question the notion that science is the exclusive preserve of people with academic degrees and established positions.[17]

16. An early post-Máo article that discussed "Labor created humanity" in the context of defining human nature was Wáng Ruìshēng, "Guānyú rénxìng gàiniàn de lǐjiě," in Zhōngguó shèhuì kēxuéyuàn zhéxué yánjiūsuǒ, *Rénxìng, réndàozhǔyì*, 207–218 (originally published in *Zhéxué yánjiū*, no. 3 (1980). For more recent examples, see Hú Jiāxiáng, "Rénxìng sān tíyì" [Three Proposals on [the Question of] Human Nature], *Jiāngxī shèhuì kēxué*, no. 4 (1997): 28–32; Sū Fùzhōng, "Rénxìng yánjiū de fāngfǎ lùn sīkǎo" [Thinking Deeply on the Methodology of Research into Human Nature], *Shānxī dàxué shīfàn xuéyuàn xuébào*, no. 2 (1999): 19–25; Lǐ Kèshí, "Guānyú rén de běnzhì de zhéxué gàikuò" [The Philosophical Summation of Human Beings' Essence (translation provided with article)], *Huánghé shuǐlì zhíyè jìshù xuéyuàn xuébào* 13, no. 3 (2001): 43–46; and Xióng Fāng, "Máo Zédōng rén de běnzhì lùn" [Máo Zédōng's Theory of Human Nature], *Jiānghàn lùntán*, no. 9 (2004): 81–83.

17. On nineteenth-century Britain, see Adrian Desmond, "Artisan Resistance and Evolution in Britain, 1819–1848," *Osiris* 3 (1987): 77–110; and Alison Winter, "Mesmerism and Popular Culture in Early Victorian England," *History of Science* 32, no. 3 (1994): 317–343. On the recent U.K. and U.S., see Irwin and Wynne, *Misunderstanding Science?* and Steven Epstein, *Impure Science: AIDS, Activism, and the Politics of Knowledge* (Berkeley: University of California Press, 1996). On postwar Japan, see Nakayama Shigeru, "Grass-Roots Geology: Ijiri Shōji and the Chidanken," in *Science and Society in Modern Japan: Selected Historical Sources*, ed. Nakayama Shigeru, David L. Swain, and Yagi Eri (Cambridge, Mass.: MIT Press, 1974). On post-Máo China, see Nancy Chen, "Urban Spaces and Experiences of Qigong," in *Urban Spaces in Contemporary China*, ed. Deborah S. Davis, Richard Kraus, Barry Naughton, and Elizabeth Perry (Cambridge: Cambridge University Press, 1995): 347–362. The classic formulation of the thesis that artisanal knowledge contributed fundamentally to the Scientific Revolution appears in Edgar Zilsel, "The Sociological Roots of Modern Science," *American Journal of Sociology* 47 (1942): 544–562. Joseph Needham built on this thesis to suggest that China owed its failure to develop modern science to social structures that prevented the integration of artisanal and scholarly forms of knowledge. Clifford D. Conner has recently produced a synthesis of such scholarship for a general audience entitled *A People's History of Science: Miners, Midwives, and "Low*

At its height, mass science in China had the potential to be the most concerted attempt to overturn scientific elitism the world has ever seen. At the same time, however, mass science coexisted very uneasily with Máo's own depiction of the masses as superstitious, a perspective scientists found much more agreeable and which they effectively kept as the keystone of science popularization programs. The question of who may legitimately speak on scientific subjects—including human evolution—was thus a far more explicit one in socialist China than in other contexts. Mass science was, moreover, a particularly strong test of the Maoist commitment to radical populism, and as such it offers China scholars a productive way to explore the potentials and limits of this philosophy.

Superstitious or not, laypeople in China have preserved and adapted long-standing notions about humanity with respect to social relations and the place of humans in the natural world. While difficult to excavate from under the thick sediments of state propaganda, such perspectives have been present all along and are especially close to the surface in stories about half-ape, half-human creatures similar to the North American Bigfoot. In the post-Máo era, stories about these *yěrén* have circulated widely and have even received official recognition as a legitimate subject of scientific inquiry. Moreover, the widespread respect for ancestors and continued practice of ancestor worship represent strong cultural influences that underlie much popular paleoanthropology. These and other elements from Chinese popular culture have played critical though often unacknowledged roles in the production of scientific knowledge about humans in China.

––––––––

Our guides in exploring this history are two popular characters in twentieth-century paleoanthropology: Peking Man (*Běijīng yuánrén*, 北京猿人) and Wild Man (*yěrén*, 野人). The two figures often appear between the covers of the same popular science books and magazines, and many of the scientists who study Wild Man also research Peking Man or other fossil hominids. Peking Man and Wild Man both represent the primitive human, and at the same time something not quite human. As such, they have similarly served as mirrors for Chinese people in defining human identity. There is, however, an important difference. Peking Man repre-

–––––––––––

Mechaniks" (New York: Nation Books, 2005). The late-night radio show I am thinking of is Art Bell's *Coast to Coast AM*.

sents science—the triumph of a rational understanding of human origins to replace religious creationism. Wild Man is the irrepressible legend, untamable by science. Even so, they cannot be wholly separated. Peking Man is also the stuff of stories and fantasies, and Wild Man the subject of scientific inquiry, an attempt to transform legend into science. They are not fixed entities; rather, they change to speak to different people and different times. But neither are they mere reflections of shifting political contexts: sometimes they speak against the grain.

Figure 1 shows Peking Man as she has appeared in many representations since the 1950s. Note that here and in many other places she is female: the first excavated skullcap was from a female specimen, and following scientific convention, reconstructions of Peking Man *qua* Peking Man thus often depict a female. Where depicted as male, Peking Man is still typically a gentle sort of primitive, with a soft look in the eye, engaged in a life of quiet, determined labor (figure 2). Only occasionally is a wilder, more aggressive side shown, and even then the emphasis is on triumph over natural hardships and not on conflict within the species. Although she stands on the border between ape and human, it is Peking Man's human side that receives the greater emphasis: portraits evoke affection for her as an ancestor rather than aversion to her remaining bestial qualities.[18]

Since the 1950s—although arguably not before—Peking Man has been drenched in nationalist symbolism. The growing (but by no means complete) consensus among scientists internationally is that today's humans all share a common ancestor who lived in Africa as recently as 100,000 years ago. Dating from 500,000 years before the present, Peking Man would thus be an offshoot, a sideshow to the main attraction: the lineage leading to modern humans. Most Chinese paleoanthropologists (and some Western ones as well), however, support a rival theory, known as multiregionalism, in which multiple areas contributed to the emergence of modern humans. In this model, Peking Man is preserved as a Chinese ancestor, if not an ancestor of all people.

Nationalist influences and rhetoric in paleoanthropology should not, however, make us insensitive to the wealth of other meanings Peking Man has embodied. During the Máo era Peking Man was a chief protagonist in a story that emphasized not only the socialist moral that "labor created humanity" but also the *international* socialist moral that "all the

18. In 1979, paleoanthropologist Zhōu Guóxīng oversaw the construction of a new bust that appears still more human (and prettier). Scientists now see less anatomical difference than they once did between *Homo erectus* (including Peking Man) and *Homo sapiens*. The bust is on display at the Běijīng Natural History Museum.

1 Peking Man qua Peking Man. This bust sits in the anthropology wing of the Institute for Vertebrate Paleontology and Paleoanthropology's paleontology museum. It is modeled after a
bust created in the 1950s by Wáng Cúnyì under the direction of Wú Rǔkāng and Wú Xīnzhì.

2   Painting by the IVPP artist Lǐ Róngshān depicting Peking Man engaged in tool manufacture. Executed in the early 1970s, it was one of the paintings IVPP took to Tibet in 1976 for use in a traveling exhibition (see chapter 5). It is very similar to the statue that currently sits in the front hall of the Peking Man site exhibition, but the standing figure in the statue is more recognizably female. Representations of a standing female Peking Man manufacturing tools have been standard fare at least since 1949. See, for example, Yáng Háotīng's drawing (based on an earlier statue) in Péi Wénzhōng, *Zìrán fāzhǎn jiǎnshǐ* [A Short History of Natural Development] (Shànghǎi: Gēngyún chūbǎnshè, 1950), 56. (Courtesy of Lǐ Róngshān.)

world is one human family." Both these themes remain in contemporary Chinese popular writings on human evolution. Thus, in addition to the narrowly nationalistic devotion to Peking Man as an ancestor, there is a more broadly human one.

A poem published in 1979 in the popular magazine *Science Hobbyist* captures yet another kind of sentiment that some have harbored for Peking Man.

Who brushed from your brow
The accumulated yellow mud of the ages?
Whose great hand warmly and softly
Woke you from a primeval dream?

Painstaking as a young lady embroidering flowers,
Grueling as panning for gold in the sand,
On the second of December in 1929,
A Chinese scholar's hands, trembling with excitement,

From layers of earth more than thirty meters deep,
Oh so lightly, oh so carefully raised you . . .[19]

Scientists and others involved in science dissemination continue to recall the excitement of the original discoveries, their international significance, and the scientific culture that grew around the excavations. For them, Peking Man represents love for science itself.

The portrait of Wild Man (*yěrén*) in figure 3 is also female and primitive, but here the similarity with Peking Man seems to end. Whereas Peking Man is soft and gentle, Wild Man is passionate and savage. Whereas Peking Man lives a sociable, cooperative existence with her fellow Peking Men, Wild Man lives outside law and custom, surviving by kidnapping humans of the opposite sex and forcing them to serve as mates. Wild Man's wildness is terrifying but also exciting, especially for people who feel too much the constraints of civilization.

Like the North American Bigfoot, the modern *yěrén*'s ancestry is in legends about creatures that straddle the boundary between human and animal. The scientists who pursue them today theorize on their likely relationship to modern humans: some suggest they are descendants of the fossil ape *Gigantopithecus*, others of the (probable) human ancestor *Australopithecus*. They hope to bring the *yěrén* out of the realm of legend and into the realm of science. This has proved difficult. As a poem from 1985 entitled "I Am *Yěrén*" puts it,

You want to see but I'm difficult to find, want to capture but I'm difficult to catch;
I am the living specimen you need to explain, you want to explain.[20]

Not only have numerous scientific expeditions failed to produce a specimen, but the popularity of *yěrén* can be attributed in part to their very elusiveness: they are a testament to the limitations of science. As such they "strike a blow" against scientific elitism, but in the process they strike another for something that smacks of superstition. Where they stand in the class politics of knowledge is an open question.

If Peking Man's significance rests on her accepted position as an ancestor, Wild Man's relationship to modern people is somewhat more complicated. Peking Man represents the root of humanity, a primitive state that has left its mark but to which there is no return. She exists at

19. Zhāng Fēng and Xiāo Yán, "'Běijīng rén' zhī gē" [Ode to Peking Man], *Kēxué àihàozhě*, no. 1 (1979): 22.

20. Zhōu Liángpèi, *Yěrén jí* [*Yěrén* Collection] (Běijīng: Huáxià chūbǎnshè, 1992), 96–99.

野人木偶记

紫枫编

3   Cover of Zǐ Fēng, ed., *Yěrén qiú'ǒu jì* (Běijīng: Zhōngguó mínjiān wényì chūbǎnshè, 1988), which included Sòng Yōuxīng's story "A *Yěrén* Seeks a Mate." Note the similarity between the face at lower left and the bust of Peking Man in figure 1.

a fixed node in a linear scientific temporality. Wild Man, on the other hand, comes from a world of legends where animals sometimes become human, and humans easily revert to beasts. This potential for two-way movement places her in a powerful position from which to speak on modern humanity. From the same poem,

If we look at one another, we'll see who still retains a tail;
You are my bright mirror, and I am yours.

Sometimes the reflection shows the human to be more barbaric than the beast.

When you fell trees and hunt wild animals, you acquire in your greed a wildness you cannot cast off.

In such cases Wild Man becomes a powerful voice for those crushed by the inhumanity of modern society.

––––––

The narrative that follows will cover slightly more than one century of Chinese history, from 1898 to 2005. It is a serious investigation of two possibilities that are too often dismissed in our cynical times. The first is the possibility of human identity. Scholars have rightly pointed to all the ways in which constructions of the human are wrapped up in constructions of race, nation, and gender. I hope to demonstrate, however, that human identities are not reducible to specific social identities, but have themselves played significant historical roles in twentieth-century China and as such are worthy of more attention. In the Máo era, when scholars of literature could not freely discuss "human nature," humanism survived in the realm of science.

The second is the possibility of mass science. I suggest that Máo was right to emphasize the class politics of knowledge and right to think that laborers had something to offer science. Yet Máo's own conception of "the masses" as superstitious blinded him to some of the most important contributions they have made. To see the true promise of mass science, we will have to hold Máo accountable for the limitation of his vision and pull away the curtain of "superstition" that cloaks the richness of Chinese popular culture.

ONE

"From 'Dragon Bones' to Scientific Research": Peking Man and Popular Paleoanthropology in Pre-1949 China

Celestial Clouds and Zip Wires

In 1934, a dynamite blast during the excavation of the Peking Man fossils destroyed the Temple of the Hill God (山神廟) at Zhōukǒudiàn.[1] The scientists involved in the research arranged for a new temple to be built, but in place of the original mural portraying a deity they installed one that vividly evoked the complex social and cultural milieu of the time (figure 4). Composed in a recognizable style of Chinese landscape painting, the new mural retained the swirling clouds that connote the divine in Daoist representations and a narrative element commonly seen in Chinese art. Yet the subject matter was thoroughly modern: it depicted the scientific excavation work then in progress at the site. Prominently featured was the zip wire installed to facilitate the transport of baskets of excavated rock and

1. The destruction may have occurred in 1933 or 1935. A photograph of the old temple exists from 1933, and one of the new temple exists already from November 1935. See Jiǎ Lánpō, *Zhōukǒudiàn jìshì* [Chronicle of Zhōukǒudiàn] (Shànghǎi: Shànghǎi kēxué jìshù chūbǎnshè, 1999), 71, 104.

4   Mural created for a new temple at Zhōukǒudiàn built to replace one destroyed during excavations. From Jiǎ Lánpō, *Zhōukǒudiàn jìshì* [Chronicle of Zhōukǒudiàn] (Shànghǎi: Shànghǎi
kēxué jìshù chūbǎnshè, 1999), 137.

gravel, of which participants were duly proud.[2] Also visible in the painting were the many types of people who participated in the excavation
and in daily life in the surrounding community. Workers dug in the excavation pit, which was marked in sections with Roman letters, and filled
baskets with material to be sent down the wire. Scientists in Chinese and
Western clothes—although interestingly no obvious examples of Western scientists—observed the work. Women in Chinese dress walked up
the road to the site, while a man at the bottom sold food out of a basket
to a customer. Two other women stood reading a sign that identified the
site and the scientific work underway.

Through its inclusions, and also its omissions, the mural captured
much of what defined the work at Zhōukǒudiàn and what it meant for
Chinese society at large. It expressed the hope that China could become
modern without losing its Chinese "essence," and that Chinese and
Westerners—or at least their clothing—could work together toward a

2. Jiǎ Lánpō, interview with Dennis Etler, 17 September 1999. Péi Wénzhōng, *Zhōukǒudiàn dòngxué céng cǎijué jì* [Record of Excavations of the Zhōukǒudiàn Cave Strata] (Běijīng: Dìzhàn chūbǎnshè,
2001 [1934]), 14–17.

common goal while respecting Chinese leadership in China. It was significant that the goal in question, along with the means for achieving it, was science—an international endeavor in which Chinese people were struggling to participate as equals. But not all Chinese could be equals: whether in Chinese robes or Western suits, scientists could be distinguished from the workers who handled the heaviest and least rewarding jobs. Finally, the painting appeared to symbolize simultaneously the replacement of religion with science and also the accommodation of science to existing Chinese ideas and practices, including spiritual ones. Having blown up the original temple, scientists replaced it with an altar to science and a depiction of themselves as the principal agents in the transformation. The human power of science represented by the zip wires, however, appeared nonetheless as an extension of the power of heaven coming to earth through the ribbons of clouds on which immortals were often seen to descend. These core relationships—nationalism and internationalism, elite and worker, tradition and modernity, science and "superstition"—were to continue to frame popular paleoanthropology throughout the twentieth century.

A Willingness to Change

An adage from the ancient classic, the Book of History (尚書), has often been called upon in China to express the nature of the human: 惟人萬物之靈.[3] Pithy and potent, this quotation (which serves as the epigraph to this volume) is also open to diverse interpretations—qualities that help explain its continued cultural significance. 人 is "human"; 萬物 is "the ten thousand things." The difficulty is with 靈 (líng). It typically refers to a spiritual essence or quality. It may denote a female shaman, a divinity, the soul, the austerity of a leader, the quality of being efficacious, innate intelligence, fortune, or excellence.[4] The adage has been translated into English as "man is the soul of all creation," "man is the best of the ten thousand things," and "of all creatures, man is the most highly endowed."[5] An equally plausible translation would be "humans are the intelligent ones of the ten thousand things." The connotations

3. From the Upper Tàishì (泰誓上) section. This section appears only in the "old text" version of the classic, commonly considered a "forgery" from the Hàn dynasty.
4. Wú Zéyán, Huáng Qiūyún, and Liú Yèqiū, eds., *Cí yuán, dàlù bǎn* [Origin of Words, Mainland Edition] (Táiběi: Táiwān shāngwù yìnshūguǎn, 1993 [1989]), 2:3345.
5. Stuart Schram, *Chairman Mao Talks to the People: Talks and Letters: 1956–1971* (New York: Pantheon Books, 1974), 220; James Reeve Pusey, *China and Charles Darwin* (Cambridge, Mass.: Harvard University Press, 1983), 195; James Legge, trans., *The Chinese Classics III: The Shoo King* [Shū jīng] (Hong Kong: Hong Kong University Press, 1970 [1861]), 283.

of intelligence and leadership led Japanese and Chinese intellectuals to use 靈長 (efficacious leader) to translate the term "primate" in their writings on biology and evolution. This was probably the sense that Máo Zédōng had in mind when in a 1964 speech he said: "In the past, they said . . . '人為萬物之靈.' Who called a meeting to elect [humans to this position]? They appointed themselves."[6]

Darwin is often credited with promoting the radical principle that humans are animals. In the century since Darwinism was first introduced to China, the ancient adage from the Book of History has appeared to different people either as prescient or as old fashioned, depending on whether they have understood it to acknowledge the place of humans among other living things or to emphasize their exceptional qualities. For one famous advocate of science in 1921, the spiritual quality of the term 靈 was problematic, and he contrasted it with the evolutionary perspective then taking Chinese intellectuals by storm: "People who consider themselves 'the spirit of the ten-thousand things' have forgotten that they are still biological organisms." It was only by "knowing themselves" as such that the Chinese people could achieve a "true self-liberation movement."[7]

Actually, it is an ancient notion, in China and elsewhere, that the boundary between humans and animals is porous. Chinese writings throughout imperial times abounded with tales of creatures part human, part animal, and others—like fox spirits and snakes—that could change into human form and back again. The "hairy people" (毛人) are a late example: they appear in a Qīng dynasty (1644–1911) gazetteer and in the writings of the eighteenth-century folklorist Yuán Méi. Legend has it that they descended from people who had fled conscript labor on the first Qín emperor's Great Wall; removed from the civilizing influences of society, they had reverted to beasts.[8] According to classical Chinese texts, humans were originally not separate from the animals; they achieved humanity only through the actions of the sage kings.[9] If the classics told that humans had originated among the beasts, they also suggested that

6. Note that modern allusions to this adage typically adopt the more modern grammar that Máo uses here. I have slightly altered the translation from Schram's in *Chairman Mao Talks*, 220. For a more explicit connection of the adage to the term for "primate," see Wáng Xiǎoshí, *Cóng yuán dào rén: Tōngsú jiǎnghuà* [From Ape to Human: A Simple Account] (Shànghǎi: Xīnyà shūdiàn, 1950), 17.

7. Zhōu Zuòrén, quoted in Liú Wéimín, *Kēxué yǔ xiàndài Zhōngguó wénxué* [Science and Modern Chinese Literature] (Héféi: Ānhuī jiàoyù chūbǎnshè, 2000), 41.

8. Yáng Yánliè, ed., *Fāng xiàn zhì* [Fāng County Gazetteer] (Táiběi: Chéngwén chūbǎnshè, 1976 [1865]), 847–848; Yuán Méi, *Zhèng xù Zǐ bù yǔ* [What Confucius Did Not Discuss, Corrected and Continued] (Táiběi: Xīnxìng shūjú, 1978), 5240–5241.

9. Mark Edward Lewis, *Sanctioned Violence in Early China* (Albany: State University of New York Press, 1990), 169–170.

humans could slip back across the boundary if they abandoned that which distinguished them—namely, their moral sensibilities and adherence to ritual.[10] By the same logic, members of ethnic groups whose customs differed were often classified along with the "wild beasts."[11] What made Darwinism as introduced to China such a powerful new idea was not simply the placement of humans within the natural world, but the law of progressive change that went with it.[12] When Peking Man arrived on the scene, she would be understood as part of this forward momentum and not, like the "hairy people," as an example of slipping back across the boundary into the realm of animals. While in Chinese folklore the possibility of transformation between animal and human was two-way, in modern evolutionary theory, movement is one-way. Humans *were* like the rest of the animals but have evolved and cannot go back.

The first to write on human evolution in China were not scientists. They were officials and intellectuals who sought to make sense of the related experiences of a declining Qīng empire, encroaching imperialist powers, and an emerging Chinese nationalism. In the writings of Darwin, Thomas Henry Huxley, Herbert Spencer, Pyotr Alekseyevich Kropotkin, and others, Chinese writers in the first half of the twentieth century hoped to find answers to the questions Why is China weak? and How can China be made strong?

The man most responsible for introducing Darwinian thought to China was Yán Fù (1853–1921), an educator and Westernizer who in 1898 published a paraphrastic and heavily annotated translation of Thomas Huxley's *Evolution and Ethics* (1893), entitled in Chinese *On Evolution* (天演論).[13] Yán's motivation for writing the text and the reason for its widespread influence was the apparent relevance of the subject to "self-strengthening and the preservation of the race."[14] It was not Darwinism per se, but social Darwinism, that interested Yán and his many followers,

---

10. "If the sense of duty or knowledge of [correct hierarchical relations] were lost, then he would revert to the level of the beasts." Lewis, *Sanctioned Violence*, 172.

11. Lewis, *Sanctioned Violence*, 173. Those who introduced Darwinism to China were aware of this history and sometimes used it to highlight both the existence of historical precedents and the difference between such ideas and those of Darwinists. Qiáo Fēng, "Rénzhǒng qǐyuán" [Theories of Human Origins], *Dōngfāng zázhì* 16, no. 11 (1919): 93–100.

12. Some might object that Darwin did not espouse a "law of progressive change," which is why I say "Darwinism as introduced to China." Nonetheless, an excessively protective attitude toward Darwin is unwarranted. There is much in Darwin's writings that celebrates progressive evolution, and also much that (again, contrary to his celebrated image) encourages application of evolutionary principles to human societies.

13. Benjamin Schwartz, *In Search of Wealth and Power: Yen Fu and the West* (New York: Harper and Row, 1964), 42–90.

14. Yán Fù quoted in Schwartz, *In Search of Wealth*, 100.

including the notable reformer Hú Shì (1891–1962), who adopted the character 適 for his given name to evoke the phrase "survival of the fittest" (適者勝存).[15]

Yán drew selectively from both Herbert Spencer and Thomas Huxley to generate a vision of evolution suitable for a Chinese reformer of his era. Yán adopted Spencer's optimism with respect to evolution as a progressive force while rejecting his individualism and his willingness to forego human action for the betterment of society. Huxley proved the perfect complement: eschewing Huxley's pessimistic view of evolution as a necessarily brutal force, Yán nonetheless embraced Huxley's call for people to organize to improve themselves and their societies physically, intellectually, and morally.[16]

Yán's synthesis, like the new painting in the Temple of the Hill God, marked both an accommodation to long-standing Chinese beliefs and a reorientation to a modern notion of progressive change. Considering human ethical action an essential part of the struggle for existence left room for Confucian activism. It also preserved what had long been held in China to be the crucial distinction between humans and animals, their understanding of morality and their ability to manifest it through correct practice—in short, their humaneness (仁). Moreover, Yán's sense of a cosmic pattern governing nature and human society strongly resonated with Daoist ideas.[17]

At the same time, Yán's Darwinism was progressive. Unlike the two-way movement between human and animal found in Chinese legends, in the evolutionary theory that Yán introduced the position of animals became fixed in the human past as "ancestors"—a concept bound to make many filial Chinese somewhat uncomfortable. Looked at in the other direction, however, the struggle for existence moved nature and society ever closer to perfection. Yán found in evolutionary theory the lesson he thought China most needed to learn. Change was both unavoidable and beneficial. To survive and prosper, China had to embrace change and release the progressive energies of the people that had been inhibited by ignorance, poverty, and weakness.[18]

Thus, Yán Fù and his followers were also voluntarists. For the problems that concerned late Qīng-dynasty scholars—national strength in

15. Among his classmates there were also "Natural-Selection Yáng" and "Struggle-for-Existence Sūn." Jerome B. Grieder, *Hu Shih and the Chinese Renaissance: Liberalism in the Chinese Revolution, 1917–1937* (Cambridge, Mass.: Harvard University Press, 1970), 27.

16. Pusey, *China and Charles Darwin*, 158–175.

17. Schwartz, *In Search of Wealth*, 106–110; Pusey, *China and Charles Darwin*, 164–165.

18. Schwartz, *In Search of Wealth*, 49, 56.

the face of imperialism and above all the need for modernizing change—continued to occupy their successors through the disunity, war, and revolution that filled the republican era (1912–1949). In a 1927 speech at Whampoa Military Academy, the famous Chinese writer Lǔ Xùn offered this interpretation of human origins: "Why did men become men, and monkeys remain monkeys? This was simply because monkeys were unwilling to change—they *liked* to walk on all fours."[19] This was the dominant understanding of evolution in China when Peking Man stepped on the scene in the late 1920s, and it remained powerful for many decades to come. Significantly, the same voluntarism was also at play in Chinese interpretations of Marxist theory, which emphasized the power of human will in revolutionary transformation more than the notion that society follows set laws of development.[20] In 1964, Máo was to say: "I don't believe that men alone are capable of having two hands. Can't horses, cows, sheep evolve? Can only monkeys evolve? And can it be, moreover, that of all the monkeys only one species can evolve, and all the others are incapable of evolving?"[21]

As powerful as this sense of progressive voluntarism was, another very different but equally seductive evolutionary paradigm had its own niche in early twentieth-century Chinese books on anthropology.[22] Ernst Haeckel's theory of recapitulationism vividly represented the developmentalist bent of nineteenth-century Western philosophy. It saw in the development of the individual organism a repetition of the evolutionary history of its ancestors. This contributed to a sense of deterministic evolution somewhat at odds with the voluntarism described above. Perhaps even more important, the recapitulationist paradigm also encouraged the dissemination in anthropological texts of examples of atavism (the Chinese, 返祖, literally means "returning to an ancestral [state]"), such as human babies born with tails or with thick body hair. Their primary function in such texts was to reinforce a progressive and deterministic understanding of evolution. They served as evidence that each new organism passed through the primitive stages on its way to more advanced states. Yet examples of atavism could not but also evoke that same slippery boundary between humans and animals that was found in

19. Quoted in James Reeve Pusey, *Lu Xun and Evolution* (Albany: State University of New York Press, 1998), 135. On Wú Zhìhuī's volutarist interpretation of evolution, see Pusey, *China and Charles Darwin*, 386–389.

20. Maurice Meisner, *Li Ta-chao and the Origins of Chinese Marxism* (Cambridge, Mass.: Harvard University Press, 1967).

21. Schram, *Chairman Mao Talks*, 220–221.

22. Frank Dikötter, *The Discourse of Race in Modern China* (London: Hurst and Company, 1992), 138–142.

Chinese classical origin stories and accounts of half-animal, half-human creatures. Similarly, a newspaper article in 1935 described a Californian case of Paget's disease (which can cause bowing of the limbs) and concluded, "While Darwin said that humans evolved from apes, this strange account shows that humans can also revert back into apes."[23] Such latent symbols of degeneration or human bestiality were to survive through the decades until they experienced an explosion of popular interest beginning in the late 1970s.

Nationalism and Internationalism

The commitment to change and fascination with evolutionary theories as guides to such change were inextricable from a concern for China as a nation in an international system defined by conquest and survival. Yán Fù's brand of social Darwinism differed significantly from that of Herbert Spencer and others in the West, for whom the "struggle for existence" took place primarily among individuals. Yán and other writers in China focused instead on the struggle among groups (群), a general term that easily accommodated the more specific idea of nations. Yán found meaning in the words of the ancient Chinese philosopher Xúnzǐ: "Animals cannot organize as groups; only men can organize as groups."[24] While this was not the only difference between humans and other animals emphasized in early twentieth-century China—others included intelligence, consciousness, and the ability to make tools and other objects—it was perhaps the most important.[25] It lent weight to the notion that in their struggle to survive, Chinese people must focus on strengthening the collective body, increasingly understood as the Chinese nation. As one biographer summarizes Yán's position, "Values, institutions, ideas— the whole content of culture—must be judged in terms of one criterion: will it preserve and strengthen the nation-state?"[26]

A 1918 book, *Vista on Human Evolution*, vividly expressed a similar concern for the survival of the Chinese nation. In a section on "inherent

23. "Rén biàn hóu zhī qíwén" [Strange News of a Person Transforming into an Ape], *Běipíng chénbào*, 6 June 1935. Cited also in Frank Dikötter, "Hairy Barbarians, Furry Primates, and Wild Men: Medical Science and Cultural Representations of Hair in China," in *Hair: Its Power and Meaning in Asian Cultures*, ed. Alf Hiltebeitel and Barbara D. Miller (Albany: State University of New York Press, 1998), 57.

24. Pusey, *China and Charles Darwin*, 58. Other important thinkers who highlighted struggle among groups included Liáng Qǐchāo, Zhāng Bǐnglín, and Kāng Yǒuwéi.

25. For some of these other differences, see Guō Mòruò, *Gémìng jīngshén rénlèi jiqiǎo zìrán* [With Revolutionary Spirit, Humanity Constructs Nature] (Shànghǎi: Kāimíng shūdiàn, 1928), 4–5, 17. This appears to be a rare book. It is not available in United States libraries, and the copy in the Chinese National Library is a facsimile.

26. Schwartz, *In Search of Wealth*, 50.

faculties" (本能), the author first described various animals: birds have feathers for flying; beasts have limbs for running; lions and tigers have claws and teeth for fierceness. "Those who attain them will live, and those who do not will die." He then turned not to humans in general, but to "our race [我种民] with [our] feet planted in east Asia." The historical annals related that over the ages, the Chinese had been pressed by outsiders on all sides: "Up until today, our race has not been vanquished, but is able to rise from the ashes to test its power and measure its strength against the other races. This is [the meaning of] inherent faculties."[27]

The integration of biological principles of evolution with the political category of nations inevitably produced powerful and dangerous conceptions about "racial" identity. Discussions of race abounded in early twentieth-century China and drew much force and meaning from the discussions of nations and imperialism in which China, along with most of the rest of the world, was thoroughly entangled.[28] Such notions fed into and in turn were enriched by the growth, beginning in the 1920s, of eugenics (優生學, the "science of superior births"), the subject of a full chapter in the influential 1918 Chinese introductory text *Anthropology*.[29]

Eugenics has a deservedly ugly reputation arising from its frequent implication in racist, classist, and sexist attempts to purify populations at the expense of individual liberties. Nonetheless, it is important to place eugenics theories and movements in their proper historical contexts. Chinese intellectuals and scientists advocated eugenics because they believed it was at least a compassionate alternative to racial extinction, and at best a positive way for China to become a nation of healthier, stronger people. Moreover, solutions were not limited to ones we might term "Mendelian," for example, prohibiting reproduction by "unfit" people and encouraging intermixing with "new blood" of neighboring ethnic groups. Rather, they also included "Lamarckian" solutions for improving prenatal care, education, and the natural and social environment. Above all, eugenicists saw themselves engaged in a "struggle for national survival."[30]

27. Chén Ānrén, *Rénlèi jìnhuà dàguān* [Vista on Human Evolution] (n.p., 1918), 2.

28. Dikötter, *The Discourse of Race*.

29. Chén Yīnghuáng, *Rénlèixué* [Anthropology] (Shànghǎi: Shāngwù yìnshūguǎn, 1923 [1918]), 236–247.

30. Mendelian inheritance is limited to what is present in immutable genes; Lamarckian inheritance includes acquired characteristics. For a very sensitive reinterpretation of eugenics in China and Japan, see Yuehtsen Juliette Chung, *Struggle for National Survival: Eugenics in Sino-Japanese Contexts, 1896–1945* (New York: Routledge, 2002), especially 71–86, 169–171. Among the intellectuals and scientists she cites are Liáng Qǐchāo, Lǔ Xùn, Zhōu Jiànrén, and Pān Guāngdàn.

Eugenics in China, as in some other countries, was largely predicated on a self-diagnosis of racial degeneration.[31] Many Chinese intellectuals accepted the nineteenth-century characterization of China as the "sick man of Asia." Yet degeneration implied a previous state of strength originating in antiquity. In the second and third decades of the twentieth century, Chinese intellectuals approved and recirculated a theory published by a French philologist in 1894 that the origins of Chinese civilization lay in Babylon.[32] Such claims resonated with social Darwinism in the portrayal of the Chinese race as a formerly strong one that had come to China in ancient times and ousted the weaker Miáo people then living there.[33] During the Nánjīng decade of 1928–1937 (the height of Chiang Kai-shek's power and the most unified period of the republican era), the dominant opinion shifted to rooting Chinese civilization in China proper straight back to the Paleolithic era—an opposite position that nevertheless continued to serve nationalist ends.[34]

The influence of nationalism in science went beyond abstract questions of race and prehistory to play an important role in the social relations of Chinese and foreign scientists active in the republican period. For many years, Chinese archaeologists fumed over the actions of foreign scientists in China who removed cultural artifacts for their own research and for display in their nations' museums. With the consolidation of its power in the Nánjīng decade, the Nationalist government gained new means to establish and enforce laws protecting Chinese artifacts.[35] In 1928, the Society for the Preservation of Cultural Objects, with government support, seized the collections of several foreign explorers who had planned to ship them out of the country.[36] Following this dramatic success, in the 1930s a series of laws and regulations strictly governed foreign participation in archaeology.[37]

Nationalism, and indeed the nation itself, does not exist but in a context of internationalism. Chinese reformers in the very late Qīng dynasty

---

31. Chung, *Struggle for National Survival*, 62.

32. Terrien de Lacouperie, *Western Origin of the Early Chinese Civilization* (Osnabrück: Otto Zeller, 1966 [1894]).

33. Fa-ti Fan, "How Did the Chinese Become Native? Science and the Search for National Origins in the May Fourth Era," in *In Search of Modernity: Re-examining the May Fourth Movement*, ed. Kai-wing Chow and Tze-ki Hon (Lanham, Md.: Lexington Books, forthcoming).

34. Fan, "How Did the Chinese"; Guolong Lai, "Digging Up China: Nationalism, Politics and the Yinxu Excavation, 1928–1937" (unpublished paper).

35. Amy Hwei-shuan Feng, "Chinese Archaeology and Resistance to Foreign Participation in the 1920s" (unpublished paper).

36. Fan, "How Did the Chinese." The explorers were Roy Chapman Andrews, Sven Hedin, and Aurel Stein.

37. Feng, "Chinese Archaeology."

learned to conceive of China as a sovereign nation in order to advocate China's interests within the system of international politics. In enacting laws to protect artifacts, "China actually participated in a process of defining national properties and rights in international scientific activities."[38] The goal of strengthening China was to protect it from imperialism, and also to allow it to take its place in the modern world of nations. For intellectuals—and perhaps especially scientists, whose work was so often collaborative—this was often manifested on a personal level. Studying abroad and forming professional relationships with people of other countries, they hoped to engage with their colleagues on equal footing. To do so required not only a cosmopolitan approach but also a strong sense of national identity. Such was the challenge facing Chinese scientists involved in the excavation of Peking Man.

Tradition, Superstition, Science

If China's leading citizens saw it as a country in need of change, science was to play a central role in the proposed transformations. Science represented what was new, modern, liberating, and above all necessary for national survival. So convincing was this idea that the commitment to spreading scientific knowledge transcended the deep political divides of a country engaged in multiple civil wars. It was perhaps inevitable that in identifying science in these ways reformers should find a corresponding negative in the Chinese tradition, as they understood it. The dominant narrative of the early twentieth century in China, centering on the great cultural transformation of the May Fourth Movement (1915–1925), indeed paints the period as one of youthful energy embracing science and overturning tradition and superstition.[39] Yet reformers and revolutionaries were by no means united in a wholesale acceptance of science as a replacement for Chinese tradition. Questions of how long-standing Chinese ideas and values were to fare in the modern world resurfaced repeatedly.[40] And just as Yán Fù had understood evolution in the moral

38. Fan, "How Did the Chinese." John Schrecker made a similar point with respect to Qīng responses to German colonialism in Shāndōng Province. See his *Imperialism and Chinese Nationalism: Germany in Shantung* (Cambridge, Mass.: Harvard University Press, 1971).

39. The May Fourth Movement is named after the great May Fourth Incident of 1919, in which three thousand students at Běijīng University marched in Tiānānmén Square to protest the Treaty of Versailles, which handed the German concessions in Shāndōng Province to the Japanese. The movement as a whole embraced broader cultural, social, and political changes and is perhaps best considered to have lasted from the founding of *New Youth* magazine in 1915 to the May Thirtieth Movement in 1925. The most thorough account of scientism in and around the May Fourth Movement is D. W. Y. Kwok, *Scientism in Chinese Thought, 1900–1950* (New Haven, Conn.: Yale University Press, 1965).

40. Examples include the famous 1923 debate over "science and metaphysics" and the prolonged

terms of Confucian thought, even those science promoters most rabidly opposed to tradition borrowed more than they knew from long-standing Chinese philosophical perspectives.[41]

These enduring ambiguities complicated but did little to stanch the overwhelming social energy behind the movement to make China a "scientific nation" by transforming the knowledge and worldviews of the general population. The notion of superstition (迷信) served as a useful catchall term that denoted the opposite of science. As one Chinese educator neatly put it, "The world of the twentieth century is a scientific world. We need a scientific China for the scientific world. A superstitious nation will find it difficult to survive in a scientific world."[42] Superstition was a powerful and viable concept in large part because it was a nebulous one. For serious antitraditionalists, it could stand in for the whole of premodern Chinese thought.[43] For more selective cultural critics, it could refer, for example, to Daoist mysticism in contrast with the Confucian precept of not concerning oneself with ghosts.[44] For rural reformers and revolutionaries who sought to bring modernity to the Chinese countryside, it could represent specific folk beliefs and practices. For the great pragmatist reformer Hú Shì, what was at issue was a proper critical attitude for evaluating the truth and worth of all ideas, old and new.[45] It was perhaps in a similar spirit that an influential anarchist spoke out against what he saw as an uncritical acceptance of Darwinism then in vogue: "Those who look towards evolutionary progress are as superstitious as those who pray to gods and spirits."[46] These last two examples, however, were exceptional: for most, "superstition" referred unproblematically to any idea or practice that appeared unscientific or otherwise impeded China's entry into the modern world.

---

arguments over the legitimacy of traditional Chinese medicine. Charlotte Furth, *Ting Wen-chiang: Science and China's New Culture* (Cambridge, Mass.: Harvard University Press, 1970); Ralph Croizier, *Traditional Medicine in Modern China: Science, Nationalism, and the Tensions of Cultural Change* (Cambridge, Mass.: Harvard University Press, 1968).

41. Wang Hui, "The Fate of 'Mr. Science' in China: The Concept of Science and Its Application in Modern Chinese Thought," *positions* 3, no. 1 (1995): 1–68.

42. Táo Xíngzhī (1891–1946, like Hú Shì an educator and follower of John Dewey), quoted in Zhōngguó kēpǔ yánjiūsuǒ, ed., *Zhōng wài kēpǔ chuàngzuò bǐjiào yánjiū jiēduànxìng yánjiū bàogào, guónèi kēpǔ chuàngzào bùfēn* [Preliminary Research Report on Research Comparing Chinese and Foreign Science Dissemination Materials, Section on Chinese Science Dissemination Materials] (Běijīng: Zhōngguó kēpǔ yánjiūsuǒ, 2002), 13.

43. Chén Dúxiù is a good example. Kwok, *Scientism in Chinese Thought*.

44. Shěn Tiānháo, "Rén guǐ màn tán," in *Zhōngguó kēxué xiǎopǐn xuǎn, 1934–1949* [Selected Chinese Short Writings on Science, 1934–1949], ed. Yè Yǒngliè (Tiānjīn: Tiānjīn kēxué jìshù chūbǎnshè, 1984), 417.

45. Grieder, *Hu Shih*, 111. Hú Shì was a student of John Dewey.

46. Zhāng Bǐnglín quoted in Pusey, *China and Charles Darwin*, 414.

Intellectuals and political leaders of all stripes throughout the early twentieth century largely agreed on the need to eliminate superstitious beliefs and foster in China an appreciation for "science," sometimes conceived narrowly as scientific research and methodology but more often broadly as a scientific worldview. In the 1930s, in the midst of war and revolution, influential people from across the political spectrum created specific programs designed to bring science to the people. Rural reformers took their dreams for a modern China down to the countryside and included in their comprehensive reform packages efforts to educate rural people about science.[47] Meanwhile, communist activists pursued their own science education projects in the rural base areas. In the "revolutionary cradle" of Yán'ān, they created a Science Popularization Movement (科學大眾化運動), which included a major campaign against superstition in 1944–1945.[48]

In addition to such intensive rural efforts, published materials worked to spread scientific knowledge in urban areas. Throughout the early twentieth century, general-interest magazines like *Eastern Miscellany* (東方雜誌) and *Women's Magazine* (婦女雜誌), and political ones like *New Youth* (新青年) and *New Tide* (新潮), regularly printed articles on scientific topics, but the 1930s saw the first real explosion of books and magazines dedicated to disseminating scientific knowledge. For example, the Science Society of China (中國科學社), founded by Hú Shì and his Chinese classmates at Cornell University, launched a magazine for intellectuals entitled *Science* (科學雜誌) in 1915 but waited until 1933 to publish a magazine for broader audiences, *Popular Science* (科學畫報).

In 1931, the reformer Táo Xíngzhī obtained funding from the owner of *Shēn bào* (one of China's most important newspapers) to found the Science Dissemination Movement (科學下嫁運動, literally "Hand Down Science Movement"). A former student of John Dewey, Táo chose to focus on children's education. Under his leadership, the Science Dissemination Movement published a series of short books for children on scientific subjects—for example, *Exploring the Life of the Praying Mantis*. The proceeds from these publications helped found the Children's Science Correspondence School (兒童科學通訊學校), aimed at both children

---

47. Examples include Yàn Yángchū (James Yen), Táo Xíngzhī, and Liáng Shùmíng. James C. Thomson, *While China Faced West: American Reformers in Nationalist China, 1928–1937* (Cambridge, Mass.: Harvard University Press, 1969); Charles W. Hayford, *To the People: James Yen and Village China* (New York: Columbia University Press, 1990).

48. Yuán Qīnglín, "Dǒng Chúncái kēpǔ chuàngzuò de shíjiàn hé lǐlùn" [The Practice and Theory of Dǒng Chúncái's Science Dissemination Writings], in Zhōngguó kēpǔ yánjiūsuǒ, *Zhōng wài kēpǔ chuàngzuò*, 34; James Reardon-Anderson, *The Study of Change: Chemistry in China, 1840–1949* (Cambridge: Cambridge University Press, 1991), 359–363.

and elementary school teachers. Gāo Shìqí—who later became China's most celebrated popular science writer—contributed to the course materials.[49] Although the Science Dissemination Movement did not have an explicit political affiliation, its participants were in various ways clearly estranged from Chiang Kai-shek's Nationalist Party. The owner of the *Shēn bào* was assassinated in 1934 by Chiang Kai-shek's Blue Shirts. Táo was a vocal critic of Chiang's policies and cooperated with the Chinese Communist Party on a number of projects.[50] Other members of the Science Dissemination Movement, including Gāo Shìqí, moved to Yán'ān in the late 1930s, contributed to the communists' Science Popularization Movement there, and went on to distinguished careers in the People's Republic after 1949.

The Nationalists have often been criticized for neglecting science both in their policies and in their very ideological foundation. However, scholars have brightened the picture a bit by recognizing overlooked efforts and accomplishments.[51] A modest revision of this type is certainly deserved in the field of science dissemination as well. In 1932, the Guómíndǎng ideologue Chén Lìfú spearheaded the Chinese Association for the Scientization Movement (中國科學化運動協會). Known for his conservative cultural perspectives, Chén was exactly the type left-leaning intellectuals saw as an impediment to scientific and cultural progress. Yet Chén gathered more than fifty people, mainly political elites and scientists educated at American universities, to launch the association. Its eleven branches created radio lectures, organized exhibitions, and published several magazines beginning with their core publication *Scientific China* (科學的中國).[52] Although it claimed to target people without

49. Yuán Qīnglín, "Dǒng Chúncái kēpǔ chuàngzuò," 32–33. Some of the Children's Science Correspondence School's activities proceeded under the name Natural Science Garden (自然科學園). The noted educator, writer, and revolutionary Dǒng Chúncái was a key member of the organization.

50. Yusheng Yao, "The Making of a National Hero: Tao Xingzhi's Legacies in the People's Republic of China," *Review of Education, Pedagogy, and Cultural Studies* 24 (2002): 259.

51. William C. Kirby, "Technocratic Organization and Technological Development in China: The Nationalist Experience and Legacy, 1928–1953," in *Science and Technology in Post-Mao China*, ed. Denis Fred Simon and Merle Goldman (Cambridge, Mass.: Council on East Asian Studies, Harvard University, 1989); J. Megan Greene, "GMD Rhetoric of Science and Modernity (1927–70): A Neo-traditional Scientism?" in *Defining Modernity: Guomindang Rhetorics of a New China, 1920–1970*, ed. Terry Bodenhorn (Ann Arbor: Center for Chinese Studies, University of Michigan, 2002).

52. Péng Guānghuá, "Zhōngguó kēxuéhuà yùndòng xiéhuì de chuàngjiàn, huódòng jíqí lìshǐ dìwèi" [The Establishment, Activities, and Historical Significance of the Chinese Association for the Scientization Movement], *Zhōngguó kējì shǐliào* 13, no. 1 (1992): 60–72; Huáng Dàoxuàn and Zhōng Jiàn'ān, "1927–1937 nián Zhōngguó de xuéshù yánjiū" [Chinese Academic Research, 1927–1937], *Jìndài Zhōngguó yánjiū*, 30 March 2006, http://jds.cass.cn/Article/20060330092444.asp (viewed 2 August 2006).

formal educational opportunities, *Scientific China* assumed quite a bit of knowledge and was not visually inviting. Nevertheless, the lead article expressed the importance of science dissemination work in no uncertain terms: "Bringing scientific knowledge to the common people [民間]" was a necessary component of strengthening them and helping the Chinese ethnicity and culture survive the "life and death" crisis they faced.[53] Thus, cultural nationalism did not preclude commitment to science dissemination; it just offered a slightly different rationale for it.

In 1934, a new popular science magazine unabashedly opposed to Chiang Kai-shek's conservative regime emerged. Chén Wàngdào organized others in the League of Left-Wing Writers to create the intellectual but highly vernacular magazine *Tàibái* (太白), which published short pieces on scientific subjects along with essays on the art of popular science writing itself. The name of the magazine reflected its founders' political agenda. *Tàibái* denoted the planet Venus, signifying the dawning of a new day after the dark of Nationalist rule. Moreover, in the face of Nationalist calls to return to the old, literary Chinese writing language, *Tàibái* was a pun that meant "extremely vernacular." The magazine was thus a resurrection of the May Fourth spirit, which had promoted (and achieved) the widespread adoption of the vernacular in Chinese writing, and which moreover had celebrated science and democracy as twin liberators from the evils of the old society.[54]

Language and audience were thus political issues in addition to practical ones. Some who wrote science dissemination materials acknowledged that they too often served only as "little dainties" to be consumed "over a few cups of tea and glasses of wine"—a trope for an elite leisure activity.[55] As with science educators in so many other times and places, the struggle was to resist the urge to write interesting prose accessible only to a privileged few, and to focus instead on the more important task of "eradicating bit by bit the traditional, antiscientific spirit and

53. "Zhōngguó kēxuéhuà yùndòng xiéhuì fāqǐ zhǐ qù shū" [The Establishment and Purpose of the Chinese Association for the Scientization Movement], *Kēxué de Zhōngguó* 1, no. 1 (1933): 1–3. See also Laurence Schneider, *Biology and Revolution in Twentieth-Century China* (Lanham, Md.: Rowman and Littlefield, 2003), 95–96.

54. Máo Dùn, cited in Liú Wéimín, *Kēxué yǔ xiàndài Zhōngguó wénxué*, 159. Liú provides an entire chapter on the "Tàibái phenomenon," which he considers enormously significant in the history of science writing not only because of its literary merits but also because it spanned such boundaries as literature versus science, left-wing writers versus democratic writers, and so on (p. 178). He also provides analysis of several of the other magazines mentioned below, with particular attention to their literary qualities (pp. 179–203).

55. Chéng Shíkuǐ, "Zěnyàng shǐ mínzhòng kēxuéhuà" [How to Make the People Scientific], *Kēxué huàbào* 1, no. 10 (1937); and Liǔ Shí, "Lùn kēxué xiǎopǐn wén" [On Science Writing], originally published in a 1934 issue of *Tàibái*, collected in Yè Yǒngliè, *Zhōngguó kēxué*, 750.

thoughts from the brains of the people," in other words of "stamping out popularly held antiscientific superstitions."[56]

Like *Scientific China*, *Science World* (科學世界) and *Scientific Life* (科學生活) were written for relatively well educated people and had limited popular appeal. Another magazine, also called *Scientific Life* (科學生活), similarly assumed a fairly high level of education but used lively images and a more friendly writing style to draw readers in. *Wonders of the Universe* (宇宙奇觀) was more self-consciously popular and included articles on such subjects as strange animals, the social organization of bees, the effects of sunspots on humans, the world's largest shovel, and H. G. Wells's science fiction. At least one magazine, *Everyday Science* (家常科學), targeted children and focused on providing scientific explanations for commonly encountered phenomena.[57]

Far more dedicated than any of these to popularizing science on a truly mass level, however, was a socialist magazine called *Mass Science* (大眾科學), which began publication in 1946. The magazine emphasized articles related to agriculture, provided concrete and useful information, and catered to people with limited education. The same year, another socialist magazine titled *Science and Life* (科學與生活) briefly appeared in the wartime Nationalist capital of Chóngqìng. The editors (members of an underground communist cell) proclaimed their mission to be "spreading some seeds of science across this scientifically backward land," and they declared, using Maoist language, their intention to strive for increasingly "popular" (通俗) and "lively" (生動) writing to suit their audience.[58]

This was the world into which Peking Man emerged. Chinese intellectuals of the late nineteenth and early twentieth centuries were painfully aware of China's weak position internationally and the need for China to change in order to survive. Evolutionary theories that emphasized progressive, unidirectional change appeared to offer guidance in this respect. The teachings of social Darwinism, combined with the realities of imperialism, suggested that to change meant first of all to become a nation, and second to become a *scientific* nation. Yet this attempt at transformation was fraught with enormous challenges. What place

56. Chéng Shíkuǐ, "Zěnyàng shǐ mínzhòng kēxuéhuà."

57. *Science World* began in 1932 and was published by the Nánjīng Society for the Natural Sciences. *Scientific Life* began in 1936 and was published in Běipíng. *Scientific Life* began in 1935 and was published in Shànghǎi. *Wonders of the Universe* began in 1936, and *Everyday Science* in 1937.

58. "Jīnhòu de běnkān" [This Magazine's Future], *Kēxué yǔ shēnghuó* 1, no. 8 (1946), inner cover. The editors included Chén Rán, Jiàng Yīwěi, and Liú Róngzhù, and the publishers were the sympathetic Life Books, Reading Press, and New Knowledge Books, which merged two years later to become Life, Reading, and New Knowledge Books (生活、讀書、新知書店). Gān Lí, "Wú Yùzhāng hé 'Lùn qìjié'" [Wú Yùzhāng and "On Morals"], *Hóng yán chūnqiū*, no. 3 (2004): 14.

would there be in modern China for Chinese traditions? How could the Chinese population be transformed from a superstitious to a scientific one? While the first question was never resolved, and the second never adequately addressed, both consumed the work of many of China's most prominent citizens.

First Contacts

The late 1920s were event-filled years for China. After cooperating with the Chinese Communist Party in a united front against the warlords and imperialists, in 1927 Chiang Kai-shek turned on his erstwhile allies and sent the communists scrambling for a foothold in the countryside. Nineteen twenty-eight marked the beginning of the Nationalist Decade but not the end to Chiang's struggles to maintain control over what was still a land bubbling with strife and military opportunism. For the next ten years, Chiang would focus on exterminating his communist rivals while the Japanese imperialist threat mounted ever higher and warlords continued to rule outlying provinces. Only in 1937 would he agree to a second united front with the communists, but by then the war with Japan had reached new levels.

Amid the turmoil and contest, a series of less deadly but nonetheless historic explosions in the hills southwest of Běijīng rocked the international scientific community. In 1926, scientists announced that three teeth discovered among fossils excavated in the mining community of Zhōukǒudiàn were human; they named the finds "Peking Man." Around the world, eyes turned toward China to see what these clues would reveal about the origins of humanity.

Peking Man—then *Sinanthropus pekinensis*, now *Homo erectus pekinensis*—emerged into a scientific field full of theories, but hungry for empirical evidence, about where, how, and from what humans had evolved. Nineteenth-century writers on human evolution based their claims more on comparative anatomical and embyrological studies and on general principles derived broadly from natural history than on fossil evidence, which was relatively scarce and difficult to interpret. The discovery of Neanderthal fossils was first announced in Europe in 1857, but not until the turn of the twentieth century did they become fairly well accepted as links between apes and humans.[59] In the early 1890s, the Dutch scientist Eugene Dubois discovered human fossils of a much more

59. This paragraph is largely drawn from Peter Bowler, *Theories of Human Evolution: A Century of Debate, 1844–1944* (Baltimore: Johns Hopkins University Press, 1986), especially 31–40.

primitive type in Java, where he had traveled following Haeckel's suggestion that humans had evolved from Asian apes. (Here Haeckel differed with Darwin, who saw Africa as the most likely place of human origins.) The fossils, called *Pithecanthropus erectus* (literally, the ape-human that stands upright), or "Java Man" in the vernacular, were deeply controversial, with some scientists insisting that the bones belonged to modern humans and others convinced they were of apes. Adding to the confusion were the Piltdown fossils, which were reported in 1912 and only in 1954 proved to be fraudulent, doctored fragments of an ape jaw and a modern human skull. "Piltdown Man" conveniently provided powerful evidence for the contemporary theory that the enlargement of the brain had been the first important transformation in the emergence of the human species. Nonetheless, the fossils' inconsistencies left much room for doubt, and by the 1920s they were considered "simply too controversial for anyone to feel comfortable using them as important evidence for the course of human evolution."[60]

The Peking Man fossils, in contrast, were relatively free from controversy. They bore striking resemblances to the Java Man fossils but were different in certain key ways. Perhaps most important, Peking Man's remains were discovered in a cave, while the bones of Java Man were found scattered in an alluvial deposit, leading some critics to question whether they belonged to the same individual or even the same species.[61] Here was new evidence of primitive humans without many of the problems posed by Dubois's find. And with the generous international funding generated by the enthusiasm over the first few teeth, the collection grew to include the remains of dozens of individuals, numerous stone tools, and evidence of the control of fire—truly, in the words of one foreign scientist, an "embarrassment of riches."[62]

Further contributing to Peking Man's ready acceptance among scientists internationally, the appearance of the Peking Man fossils coincided with a spreading conviction among scientists that Asia rather than Africa was the site of early human evolution.[63] The theory of Asian origins

60. Bowler, *Theories of Human Evolution*, 37. Nonetheless, professional and especially popular literature still often discussed Piltdown Man.

61. Grace Shen has argued in an unpublished paper that it was Peking Man's "home," that is, the existence of the cave itself as a likely residence, that provided the necessary support both for the acceptance of the Peking Man fossils as scientific evidence and for the concentration of resources for the intensive study of that evidence. Grace Shen, "Mining the Cave: Global Visions and Local Traditions in the Story of Peking Man" (paper presented at the annual meeting of the History of Science Society, Vancouver, 3 November 2000).

62. Roy Chapman Andrews, "An Embarrassment of Riches," in *Leader Reprints*, no. 51 (Peiping [Běijīng]: Leader Press, 1930), 19–21.

63. This issue was different from the current debate over the recent out-of-Africa hypothesis,

inspired a number of well-positioned scientists to organize excavations in China and Mongolia.[64]

The discovery of Peking Man did not happen in one day, or even in one year. Discoveries seldom do. Rather, a host of significant encounters and unearthings accumulated gradually over years of dedicated work, leading to the naming of Peking Man as *Sinanthropus pekinensis* in 1927 and the celebrated finding of the first complete skullcap in 1929. The steps in this process involved long-established Chinese forms of knowledge in addition to modern science as practiced by Westerners and, increasingly, Chinese people as well. The story begins with dragon bones.

In 1899, a German naturalist, prevented by the Boxer Uprising from exploring China's interior as he had intended, encountered a number of curious objects called "dragon bones" (龍骨) and "dragon teeth" (龍牙) in apothecary shops in Běijīng and other accessible cities. Upon returning to Germany, he delivered them to a paleontologist in Munich, who identified them as the fossilized bones of various mammals. In the words of the Swedish geologist and archaeologist J. Gunnar Andersson, "The veil of mysticism over these objects was removed by the exhaustive and perfected investigations of the learned Munich paleontologist, while at the same time these once so mysterious objects gained greatly in scientific interest."[65] Thus began a long and productive, though not unproblematic, relationship between dragon bones and modern science, and so also between scientists in the international scientific community and the local Chinese people who knew where to find what they sought. Out of these exchanges came the discovery of the Peking Man fossils.[66]

The history of fossil collecting is strikingly different in China compared with Europe because these objects have long been valued for a reason other than science or aesthetics. Dragon teeth and dragon bones

---

discussed fully in chapter 8. The recent out-of-Africa hypothesis deals with the origins of modern *Homo sapiens*, while the question discussed here is the much earlier evolution of the first humans from apes.

64. For example, it strongly influenced Davidson Black in his decision to move to China in 1919, and it lay behind the Central Asiatic Expeditions to Mongolia led by Roy Chapman Andrews in 1921.

65. J. Gunnar Andersson, *Children of the Yellow Earth* (London: Kegan Paul, 1934), 76. Western naturalists began collecting fossils in China at least by the mid-nineteenth century. See Fa-ti Fan, *British Naturalists in China, 1760–1910* (Cambridge, Mass.: Harvard University Press, 2004), 111.

66. The story of the discovery of Peking Man has been told in many places. See Andersson, *Children*; Harry L. Shapiro, *Peking Man* (New York: Simon and Schuster, 1974); Noel T. Boaz and Russell L. Ciochon, *Dragon Bone Hill: An Ice-Age Saga of Homo Erectus* (New York: Oxford University Press, 2004); and three versions of a book by Jiǎ Lánpō and Huáng Wèiwén, a Chinese version and an English version published in mainland China and an illustrated edition in Chinese published in Táiwān (see bibliography under "Other Primary Sources in Asian Languages" and "Other Primary Sources in English").

have been part of Chinese pharmaceutics probably since the first century A.D. The oldest Chinese medical text, the *Divine Husbandman's Materia Medica* (神農本草經, circa A.D. 100), discussed using dragon teeth to treat various spasms and convulsions. A fifth-century herbal reference recommended pulverized dragon bones as a remedy for strengthening the kidneys.[67] Today, dragon bones are still used in many parts of China—for example, as a topical treatment on cuts. Their widespread use for a great variety of ailments is worthy of the powerful associations they have with dragons—bringers of rain, masters of rivers, and symbols of imperial authority.

The importance of dragon bones to Chinese people has been not only medical, but also economic. People often encounter dragon bones in the course of their regular agricultural and industrial labor, and in places with especially rich deposits, local people have established elaborate mining operations specifically to extract dragon bones. Whether or not they believe them to be literally the bones of dragons, the cash obtained from the sale of dragon bones to apothecaries is always a welcome addition to a family's income.[68]

Chinese apothecaries offered easy access to dragon bones for foreign scientists eager to chart China's natural history. But fossils obtained in this manner were limited in what they could tell the scientists. The pharmacists did not know precisely where they had been collected, in what strata they were found, or what other fossils had lain beside them. For such information—critical in the determination of age, not to mention as clues for further excavation—scientists had to move upstream to the people, typically farmers and other rural laborers, responsible for excavating the dragon bones and selling them to apothecaries or wholesalers. Such was the case in the discovery of Peking Man.

In 1918, J. Gunnar Andersson heard from another foreign scientist

---

67. On dragon bones, Lǐ Shízhēn's *Classified Materia Medica* (本草綱目, completed in 1578) cites the text *Bié lù*, compiled by Liú Xiàng (77?–6 B.C.). It is unclear from this text, however, whether the dragon bones served medical purposes. The *Classified Materia Medica* further cites Léi Xiào's fifth-century *Duke Léi's Treatise on the Processing of Herbs* (雷公炮炙論). Lǐ Shízhēn, *Běncǎo gāngmu* [Classified Materia Medica] (Shànghǎi: Shànghǎi gǔjí chūbǎnshè, 1991 [1578]), 3:255–256. The *Divine Husbandman's Materia Medica* probably dates to circa A.D. 100, but the original text was lost. The oldest extant version includes commentary by the third-century scholar Wú Pǔ.

68. Andersson, *Children*, 82: "Workmen in these mines have a very good idea of the real nature of the skulls which they bring to light. The resemblance of a *Hipparion* to that of a horse is sufficiently striking for the nature of the animal to be approximately clear to them. . . . On the spot, therefore, it is fairly well known that, at any rate in part, there is no question of dragons, but only of recent mammals. But the teeth are struck out of the skulls and the sick Chinaman who buys from the chemist in his native town . . . is assuredly convinced that he is enjoying the help of his revered patron, the dragon."

about "Chicken Bone Hill" in the mining town of Zhōukǒudiàn, about thirty miles southwest of Běijīng. As recounted in the opening pages of the introduction, scientists began excavating there but in 1921 learned from a local man of the far richer trove of dragon bones in a nearby limestone fissure. Convinced that there was a strong possibility the site would yield hominid fossils, Andersson instructed a junior colleague, Otto Zdansky from Austria, to continue excavating, and in 1926 a scientist in Sweden identified two of the teeth Zdansky had collected as *Homo* specimens. The finds, informally called "the Peking Man," were not immediately recognized by all as unequivocally human; one scientist suggested they might belong to a carnivore. A colleague later asked Andersson slyly, "How are things just now with the Peking man? Is it a man or a carnivore?" Andersson's celebrated reply was, "Our old friend is neither a man nor a carnivore but rather something half-way between the two. It is a *lady*." For several months, the joke stuck, and the finds took on the name "Peking Lady" in English.[69]

Despite some doubts, the finds had enough potential to spur a cooperative arrangement among the Geological Survey of China (founded by geologist Dīng Wénjiāng), the Peking Union Medical College, and the Rockefeller Foundation to increase greatly the scope of the excavations, an arrangement that led to the formal establishment in 1929 of the Cenozoic Research Laboratory (新生代研究室) under the Geological Survey of China. It was agreed that Dīng would be in charge of the project and that the Geological Survey, and thus the Chinese nation, would retain property of all materials excavated.[70] In 1927, a Swedish scientist uncovered a third tooth, whose excellent preservation was enough to confirm that the three teeth together belonged to an early human species that could now be officially named *Sinanthropus pekinensis* (which might be translated as "China Man of Peking"). By this time, a growing consensus among Chinese and foreign scientists alike viewed Zhōukǒudiàn as the most promising site for researching human evolution in the world.

Who Discovered Peking Man?

In the following years, Zhōukǒudiàn became a site remarkable not only for the pursuit of knowledge about human origins, but also for the coming together of several distinct cultures. Western scientists from

69. Andersson, *Children*, 105–106. For better or worse, this joke does not work in Chinese, which lacks a term like "man" that is both gendered and used to refer to humanity as a whole.
70. Andersson, *Children*, 106.

many countries, Chinese scientists and budding scientists, Chinese technicians, and local Chinese workers all had a hand in the unearthing of Zhōukǒudiàn's "embarrassment of riches." This international and socially diverse group's efforts were all the more extraordinary for having taken place during a period of political and military upheaval: work was suspended for only three months in 1927 during the worst of the fighting between Chiang Kai-shek and his warlord allies on the one hand and the warlord Zhāng Zuòlín on the other. In 1937, the Japanese invasion brought a close to the excavations, and America's entry into the war against Japan in 1941 saw the disappearance of all the Peking Man fossils thus far excavated and virtually the end of the laboratory work in Běijīng until 1946.

Set in the swirling political currents of late republican China, Zhōukǒudiàn was a place where disparate social actors nevertheless managed to work together with extraordinary results for science and for the Chinese nation. This is not to say, however, that cooperation was always easy or that all participated on equal terms. The complex relationships among these actors would be examined and reexamined in the light of future political developments for decades to come.

Chinese scientists in the early twentieth century groped for footing in two widely separated worlds. Most were very highly educated by Chinese standards of the time but also painfully aware of how low those standards were. To be considered truly well educated, one had to seek a diploma overseas. China's insecurity as a nation thus played out in the insecurity of Chinese scientists among their foreign colleagues. At the same time, while they often faced economic instability along with the rest of the country, their advanced education, knowledge of foreign languages, and associations with Western scientists separated them from the vast majority of the Chinese population.

Of the Chinese scientists, the geologist Dīng Wénjiāng (V. K. Ting) is the most widely known in the West, as he was one of the most prominent scientists of the republican era as well as an important political figure.[71] He served as the director of the Geological Survey until 1926, when the geologist Wēng Wénhào (W. H. Wong) took over. Yáng Zhōngjiàn (C. C. Young) and Péi Wénzhōng (W. C. Pei) served as field directors; both became key figures in paleoanthropology in the late republican and socialist eras. Yáng (1897–1979) was born in Shǎnxī, graduated from the geology department of Běijīng University in 1922, and went on to obtain

---

71. See Furth, *Ting Wen-chiang*. Since the story of the excavation of Peking Man is told elsewhere, I will focus only on those people who will appear again in the book.

a Ph.D. in paleontology from the University of Munich in 1927. After the revolution, he became the director of the Institute of Vertebrate Paleontology and Paleoanthropology (successor to the Cenozoic Research Laboratory) and was elected an academician in the Chinese Academy of Sciences in 1955.[72]

Born in Héběi Province, Péi Wénzhōng (1904–1984) graduated from the geology department of Běijīng University in 1927, five years after Yáng. Péi's work for the Cenozoic Research Laboratory at Zhōukǒudiàn would undoubtedly have been the beginning of a prestigious scientific career in any case, but his prominence in the discovery of the first Peking Man skullcap in 1929 sealed this fortunate fate. In 1935, he was invited to attend the University of Paris, where he obtained his doctoral degree.[73] After the communist revolution in 1949, in addition to his work at IVPP, he served in the All-China Association for the Dissemination of Scientific and Technological Knowledge (中華全國科學技術普及協會) and as the director of the Běijīng Natural History Museum. He was elected to the Academy in 1955.

Like Péi, Jiǎ Lánpō (1908–2001) was an archaeologist from Héběi who got his start at Zhōukǒudiàn and went on to a prominent career at IVPP. Jiǎ, however, took a very different professional road. He was twenty-one when he graduated from secondary school in 1929, and he never pursued postsecondary formal education of any kind, not to speak of education abroad. Nevertheless, he greatly enjoyed natural science and geography, interests cultivated through subscription to such magazines as *China Traveler* (旅行家). In 1931, Jiǎ learned from Péi Wénzhōng, who managed a store in Běijīng owned by a relative of Jiǎ's, that the Geological Survey was hiring. Jiǎ went for an interview and was one of two people selected. When Jiǎ arrived at Zhōukǒudiàn, he began learning through practice how to excavate and identify fossils and artifacts. His skills quickly won him the respect of the resident scientists, including Yáng Zhōngjiàn and the German anatomist Franz Weidenreich. After 1949, he authored a great many professional and popular books and articles as a paleoanthropologist at IVPP. He was elected to the Academy in 1980.[74]

72. Wáng Hénglǐ et al., eds., *Zhōngguó dìzhì rénmíng lù* [Biographical Dictionary of Chinese Geology] (Wǔhàn: Zhōngguó dìzhì dàxué chūbǎnshè, 1989), 127. See also Lǐ Èróng, ed., *Yáng Zhōngjiàn huíyì lù* [The Memoirs of Yáng Zhōngjiàn] (Běijīng: Dìzhì chūbǎnshè, 1983).

73. Wáng Hénglǐ et al., *Zhōngguó dìzhì*, 251–252. See also Liú Hòuyī and Liú Qiūshēng, *Fāxiàn Zhōngguó yuánrén de rén, Péi Wénzhōng* [Péi Wénzhōng, the Person Who Discovered Peking Man] (Kūnmíng: Yúnnán rénmín chūbǎnshè, 1980). Péi studied under Henri Breuil.

74. Jiǎ Lánpō, interviews with Dennis Etler, 15 September 1999 and 29 September 1999; "Kēxuéjiā zhuànjì dà cídiǎn" biānjí zǔ, ed., *Zhōngguó xiàndài kēxuéjiā zhuànjì* [Biographies of Modern Chinese Scientists], 6 vols. (Běijīng: Kēxué chūbǎnshè, 1991–1994), 2:334–335.

Beginning in the 1930s, both Jiǎ and Péi supplemented their salaries by writing articles for newspapers and magazines.[75] Like many of their contemporaries, they saw themselves not narrowly in terms of their scientific work alone, but as intellectual contributors to China's changing society. Péi even wrote a short story, published in a newspaper supplement, that drew the attention of China's most influential writer, Lǔ Xùn.[76] A sharp awareness of the political struggles that rocked China during this period further drove Péi's extrascientific endeavors. As a student at Běijīng University, Péi had participated in political activities organized by the early communist leader Lǐ Dàzhāo.[77] In 1947, he wrote an article on Nationalist campaigns to "mop up" (掃蕩) communists in his home village. The article, which condemned the campaigns for actually targeting the common people, was printed in the newspaper Dàgōng bào, and an excerpt was adapted and printed in the American magazine Time.[78]

In addition to such intellectual and political reasons for writing in popular forums, Péi and Jiǎ were also compelled by their need to make ends meet. When Jiǎ first met the geologist Wēng Wénhào, then director of the Geological Survey, Wēng asked him why he wanted to do this kind of grueling work. Jiǎ answered, "I have to eat!" to which Wēng replied, "Excellent, that's the truth."[79]

The need to eat was a much greater incentive for local people who worked as day laborers at the site. Depending on the stage of the work, the project employed anywhere from ten to more than one hundred such workers. Their pay was five or six jiǎo (dimes) per day. According to Péi's 1934 account, the usual employment around Zhōukǒudiàn was far less attractive. Péi described Zhōukǒudiàn's two different industries, the products of which dirtied local workers with two distinct colors. Those who excavated limestone were white, while those who excavated coal were black. The limestone quarriers were mainly local people, and they made an average of one and a half to three dimes a day. The coal

---

75. Péi Shēn, interview with the author, 29 October 2002; Péi Wénzhōng, "Fǎguó shǐqián yízhǐ tànfǎng jì" [Notes on a Visit to a Prehistoric Site in France], Lǚxíng zázhì 10, no. 6 (1936): 45–68; and Jiǎ Lánpō, "Yóu Jiǔquán dào Jīntǎ" [From Jiǔquán to Jīntǎ], Xībĕi tōngxùn 3, no. 9 (1948): 24–27.

76. Péi's interest in literature continued in his later years. In 1962, he was inspired by his experiences exploring caves to write a beautiful essay for the magazine People's Literature on the compatibility of scientific and poetical views of the natural world. Péi Wénzhōng, "Shāndòng de tànchá hé shīrén de huànxiǎng" [Cave Exploration and the Poet's Imagination], Rénmín wénxué, no. 8 (1962): 72–74.

77. "Kēxuéjiā zhuànjì dà cídiǎn" biānjí zǔ, Zhōngguó xiàndài kēxuéjiā zhuànjì, 4:281.

78. Péi Wénzhōng, "Jīnrì zhī xiāngcūn" [The Village Today], Dàgōng bào, 23 May 1947, continued on 24 May 1947; "Mopping Up the People," Time,14 July 1947, 29–30.

79. Jiǎ Lánpō, interview with Dennis Etler, 15 September 1999.

miners, on the other hand, were brought in from the northern part of the province. They or their families were given a certain amount of money up front, and then given the rest after they had worked a year. They made only forty or fifty yuán (dollars) per year and were virtually slaves: if caught running away they were beaten to within an inch of their lives.[80] While difficult and sometimes dangerous, employment at Dragon Bone Hill generated significantly better income and, for some at least, was interesting work. The research also undoubtedly benefited from this arrangement: many of the skills used in mining—from digging to detonating dynamite—transferred readily to scientific excavation.

Between the scientists and workers in rank were the technicians. Jiǎ Lánpō and Péi Wénzhōng themselves began their employment at Zhōukǒudiàn in this capacity. The dozen or so technicians earned twelve to eighteen yuán per month. Unlike the workers, who were required to move far away whenever a fossil was located, the technicians were trusted with the delicate work, and they were largely responsible for excavating fossils under the direction of the scientists.[81] In addition to Jiǎ and Péi, at least two other technicians went on to rewarding careers with IVPP after the revolution. One was Wáng Cúnyì. Wáng's thorough work and devotion to archaeology over the decades earned him a reputation as "China's notable expert collector and excavator of fossils."[82] He has also been acclaimed for his sculpted reconstructions of humans and animals from bones, both fossil and modern. The other was Liú Yìshān. Liú was hired after the revolution to provide interpretation to people visiting the exhibition hall, and he gained much recognition for his talents in this area.[83]

The Chinese participants in the early Zhōukǒudiàn excavations were thus a diverse group and brought rich sets of experiences and disparate worldviews to the work. While scientists with their commitment to modern scientific thought dominated, local people with their all-too-intimate experiences in mining and their knowledge of dragon bones also made their marks on the site. Moreover, these worlds did not remain separate. Rather, some workers became wedded to the scientific project, and the experiences and knowledge of local workers contributed to the scientific culture of Zhōukǒudiàn. Scientists were aware, even painfully

80. Péi Wénzhōng, Zhōukǒudiàn dòngxué, 12, 4–5.

81. Péi Wénzhōng, Zhōukǒudiàn dòngxué, 11–12.

82. Jiǎ Lánpō and Zhēn Shuònán, Qiān lǐ zhuīzōng liè huàshí [A Thousand-Mile Trek in Search of Fossils] (Tiānjīn: Tiānjīn kēxué jìshù chūbǎnshè, 1981), 187.

83. Visitors to the Peking Man Site Exhibition Hall in 1956 frequently expressed appreciation for Liú's interpretive services.

aware, of the plight of local workers; the excavations were intimately connected with local mining experience; and the scientists named the site Dragon Bone Hill, which honored the local knowledge of dragon bones that first led Andersson and the others to the fossil trove.

Zhōukǒudiàn was also a meeting place and even a temporary home for many of the biggest Western names in the related sciences of geology, paleontology, paleoanthropology, and archaeology. These scientists hailed from institutions to which only the most fortunate and committed Chinese could aspire. Their Western credentials gave them an automatic authority in academic circles in China. As time went on, however, they increasingly found themselves compelled to meet their Chinese colleagues partway.

J. Gunnar Andersson and several other Swedes were supported by the crown prince of Sweden, himself an amateur archaeologist. From France came not only Henri Breuil but also the Jesuit priest Pierre Teilhard de Chardin, famous for his writings on science and religious philosophy in addition to his influential research on the Piltdown Man fossils. Beginning in 1923, Teilhard had conducted fieldwork in the Ordos region of Inner Mongolia, and he later excavated in the Níhéwān Basin of Héběi; in both areas he had support from missionary-established churches.

Dearest to the hearts of the Chinese colleagues was Davidson Black (1884–1934), known for his dedication to the Peking Man research and his respect for China and for Chinese scientists. Black was a Canadian who as a young man had worked for the Hudson Bay Company shipping supplies by canoe. He first studied medicine and anatomy but became fascinated by questions of human origins. Convinced that humans had first evolved in Asia, Black accepted an invitation to join the anatomy department of the Peking Union Medical College in 1919 and soon began collaborating on Andersson's research. He became deeply involved in the Zhōukǒudiàn excavations in 1926 and had the honor of naming *Sinanthropus pekinensis* in 1927. When he died in 1934 from congenital heart disease, he was at his desk in Běijīng with Zhōukǒudiàn fossils lying before him.[84]

The man chosen to replace Black was the anatomist Franz Weidenreich (1873–1948), a Jew who had fled Germany and taken a position as a visiting professor at the University of Chicago in 1934. Weidenreich's detailed and meticulous descriptions of the growing collection of human fossils from Zhōukǒudiàn remain highly valued as critical sources for

84. Jia Lanpo and Huang Weiwen, *The Story of Peking Man: From Archaeology to Mystery*, trans. Yin Zhiqi (Beijing: Foreign Languages Press, 1990), 1.

the continued study of Peking Man. His identification of morphological similarities between Peking Man and modern Chinese people has also continued to play an important role in the work of Chinese paleoanthropologists today.

The story of the Chinese scientists and their foreign colleagues has been told and retold in different ways according to shifting political priorities. Neither the Máo-era attacks on Zhōukǒudiàn as a "capitalist stronghold of cultural aggression" in which foreigners ruthlessly dominated scientific work nor the effusive descriptions of international cooperation circulated in some post-Máo and foreign accounts do justice to the complex relationships formed between individuals and the nations they inevitably represented.[85]

Personal friendships and close mentoring relationships broke down cultural and national barriers. In later years, Jiǎ Lánpō felt great fondness for and gratitude to many of the foreign scientists, from whom he learned both English and Latin. Weidenreich was Jiǎ's greatest scientific influence, and Jiǎ remembers with appreciation that in a scientific paper and a series of newspaper interviews, the German scientist explicitly gave full credit to Jiǎ for the reconstruction of the three skulls on which the reports were based.[86]

However, other evidence suggests that a certain discomfort between foreign and native scientists lingered below the surface, never fully dissipated. In a eulogy for Davidson Black, Dīng Wénjiāng touched on what he acknowledged to be a "delicate" issue:

It is frankly admitted that sometimes we find cooperation between Chinese and foreigners in scientific work rather difficult. . . . Firstly, many foreigners are suffering from a superiority complex. Subconsciously they think somewhat like this: here is a Chinese, he knows something about science, but he is a Chinese nevertheless—he is different from a European, therefore we cannot treat him in the same way. At best his manners become patronizing. On the other hand, their Chinese colleagues are suffering from an inferiority complex. They become self-conscious and super-sensitive, always imagining that the foreigner is laughing at them or despising them.

85. Jiǎ Lánpō complained about the Cultural Revolution–era "stronghold of aggression" perspective in Jiǎ Lánpō and Huáng Wèiwén, *Zhōukǒudiàn fājué jì* [The Excavation of Zhōukǒudiàn] (Tiānjīn: Tiānjīn kēxué jìshù chūbǎnshè, 1984), 2. Ironically, Jiǎ Lánpō himself was one of the writers who popularized this conception. For a portrayal of "international cooperation," see Li Chi, *Anyang* (Seattle: University of Washington Press, 1977), 44.

86. Jiǎ Lánpō, interview with Dennis Etler, 15 September 1999; "Zhōukǒudiàn xīn fāxiàn rényuán [sic., yuánrén] huàshí" [New Anthropoid Ape [sic., *Pithecanthropus*] Fossils Discovered at Zhōukǒudiàn], *Běipíng chénbào*, 25 November 1936, 9.

Black, Dīng went on to say, was free from such a superiority complex and inspired his Chinese colleagues to lose their insecurities. "In his dealings with his Chinese colleagues, he forgot altogether about their nationality or race."[87]

Before beginning collaborative work at Zhōukǒudiàn, Chinese scientists worked with skill and determination to reach a fair agreement, including the crucial provision that all excavated materials remain in China. The document, drafted by Wēng Wénhào and signed by all parties in 1927, came in the midst of the public controversy over the Western scientists who were shipping large numbers of artifacts and fossils out of the country.[88] Just as the Society for the Preservation of Cultural Objects and the Nationalist government successfully developed and enforced laws to place cultural artifacts under Chinese control, Wēng Wénhào and others worked to ensure fair play at Zhōukǒudiàn.

Some foreigners also made significant efforts to accommodate and even prioritize Chinese concerns. This included being willing to work under Chinese direction despite the criticism they received from other foreign scientists.[89] And in a letter Black wrote to Wēng Wénhào as part of the process of drafting the formal agreement, he assured his colleague, "Our desire is to frame a proposal that will not merely be acceptable to Chinese scientists but will embody what they themselves wish to see carried out. It will then be our part to see whether we can get the necessary funds."[90]

The areas of cooperation and tension between Chinese and foreign scientists on the one hand, and scientists and local workers on the other, are nowhere better illustrated than in the discovery of the first Peking Man skullcap in 1929 and the way this story has been reworked over the years. The selection of 2 December 1929—when Péi Wénzhōng's team uncovered the first Peking Man skullcap—as the date of the discovery of Peking Man itself, and even "the beginning of the study of China's Paleolithic cultures by Chinese," has resulted in "both the elevation of the role of Chinese and the downgrading of that of foreigners in the Chinese paleontological past."[91] This date has indeed been firmly estab-

87. This eulogy is quoted in Harry L. Shapiro, *Peking Man*, 54, and in all versions (Chinese and English) of Jia Lanpo and Huang Weiwen's history of the Peking Man research.
88. For a discussion of the agreement, see Jiǎ Lánpō and Huáng Wèiwén, *Zhōukǒudiàn fājué jì*, chapter 5.
89. Furth, *Ting Wen-chiang*, 53. Sven Hedin and Teilhard de Chardin are two examples. Hedin's case is particularly noteworthy, since he was one of the explorers targeted by laws to protect Chinese artifacts from export.
90. Jia Lanpo and Huang Weiwen, *The Story of Peking Man*, 38.
91. Gregory Guldin, *The Saga of Anthropology in China: From Malinowski to Mao* (Armonk, N.Y.: M. E. Sharpe, 1994), 30.

lished as *the* moment of consequence through conferences and other forms of commemoration first held on the twenty-fifth anniversary in 1954 and then typically held every ten years beginning in 1959.

Nonetheless, it is important to recognize the enormous emphasis that Western and Chinese scientists alike placed on the discovery of that first skullcap in comparison with the previous finds, which, in the words of one famous geologist, were "limited to teeth and jawbone and skull fragments usually in a more or less crushed or battered condition."[92] As another scientist put it, "The layman can hardly wax enthusiastic about that shell of brown bone. . . . But it is difficult for a scientist to discuss it without using superlatives. Science now has for the first time a very primitive human cranium which is complete; one in which the bones are in their original positions, uncrushed, not even distorted."[93] Some scientists were also quick to highlight that "the credit for the recent discoveries goes to Mr. W. C. Pei," whom Black described as a "corking field man," and that it was "entirely due to his skill and devotion that this bulky mass with its unique and fragile contents reached the Cenozoic Laboratory quite undamaged."[94]

But did Péi Wénzhōng deserve all of the credit for the find? By most accounts, yes. Few have argued with the basic story that Péi, at 4:00 p.m. on 2 December, the last day of the 1929 excavation season, "climbed down to the bottom of the pit to start a final survey of a cavity, and while digging in the softer material, uncovered a curved bone which he at once recognized as part of the long-sought skull."[95] Jiǎ Lánpō, however, has consistently told a somewhat different story based on interviews with the respected technician Wáng Cúnyì. In a 1984 book, Jiǎ first quotes Wáng: "There were four people excavating at the bottom [of the pit]. I can only remember three of their names. The three people were Qiáo Déruì, Sòng Guóruì, and Liú Yìshān." Jiǎ then continues: "Because fossils had been found [in that pit], Péi Wénzhōng often went in to investigate. After hearing someone say that down below there was a round thing exposed, he went down with [a few] technicians to excavate. It was only when more and more of it had become exposed that he cried out, 'A human skull!'"[96]

92. George B. Barbour, "The Skull of the 'Peking Man,'" *Leader Reprints*, no. 51 (Peiping [Běijīng]: Leader Press, 1930), 2

93. Roy Chapman Andrews, "An Embarrassment," 19–20.

94. Barbour, "The Skull of the 'Peking Man,'" 2; Jia Lanpo and Huang Weiwen, *The Story of Peking Man*, 68; Davidson Black, "Explanation by Dr. Davidson Black," *Leader Reprints*, no. 51 (Peiping [Běijīng]: Leader Press, 1930), 11.

95. Barbour, "The Skull of the 'Peking Man,'" 2–3.

96. Jiǎ Lánpō and Huáng Wèiwén, *Zhōukǒudiàn fājué jì*, 46.

Personal conflicts between Jiǎ and Péi probably contributed to Jiǎ's spin, which while still acknowledging Péi's important role, detracts significantly from the sense that Péi made the important discovery alone. (It must be mentioned, however, that Péi himself tended to be modest when he described the event.)[97] Nonetheless, many historians and sociologists of science will appreciate the way Jiǎ's revisionism departs from the conventionally heroic story of discovery and accounts more fully for the diverse actors who together participated in the find. In China of 1984, after decades of socialist rhetoric and policy, the revised history spoke also to the class politics of science: not just intellectual elites, but workers, too, had an acknowledged role in scientific research.

If the question of who discovered Peking Man was one of political consequence, the question of who lost, or who stole, Peking Man became positively explosive.[98] The 1927 agreement had mandated that the fossils and artifacts unearthed would remain in China. This sensible precaution was designed to prevent foreigners from pilfering China's cultural relics. But China was an increasingly unstable place, and it was not long before Chinese scientists' chief concern was their Japanese enemies rather than their Western colleagues. Still, when Franz Weidenreich returned to the United States in 1941, he requested only that replicas of the fossils be produced and sent to him; he was not in a position to take the originals with him. What happened next has been difficult to establish, though many have tried. The Chinese scientists came to an agreement to ship the fossils to the United States for safekeeping. They were packed for shipment but never arrived. In the more than sixty years since, allegations have been made against the Japanese army, Japanese scientists, the Japanese emperor, American scientists, the American Museum of Natural History, and American service members. Today, under relatively peaceful and friendly conditions, the most plausible scenario is generally forwarded: that the fossils were inadvertently blown up or sunk in transit.

97. Péi Wénzhōng, *Zhōukǒudiàn dòngxué*, 28–30. In a 1982 article, Péi gave credit to Liú Yìshān, saying: "The technician Liú Yìshān was lying on the ground carefully excavating when he suddenly exclaimed, 'I've found an important fossil.'" Péi Wénzhōng and Shí Mòzhuāng, "Dáěrwén de yùyán yǔ rénlèi de quēhuán" [Darwin's Prediction and the Missing Link], *Dà zìrán*, no. 1 (1982).

98. Many books address the search for the missing fossils. A book published in Hong Kong in 1952 disputed the recent claims from the mainland and suggested that the Soviets were responsible for the loss: Wèi Jùxián, *Běijīngrén de xiàluò* [The Whereabouts of Peking Man] (Hong Kong: Shuōwénshè, 1952). A recent mainland book is Lǐ Míngshēng and Yuè Nán, *Xúnzhǎo "Běijīngrén"* [The Search for "Peking Man"] (Běijīng: Huáxià chūbǎnshè, 2000). The English-language book on the subject best known in China is Harry L. Shapiro, *Peking Man*. Shapiro was a physical anthropologist at the American Museum of Natural History and so had a personal interest in solving the mystery.

Presenting Peking Man

In 1933, the Chinese author of a new popular book on Peking Man explained his reason for putting pen to paper. The Peking Man excavations had caused scholars from Europe and the Americas to flock to China. They wrote about the finds not only "for a small minority of specialists, but also for the masses." In contrast, "in our nation, which produced Peking Man, one sees no record in newspapers or periodicals." With the exception of a few pictures in the newspaper *Shēn bào* and some articles in the specialist journal of the Geological Survey, he said, there was precious little written in Chinese on these important finds.[99]

The situation did not remain quite as bleak as he described. Scattered newspaper and magazine articles worked to remind the reading public of the importance of the fossils and kept them updated on the progress of the excavations at Zhōukǒudiàn.[100] Museum exhibits also helped to introduce Peking Man to those who lived in the right places and had the necessary time, funds, and inclination.[101] Nonetheless, it is remarkable, especially given the widespread interest in evolutionary theory and the nationalistic climate of the time, how soft a splash the discovery made in republican China outside scientific circles. Even some of the books and articles on anthropology written after the discovery of Peking Man failed to include these finds in discussions of Java Man, Piltdown Man, and the other key fossil evidence for human evolution.[102] This relative dearth will become more striking when we turn to the extraordinary emphasis on Peking Man that erupted immediately after the communist victory of 1949.

99. Yè Wéidān, *Běijīngrén* [Peking Man] (Shànghǎi: Liángyǒu túshū yìnshuā gōngsī, 1933), 1–2.
100. Xī Wēi, "Rényuán zěnyàng biàn rén" [How Did Anthropoid Apes Become Human?], *Kēxué huàbào* 5, no. 9 (1938): 350–355; "'Běijīngrén' wéi guóshǒu rénzhǒng" ['Peking Man' Was a Species of Headhunters], *Kēxué huàbào* 5, no. 11 (1938): 435; and Lù Jīngyī, "Rénlèi de gùshì" [The Story of Humans], *Kēxué huàbào* 14, no. 3 (1948): 167–173.
101. Zhào Jǐnfú, "Shànghǎi bówùyuàn cānguān jì" [Notes on a Visit to the Shànghǎi Museum], *Kēxué qùwèi* 3, no. 2 (August 1940): 105. On other exhibits, see Pān Jiāng, "Qián dìzhì diàochásuǒ dìzhì kuàngchǎn chénlièguǎn gàikuàng" [The Former Geological Survey's Geological and Mineralogical Exhibition Hall], in *Qián dìzhì diàochásuǒ de lìshǐ huígù: Lìshǐ píngshù yǔ zhǔyào gòngxiàn* [The History of the Former Geological Survey: Historical Narratives and Principle Contributions], ed. Chéng Yùqí and Chén Mèngxióng (Běijīng: Dìzhì chūbǎnshè, 1996), 86–87; museum pamphlet from Dìzhì diàochásuǒ (Geological Survey) in Nánjīng (title unavailable), Number Two Archives in Nánjīng, file 375:483; *The First Temporary Exhibition of the National Central Museum* (Chungking: National Herald Press, 1943).
102. Yóu Jiādé, *Rénlèi qǐyuán* [Human Origins] (Shànghǎi: Shìjiè shūjú, 1929); Chén Yīnghuáng, *Rénlèixué* [Anthropology] (Shànghǎi: Shāngwù yìnshūguǎn, 1934 [1918]); Lín Tāo, "Rénlèi chūxiàn zhī mí" [The Mystery of Human Origins], *Zhīshí yǔ qùwèi* 1, no. 6 (21 December 1939): 285–288. A 1930 book did not include Peking Man in its survey of human fossils but did discuss the fossils in a section on the geographic origins of humans: Zhāng Zuòrén, *Rénlèi tiānyǎn shǐ* [The History of Human Evolution] (Shànghǎi: Shāngwù chūbǎnshè, 1930), 104–139, 177–183.

Take, for example, the presentation of Chinese history in school textbooks. While history textbooks of the socialist period typically integrated Peking Man into the historical narrative of China, few textbooks of the republican period did. Of fourteen surveyed history textbooks, all published between the naming of Peking Man in 1927 and the founding of the People's Republic in 1949, only two included a discussion of Peking Man.[103]

Republican-era history textbooks continued to present earliest Chinese history as it appeared in the Chinese classics. First, Pángǔ split heaven and earth and his body formed the mountains, rivers, and other natural topography. Nǚwā then created humans from clay. In the beginning, people ate wild fruits and meats, drank water and blood (the classical trope for primitive eating and drinking was "ate hair and drank blood"), lived in the wild in holes in the ground, and wore leaves and animal hides. Then came a series of legendary figures who taught the people key civilizing inventions. Nest Dweller (有巢氏) taught them to build homes in the trees. Flint Maker (燧人氏) showed them to make fire by drilling wood, thus allowing them to cook their food (a long-standing Chinese mark of civilization, in contrast with "savages," who allegedly continued to eat their food raw). Fúxī (伏犧) invented snares for trapping fish and game, and the Divine Husbandman (神農) taught the people to domesticate animals. Textbooks then often turned to the story of the Yellow Emperor (黃帝) and his battle with the Miáo ethnic group to recount the early formation and flourishing of the Hàn people.

These legends, which had also found their way into early Chinese writings on human evolution, were rarely presented as received wisdom. How could they have been in an age when the contrast between "science" and "superstition" was perceived as central to intellectual life and social reform? Indeed, the move from mythical origin stories to scientific explanations of evolution served as a common narrative structure in the introductions to writings on human evolution. In part, these were of the familiar genre of science history "prologues" that introduced a scientific field by giving a brief overview of its development. In the West, this history progressed, as one would expect, from the Bible's account of Genesis through Georges Cuvier's (1769–1832) theories of catastrophism to Lamarck and finally Darwin. Chinese writers in the republican period sometimes added material from Chinese intellectual and religious

103. Chén Dēngyuán, *Shìjiè zhōngxué jiàoběn gāozhōng běnguó shǐ* [World Secondary School Textbooks: Chinese History], vol. 1 (Shànghǎi: Shìjiè shūjú, 1933); Guólì biānyì guǎn, *Chūjí zhōngxué lìshǐ* [Lower Secondary School History], vol. 1 (Shànghǎi: Shāngwù yìnshūguǎn, 1948).

48

history: from Pángǔ creating the world and Nǚwā creating humans to Daoist theories of *qì*. In this way, China too found a trajectory from superstition to science.[104]

Some of the textbook authors tackled the issue of legend versus science head-on. The point as they saw it was not to believe the legends but rather to obtain from them a feeling for how ancient people lived. As one textbook put it, "If you take preposterous myths and subject them to selective competition, then lay them out and analyze them according to the general principles of human evolution, these ancient legends will provide a glimpse of the first people's [初民] lives."[105] Many of the textbooks (eight of the fourteen surveyed) provided a discussion of archaeological evidence alongside the legends. Archaeology lent the impression of scientifically dependable facts, and the legends preserved a role for Chinese tradition while affording readers a more intimate sense of how their early ancestors lived. But then why was Peking Man not a part of this story?

One likely answer relates to disciplinary infrastructures. Archaeology in early twentieth-century China, as did its intellectual predecessor antiquarianism (金石學), served in large part as a means of verifying or discrediting historical narratives as they appeared in the classics.[106] The Peking Man excavations, however, were conducted by members of the Geological Survey, and were thus associated more with the sciences of geology and paleontology. It is possible that when textbook authors browsed archaeological writings for materials to include in their overviews of early Chinese history, they were simply more likely to find useful descriptions of Ānyáng and other Chinese archaeological excavations than of Peking Man. These archaeological excavations, moreover, provided plenty of excitement and even glory for writers seeking to expound upon the antiquity of Chinese civilization.

I would suggest, however, that another important reason Peking Man was not readily assimilated into history textbooks was the uncertainty about Peking Man's status as a direct ancestor of modern humans. At the time Peking Man was discovered, and lasting through the 1940s, scientists were less inclined than they later became to interpret fossil evidence in a simple hereditary line from ape to human. Rather, they

---

104. See, for example, Yóu Jiādé, *Rénlèi qǐyuán*, 1–15; and Qiáo Fēng, "Rénzhǒng qǐyuán."

105. Fù Wěipíng, *Chūjí zhōngxué yòng fùxīng jiàokēshū běnguó shǐ, 1* [Revived Textbook on Chinese History for Lower Secondary School], vol. 1 (Shànghǎi: Shāngwù yìnshūguǎn, 1938), 3.

106. K. C. Chang, "Archaeology and Chinese Historiography," *World Archaeology* 13, no. 2 (1981): 156–169. On antiquarianism, see Shana Brown, "Pastimes: Scholars, Art Dealers, and the Making of Modern Chinese Historiography, 1870–1928" (Ph.D. diss.: University of California, Berkeley, 2003).

often described Peking Man and other finds as offshoots, that is, as early humans closely related to our ancestors but not quite on the main trunk of evolution leading to modern humans. Authors who did discuss Peking Man often either specifically stated that the question of whether Peking Man was a direct ancestor was still undetermined or, as in the case of the 1933 book *Peking Man* that began this section, explicitly denied that the fossils were ancestral to modern humans, were "proto-Mongoloid," or had "anything at all to do" with the taxonomy of modern racial types.[107] Even the author of a nationalistic 1947 book entitled *The Origins of the Chinese People* was reserved in his evaluation of Peking Man, noting the need for further evidence before deciding whether the fossils were Chinese ancestors.[108]

In addition to scientific uncertainty, cultural resistance may have played a role in Peking Man's failure to emerge as a widely embraced national ancestor in the republican era. An article published in a 1939 issue of a popular science magazine captured the issue nicely. "Are humans' ancestors monkeys?" the author asked. "Sensitive [敏感] people don't like to think this. Chinese people especially aren't willing to have this kind of ancestor. Actually monkeys are our cousins. Our fathers were ape-humans who died many years ago."[109]

The difficulty of promoting Peking Man unambiguously as a human, or specifically a *Chinese* human, ancestor removed the only real reason for including it in a history textbook. The two pre-1949 textbooks that did include Peking Man both explicitly claimed Peking Man as an ancestor of the Chinese people. A 1933 textbook for upper secondary school students presented Peking Man in the context of the debate over

107. The biologist Zhū Xǐ considered the question of Peking Man's relationship to modern Chinese unclear, although he noted that Peking Man was similar to modern humans in ways other fossil hominids were not. Zhū Xǐ, *Wǒmén de zǔxiān* [Our Ancestors] (Shànghǎi: Wénhuà shēnghuó chūbǎnshè, 1950 [1940]), 163–164. Yè Wéidān, *Běijīngrén*, 32–35.

108. Frank Dikötter states that Lín Yán "cited the discovery of Beijing Man at Zhoukoudian as evidence that the 'Chinese race' had existed on the soil of the Middle Kingdom since the earliest stage of civilization" (Dikötter, *The Discourse of Race*, 134). This interpretation is not supported by the text. What Lín Yán said was: "But was Peking Man an ancestor of our Chinese people [中華民族]? . . . According to most scholars' research today, the evidence is still not ample enough to make a determination—it still has not reached a decisive level. But at this point we can say at least: Peking Man of the Zhōukǒudiàn area was the earliest primitive human type in China during the Paleolithic period. So from this we can determine: Chinese soil did have the most ancient primitive human types living on it [周口店一帶的北京猿人, 是中國舊石器時代的最古的原始人類. 所以因此可以確定的: 中國的土地上是有最古的原始人類居住者的]." Lín Yán, *Zhōngguó mínzú de yóulái* [The Origins of the Chinese People] (Shànghǎi: Yǒngxiáng yìnshūguǎn, 1947), 27.

109. Shàng Guānshū, "Cóng biànxíngchóng dào rén" [From Amoeba to Human], *Kēxué shēnghuó* 1, no. 3 (1939), 91. For a socialist-era account of resistance to dishonoring one's human ancestors in this way, see Yuan-tsung Chen, *The Dragon's Village: An Autobiographical Novel of Revolutionary China* (New York: Penguin Books, 1980), 109.

whether the Chinese were sui generis or had migrated from the west or (as the Japanese promoted) from the east—a common topic in history textbooks as in other materials on human origins and Chinese history. "More recently," the author asserted, "progress in paleontology and geology has compelled scholars to come to the three-character conclusion 'Peking Man' [北京人]. . . . So now we can give a laugh to the theories of western and eastern origins."[110] A 1948 textbook for lower secondary school students skipped the debate over western origins and moved directly to an argument for Chinese nativity in China. "Where did the Chinese ethnicity [中華民族] originate? It was undoubtedly on Chinese soil." In the subsequent discussion of Peking Man, the authors emphasized that on the basis of the skull the "Chinese national" Péi Wénzhōng had discovered, researchers had determined that Peking Man lived 500,000 years ago and "resembled extremely closely modern northern Chinese people," a statement no doubt based on Franz Weidenreich's detailed studies.[111]

Some other writings similarly claimed ancestral status, or probable ancestral status, for Peking Man, often as a way of bolstering Chinese national identity by "demonstrating the native provenance and shared racial descent" of all the ethnic groups inhabiting China.[112] For example, one notable cultural conservative considered the discovery of the Peking Man fossils to provide needed evidence in support of the theory (of which he was already "convinced") that all the ethnic groups of China had descended from a single ancestor.[113] Another author even went so far as to argue that the Peking Man artifacts proved that human evolution was more advanced in China than in Europe of the same period.[114] However, such claims seem—perhaps surprisingly given the nationalistic

110. Chén Dēngyuán, *Shìjiè zhōngxué jiàoběn*, 28.
111. Guólì biānyì guǎn, *Chūjí zhōngxué lìshǐ*, 21–22.
112. James Leibold, "Competing Narratives of Racial Unity in Republican China: From the Yellow Emperor to Peking Man," *Modern China* 32, no. 2 (2006): 202.
113. Xióng Shílì, *Zhōngguó lìshǐ jiǎnghuà* [Talks on Chinese History] (Táiběi: Míngwén shūjú, 1984 [1939]), 34–35. The author of an article describing the Shànghǎi Museum's 1940 Peking Man display similarly called the skullcap replica "the true form of our most ancient ancestors." Zhào Jīnfú, "Shànghǎi bówùyuàn," 106. Xǔ Lìqún said Peking Man was "probably the ancestor of today's Chinese people," in *Zhōngguó shǐ huà* [Notes on Chinese History] (Běijīng: Xīnhuá shūdiàn, 1950 [1942]), 3. The book was republished in 1944, 1949, 1950, and 1952 (and possibly other years). I have not found the original 1942 Yán'ān edition. There are a few differences among the editions, but the relevant passages remained the same between 1944 and 1952, and there is no reason to believe they were substantially different in 1942. For other examples, see Leibold, "Competing Narratives of Racial Unity," 202–203.
114. Chén Jiānshàn, *Shǐqián rénlèi* [Prehistoric Humans] (n.p.: Zhōnghuá shūjú, 1936), 111. The author cited a newspaper article by Péi Wénzhōng that claimed the Zhōukǒudiàn artifacts were more advanced than those found in Europe during the same period.

incentive—to have been relatively few. And while Chinese writers on human evolution often expressed hope and even confidence that the debate over where humans originated would be settled in Asia's favor, most authors were careful to indicate that this was yet an unproven theory, and some preferred to remain agnostic on the issue.[115]

Peking Man's most acclaimed popular role in the republican era came not in a nationalistic treatise, but rather in a drama on the subject of the doomed Chinese tradition. In Cáo Yú's seminal 1940 play *Peking Man*, the older-than-ancient Peking Man became, somewhat ironically, an ideal symbol of China's need to resign itself to, or even be willing to embrace, change. The play had "a double ending, in which those who stick to tradition are doomed to suffer and die, while those capable of adjusting to the new reality are committed to seek a better future."[116] The title *Peking Man* itself had a double meaning: it represented both the early twentieth-century people of Běijīng, whose traditional attitudes threatened their very survival, and the creatures of half a million years earlier, for whom customs meant nothing and adapting to nature's laws was a way of life. As the character of an enlightened Chinese anthropologist described the fossil Peking Man, "This is the ancestor of humanity; it is also the hope of humanity. In those days, people loved, hated, cried, and shouted just as they wanted. They feared neither death nor life. All year long they gave full play to their natural feelings. . . . There was no code of custom to constrain them, and no civilization to bind them." Drawing from ancient Chinese legends, he further elaborated on Peking Man's primitive life: "They ate raw meat and drank fresh blood. The sun baked them, the wind blew on them, and the rain soaked them." He then concluded (perhaps oddly, given suspicions about Peking Man's cannibal habits), "They didn't have the kind of man-eat-man civilization we have today, and they were very happy."[117]

Cáo Yú embodied Peking Man in the character of a physically imposing but mute worker whose morphological similarities to the Peking Man fossils made him a subject of the anthropologist's research. When

115. A 1922 article in *Eastern Miscellany* on the discovery of the Rhodesia Man fossils, before Peking Man was discovered, concluded by suggesting that this find lent weight to the theory that humans originated in Africa. See Jiàn Mèng, "Rénlèixué shàng zuìjìn de fāxiàn" [A Recent Discovery in Anthropology], *Dōngfāng zázhì* 19, no. 5 (1922): 94–95. For agnostic overviews of the debate over place of origin, see Xīn Jièliù, "Rénlèi de huàchéng" [Human Evolution], *Dōngfāng zázhì* 25, no. 17 (1928): 81–84; and Yóu Jiādé, *Rénlèi qǐyuán*, 82–95. Zhāng Zuòrén's 1930 *Rénlèi tiānyǎn shǐ* emphasized that anywhere except the Americas could be the birthplace of humans, but that the discovery of Peking Man lent support to Asian origins (182–183).

116. John Y. H. Hu, *Ts'ao Yü* (New York: Twayne Publishers, 1972), 93.

117. Cáo Yú, *Běijīngrén* [Peking Man] (Shànghǎi: Wénhuà shēnghuó chūbǎnshè, 1950 [1941]), 155–156.

this stand-in for Peking Man suddenly spoke at the end of the play, the breaking of silence represented the possibility for change that allowed a young woman in the family (clearly modeled on the influential character of Nora from Ibsen's *A Doll's House*) to leave the house and its doomed inhabitants. The Peking Man in Cáo Yú's play thus represented transformation and human potential. This, more than national identity, was what Peking Man stood for in republican times.

Conclusion

The title of this chapter, "From 'Dragon Bones' to Scientific Research," comes from the first chapter title of Jiǎ Lánpō and Huáng Wèiwén's retrospective on the Peking Man excavations.[118] The implication, of course, is of progressive change from superstition to science. Not only does scientific research appear to replace dragon bones, but "dragon bones" is placed in quotation marks, a clear sign that it is a term to be treated with skepticism. The phrase reflects an attitude about culture and science dominant both at the time it was written (the 1990s) and during the era about which it was written (the early twentieth century). It is thus unconsciously ironic on several levels.

First, while the discovery of Peking Man did originate in local knowledge of dragon bones, scientific research never replaced this other form of knowledge. Rather, rural people even today continue to collect dragon bones for use as medicine, and scientists continue to seek out these dragon bone hunters to further their own research. Both as boon and as bane, the dragon bone trade has remained a part of scientific research on evolution in China. Continuity has also been the pattern on a more macroscopic cultural level. Scientific literacy remains low in China, and interest in religion and other forms of "superstition" has not been squelched even after almost a century. The goal of eliminating superstition and bringing the masses into a scientific worldview was to remain prominent among intellectuals and state agents in the socialist and postsocialist eras.

Change was the order of the day in early twentieth-century China. Learning to call one's ancestors apes, envisioning a future different from and better than the past, sending bright young scholars to the West for

---

118. It is specifically the Táiwān edition—Jiǎ Lánpō and Huáng Wèiwén, *Fāxiàn Běijīngrén* [The Discovery of Peking Man] (Táiběi: Yòu shī, 1996), published in 1996—that contains this chapter title. The full title reads "From 'Dragon Bones' and 'Dragon Teeth' to Scientific Research" (從 "龍骨"、 "龍牙" 到科學研究). See Jiǎ Lánpō and Huáng Wèiwén, *Fāxiàn Běijīngrén* [The Discovery of Peking Man] (Táiběi: Yòu shī, 1996).

education, participating as Chinese nationals in an international scientific community, advocating mass education—transformation was not just a buzzword but a lived reality, at least for some. But these changes were not the kind that come easily to completion. Calling for the people to be educated does not make them so, and even if reformers had been able to make more than a dent in this problem, new generations would still have come along needing what counted as education in their times. It was also far easier to criticize "tradition" and "superstition" than it was to define these terms in stable ways and arrange for a permanent solution. And Chinese scientists could strive alongside their nation as a whole to participate on equal footing among other scientists and other nations, but as the world changes so do international politics, and these relationships would have to be reworked again and again.

The socialist state did not simply inherit the problems of pre-1949 China. It also shared many of the perspectives of its predecessor. With respect to science dissemination, intellectual and political elites on both sides of the 1949 divide held to an understanding of the Chinese people as mired in superstition and in need of scientific knowledge to allow them to join the modern world. Moreover, the notion of humans as part of a natural order predicated on progressive change—with the possibility also for voluntarist change—was as central to post-1949 constructions of human identity as to early twentieth-century ones.

Yet, as we will see, there was also plenty of room for popular paleoanthropology to grow. In 1949, Peking Man was not yet firmly established as a national ancestor, and she had not yet taken on what was soon to become her most defining characteristic—labor. Popular science, moreover, had never had the champion it would find in the socialist state. And the meanings ascribed to the excavations at Zhōukǒudiàn were far from set in stone: the questions of who found and who lost Peking Man were dug up again and again over the decades so that the republican past could be reconstructed to fit the needs of a changing China.

TWO

"A United Front against Superstition": Science Dissemination, 1940–1971

A Role for Scientists in Revolution

The first years of the People's Republic of China saw an explosion of attention paid to another, earlier beginning: the dawn of humanity itself. Paleoanthropologist Jiǎ Lánpō linked these two transformative events in the preface to a 1951 picture book entitled *Our Ancestors 500,000 Years Ago.* "Actually," he wrote, "not only have we been transformed in revolution [翻了身], but following the victory of liberation, so has our ancestor—Peking Man." Jiǎ referred specifically to the attention the new state had begun paying to Peking Man with the purpose of teaching the "broad masses" about human evolution. His statement also had metaphorical significance: Peking Man was to begin a new life with new meaning as an icon of revolution.[1]

Disseminating knowledge about human evolution was a clear priority for the new state. The reason for this urgency was the conviction that a materialist understanding of history and social development was essential to the process of becoming a member of the new socialist society. The story of human evolution proved a powerful way to introduce a large number of people without advanced education to

1. Yáng Háonìng (art) and Jiǎ Lánpō (text), *Wǒmén de zǔxiān 1: Wǒmén wǔshí-wàn nián de zǔxiān* [Our Ancestors, vol. 1: Our Ancestors 500,000 Years Ago] ([Tiānjīn?]: Zhīshí shūdiàn, 1951), preface (no page number).

Marxist political philosophy. Scientists were essential to this process. Through the production of books, magazine articles, museum exhibits, films, and slide shows presented in factories and other places of work, they sought to familiarize "the masses" with the socialist interpretation of "human origins and development" and with the fossil evidence for human evolution in China. The goal was nothing less than to transform the worldview of every member of the new society—to rid the people of "superstition" and introduce them to science and to socialism.

The roots of post-1949 popular paleoanthropology lie in the republican era. More directly, they lie in Yán'ān, the cradle of the Chinese communist revolution. It was there, while fighting the Sino-Japanese and civil wars, that Máo Zédōng cemented his position as the leader of revolutionary China. The vision he put forward came to define the revolution, and in later years scientists along with everyone else would milk every ounce of meaning from his writings of this period. Activities in Yán'ān's cultural sphere aimed at eliminating superstition similarly set long-lasting precedents for the field of science dissemination. At the same time, party intellectuals were crafting the template for a new history of China, which began with the transition from ape to human and so offered Peking Man a leading role. Peking Man's new political importance as an agent of science dissemination in turn signaled to scientists one of the key functions they would be expected to fill in the new society.

The key document science disseminators came to cite in later years was Máo's 1942 "Talks at the Yán'ān Forum on Literature and Art." While the Yán'ān forum did not address natural science, Máo's speeches there codified an understanding of the class politics of knowledge that shaped the way science dissemination was to be conceptualized. In his concluding statements at the forum, Máo asserted that, "prior to the task of educating the workers, peasants and soldiers, there is the task of learning from them." This was necessary in order to achieve a "correct understanding of dissemination [pǔjí, 普及]." It was not that the masses held all the answers. Rather, they were "illiterate and uneducated as a result of long years of rule by the feudal and bourgeois classes" and so were "eagerly demanding enlightenment, education and works of literature and art which meet their urgent needs and which are easy to absorb." Máo thus formulated a two-way flow of knowledge. Intellectuals would disseminate art and literature to the uneducated masses but would base their work on the "language of the masses" and would plot their artistic advancement along the trajectory of the masses' own development.[2]

---

2. Mao Tse-tung, *Selected Works of Mao Tse-tung* (Peking: Foreign Languages Press, 1967–1971),

The notion that intellectuals and workers alike had parts to play in the task of building a socialist society was very much in keeping with the "New Democracy" that Máo had outlined in his 1940 article by that name, which in turn had emerged from the context of the "united front." With the pressures from Japanese invasion ever increasing, Máo had called for—and won—a second united front with the Nationalists under Chiang Kai-shek against the Japanese. New Democracy extended this principle of inclusion and cooperation to China's diverse social classes by providing for the "possibility . . . of a united front against imperialism, feudalism and superstition." Naming superstition as a target of attack suggested a special role for "natural scientists," who—so long as they were free of "reactionary idealism"—were expected to participate alongside the Chinese proletariat and progressive members of the bourgeoisie.[3]

Even in the beginning this inclusive ideal was often illusory. Whether for reasons intrinsic to the social relations or deriving from the deliberate machinations of Máo and other party leaders, the proposed harmony and reciprocity between elites and masses in the production of knowledge tended rather to split into two antagonistic strands: on one hand, an elitist attitude toward the ignorance of the masses, and on the other, populist attacks on that very elitism. Intellectuals had little control over the periodic political campaigns that targeted them. They seemed to recognize early on, however, that as long as the party perceived a need for eliminating superstition among the masses, they would have an essential part to play in the new society.

The Chinese socialist state's use of science for propaganda was not a disregard for science itself as an empirical mode of inquiry. Propagandists cared about science because it was useful as a means of turning political truths into "natural" ones, and they typically favored scientific theories that supported the state's political priorities. But they also thought of science as a mode of thought that liberated people from ideological authorities and allowed them to think critically. Science dissemination materials highlighted the importance of empirical evidence in proving the evolutionary relationship between humans and apes. They conspicuously presented fossils together with anatomical, embryological, and

---

3:80–84. Máo described this relationship again in his 1 June 1943 speech "Some Questions Concerning Methods of Leadership." See Mao Tse-tung, *Selected Works*, 3:117–122. A portion of this text was later included in the *Quotations from Chairman Mao*, widely known as the "little red book." For consistency, I have used the term "disseminate" in place of the term "popularize," which appears in the translated text for 普及.

3. Mao Tse-tung, *Selected Works*, 2:381.

behavioral comparisons of humans and apes as the evidence required to convince audiences of the fact of human evolution (figures 5 and 6). Here the values of scientific and political elites overlapped, and scientists found a place for themselves in the new order.

The place they found, however, was neither entirely stable nor without perils of its own. In marshaling science to fight superstition, scientists helped reinforce the state's position that only some forms of knowledge were worthy of study while others deserved elimination. If they expected their own scientific theories to be immune from this process, they were sorely mistaken. It would not be long before scientists with unorthodox theories faced treatment similar to what they themselves doled out to the most "superstitious" of "the masses."

Ghosts into People, Apes into Humans

The height of antisuperstition activities in Yán'ān was an intensive, two-year campaign that began in 1944. The drive particularly targeted witch doctors, who charged large fees for the performance of rituals designed to cure illness or bring rain. The means employed were creative: newly produced folk songs, dances, and dramas helped demonstrate in an entertaining way the emptiness of the witch doctors' tricks and the power of modern science and medicine.[4]

In 1945, the Lǔ Xùn Academy of Art and Literature in Yán'ān debuted an opera that demonstrated the link between social revolution and eliminating superstition. *The White-Haired Girl* (白毛女) was based on a local legend about a young concubine who fled to the mountains to escape a cruel landlord and whose suffering caused her hair and skin to turn white like a ghost. Folk stories about the "white-haired girl" strongly resemble the accounts in imperial-era texts of "hairy people" who had fled conscript labor on the Great Wall and whose isolation from society had transformed them into beasts.[5] In the opera, however, socialist revolutionaries dismiss the local villagers' sacrifices to the white-haired girl as superstition, and in the end they convince the girl to come out of hiding, denounce her abuser, and join the others in class struggle. Upon watching the opera, rural people familiar with the legends would undoubtedly have understood the message—even if they did not agree with it—that a socialist ordering of the world was to replace a mythological one. As

4. James Reardon-Anderson, *The Study of Change: Chemistry in China, 1840–1949* (Cambridge: Cambridge University Press, 1991), 359–363.

5. Yáng Yánliè, ed., *Fāng xiàn zhì* [Fāng County Gazetteer] (Táiběi: Chéngwén chūbǎnshè, 1976 [1866]), 3:847–848.

5   First of several comparisons between ape and human behavior in a picture book on human
    evolution first published in 1952. The comparisons were meant to demonstrate the evolution-
    ary kinship between humans and apes. Reproduced from Zhāng Mín, Shèng Liángxián, and
    Shěn Tiězhēng, *Láodòng chuàngzào le rén* [Labor Created Humanity] (N.p.: Huádōng rénmín
    chūbǎnshè, 1954 [1952]), 5.

图 8．人(右)和类人猿(左)四肢长短的比较。

6   Illustration from page 15 in Franz Weidenreich's influential book *Apes, Giants, and Man*
    (Chicago: University of Chicago Press, 1946). Beginning in the 1950s, many Chinese books
    and museum exhibits reproduced the image. The comparison demonstrates the evolutionary
    divergence of humans from apes: humans had evolved longer legs and shorter arms better
    suited to walking erect. Reproduced from Fāng Shàoqīng [pseud. for Fāng Zōngxī], *Gǔ yuán
    zěnyàng biànchéng rén* [How Ancient Apes Became Human] (Běijīng: Zhōngguó qīngnián
    chūbǎnshè, 1965), 38.

articulated in the play, "The old society turned people into ghosts; the new society changes ghosts back into people."[6]

There was no room in the new modernist, socialist ideology for anomalous creatures that moved freely across boundaries between animal and human or between the natural and spiritual worlds. Party intellectuals found, however, that such legends could be stripped of their superstitious elements and made to serve socialism. This was the fate of popular forms of knowledge at the hands of "dissemination" as outlined in Máo's "Talks at the Yán'ān Forum."

At the same time that performances of *The White-Haired Girl* worked to undermine a popular understanding of the transformative relationship between animal and human, party intellectuals were beginning to codify and disseminate a properly socialist and scientific account of the subject. In 1942, Yán'ān saw the publication of a new history of China that brought Peking Man into a Marxist interpretation of human origins and social development.[7] After a short explanation of the stages of history outlined by Marx and codified by Stalin, the author introduced Peking Man as the beginning of this process. In answering the question, How did apes become human? he briefly summarized the thesis that was to become the raison d'être for popular paleoanthropology for the next forty years. This was Frederick Engels's 1876 "The Part Played by Labor in the Transition from Ape to Human."

Engels's essay had been translated into Chinese and published in China by 1928.[8] The essential idea was that the "liberation" (解放) of the hands through the adoption of an erect gait made possible labor, and labor in turn promoted complex social interactions, language, the development of the brain, and all the other attributes that separate

---

6. Dīng Yī and Wáng Bīn, *Bái máo nǚ* [The White-Haired Girl] (Hong Kong: Hǎiyáng shūwū, 1948), 146.

7. Xǔ Lìqún, *Zhōngguó shǐ huà* [Notes on Chinese History] (Běijīng: Xīn huá shū diàn, 1950 [1942]).

8. In Chinese, the essay's title is 勞動在從猿到人過程中的作用, and in German it is *Anteil der Arbeit an der Menschwerdung des Affen*. In English, the title conventionally uses the word "man" rather than "human." It was published in Frederick Engels [Ēngésī], *Mǎkèsīzhǔyì de rénzhǒng yóulái shuō* [The Marxist Theory of Human Origins], trans. Lù Yīyuán (Shànghǎi: Chūncháo shūjú, 1928). The earliest Chinese version I have found is Frederick Engels [Ēngésī], *Cóng yuán dào rén* [From Ape to Human], trans. Cáo Bǎohuá and Yú Guāngyuǎn (Shíjiāzhuāng: Jiěfàng shè, 1950 [1948]). The first print run was nine thousand copies, and the 1950 print run brought the total to thirty thousand. The translators of the 1948 edition worked mainly from the 1935 *Collected Works of Marx and Engels* published in the Soviet Union in German, and they also consulted Russian translations. They included not only the full text of "The Part Played by Labor in the Transition from Ape to Human," but also an excerpt from Engels's *The Dialectics of Nature* on the process of human evolution. Later editions of the work preserved this combination. The 1948 text was also the basis for a 1959 translation into Arabic (in seventy-six thousand copies), considered one of China's other "minority nationality" languages.

humans from the apes. Engels noted that anthropoid apes already possess a degree of differentiation between hands and feet—the hands being used for such activities as collecting food and building nests in the trees. Yet, as he clarified, "how great is the difference between the undeveloped hand of even the most anthropoid of apes and the human hand that has been highly perfected by the labor of hundreds of thousands of years. . . . No simian hand has ever fashioned even the crudest of stone knives."[9] Engels insisted that labor itself had "perfected" the hand: "The hand is not only the organ of labor, *it is also the product of labor.*"[10]

Engels's thesis became the introduction to Marxism for a great many people after the revolution. As the influential anthropologist Liú Xián put it in 1950, everyone was now expected to study social development, and the period "longest in duration, murkiest in substance, and most difficult to study" was the transition "from ape to human."[11] *From Ape to Human* (從猿到人) was the title given to the thin and often-republished volume containing Engels's 1876 essay and a relevant excerpt from his *Dialectics of Nature*. It came to serve as a prologue to the Marxist stages of social development—namely, primitive society, slaveholding society, feudalism, capitalism, and socialism—that formed the skeleton of history as taught in the Soviet Union and in socialist China. In addition to materials produced specifically on human evolution, almost all books on "the history of social development" began with a chapter or at least a section on human evolution, usually entitled "From Ape to Human"—as did the great popularizer of Marxism Ài Sīqí's 1950 *Historical Materialism: The History of Social Development.*[12]

Virtually everything produced on the subject of human evolution during the first three decades of communist rule in China paid homage to Engels's treatise and its central message that "labor created humanity itself" (勞動創造了人本身). Even Cáo Yú's 1940 play *Peking Man* quickly incorporated this perspective. In place of language that celebrated Peking Man's emotional freedom and echoed ancient legends, the 1951 revision had the anthropologist character trumpet the liberating force of labor:

9. Engels, *Cóng yuán dào rén* (1950), 3. Since the Chinese does not differ significantly in meaning, I have borrowed this from Frederick Engels, *The Part Played by Labor in the Transition from Ape to Man* (New York: International Publishers, 1950), 8.

10. Engels, *The Part Played*, 9 (italics in original).

11. Liú Xián, *Cóng yuán dào rén fāzhǎn shǐ* [A History of the Development from Ape to Human] (Shànghǎi: Zhōngguó kēxué túshū yíqì gōngsī, 1950), author's preface, 1–2.

12. Ài Sīqí, *Lìshǐ wéiwù lùn: Shèhuì fāzhǎn shǐ* [Historical Materialism: The History of Social Development] (Běijīng: Shēnghuó, dúshū, xīnzhī, 1950). This was originally a radio lecture. The initial print run was forty thousand copies, and it was reprinted at least three times. On Ài Sīqí (1910–1966) as a popularizer, see Joshua A. Fogel, *Ai Ssu-ch'i's Contribution to the Development of Chinese Marxism* (Cambridge, Mass.: Harvard University Press, 1987), 59.

"Look: these hands have evolved . . . from the hardship of labor. . . . Our human ancestors underwent hundreds of thousands of years of labor before the ultimate creation of (raising his hands) this pair of hands capable of transforming life and humanity."[13] Many real-life anthropologists and other intellectuals were just as energetic in their efforts to disseminate this newly ascendent interpretation of human origins, and through it a scientific perspective on the social and natural worlds.

The Who and How of Science Dissemination

The standard curriculum in primary and secondary schools ensured that the younger generations would grow up with an exposure to ideas about human evolution. Primary school students typically encountered Peking Man and basic knowledge about human fossils in their history classes. Secondary school students also learned about biological evolution in their natural science classes and moreover learned that labor created humanity as part of the history of social development taught in courses on political science.

But what of the generations too old to attend school? A great many received their political education in party-organized lectures and seminars. "Political lectures" (政治大課) given by such important party intellectuals as Ài Sīqí introduced many to Engels's thesis on human evolution.[14] Many more learned about it through participation in the "small study groups" (學習小組) that cadres, workers, peasants, and soldiers attended with their coworkers.[15] Based on interviews conducted in Hong Kong with recent refugees, one China watcher reported that in small groups in the early 1950s, "one of the most universally used texts is a book called *The History of Social Development* . . . [and] everyone reads *From Monkey to Man*."[16] The latter text, of course, is what I have translated as *From Ape to Human*.

These institutions alone brought knowledge about human evolution

---

13. Cáo Yú, *Cáo Yú xuǎnjí* [Selected Works of Cáo Yú] (Běijīng: Kāimíng shūdiàn, 1951), 463. The revised version also eliminated the character of the mute, hulking worker, which must have risked offending the now-prominent laboring classes. Peking Man was instead represented abstractly by a shadow that appeared periodically at the back of the set. When Cáo Yú revised the play again in 1954, he returned the passage quoted above back to the original but retained the larger changes.

14. Péi Wénzhōng, preface to *Cóng yuán dào rén de yánjiū* [Research on from Ape to Human], by Lín Yàohuá (Běijīng: Gēngyún chūbǎnshè, 1951), 3.

15. Martin King Whyte, *Small Groups and Political Rituals in China* (Berkeley: University of California Press, 1974).

16. A. Doak Barnett, *Communist China: The Early Years, 1949–55* (New York: Praeger, 1964), 101. See also Robert Jay Lifton, *Thought Reform and the Psychology of Totalism: A Study of "Brainwashing" in China* (New York: Norton, 1961), 257.

to many millions. But beyond schools and study groups, human evolution was a subject of great emphasis for an extensive web of people and institutions working to disseminate scientific knowledge to the masses. Science dissemination (*kēxué pǔjí*, 科學普及) was important enough to the new state that a special bureau was created for it just one month after the establishment of the People's Republic; its catchy and well-known abbreviation (*kēpǔ*, 科普) further testified to its significance in the state's political vocabulary.[17] *Kēpǔ* had four generally recognized goals: to enable laborers to grasp scientific production technology; to disseminate knowledge of natural science so as to eliminate superstitious thought; to publicize new inventions from the laboring classes in order to cultivate patriotic spirit; and to spread knowledge of health and hygiene in order to protect the people's health.[18] The task of educating people about human evolution was one of the most powerful means of achieving the second of these four goals.

In framing their approach to science dissemination, many authors specifically cited Máo's "Talks at the Yán'ān Forum." For example, the president of the Academy of Sciences, Guō Mòruò, wrote an article entitled "Dissemination and Raising Standards in Science," which he published in the journal *Science Dissemination Bulletin* in 1950. He began with a direct quotation—"Raise standards on the foundation of dissemination; disseminate under the guidance of raising standards"—and then discussed it in relation to science.[19] Popular books on human origins similarly referenced the model of dissemination presented in the "Talks at the Yán'ān Forum."[20]

While *kēpǔ* was undeniably a top-down enterprise, there were by no means one clearly defined group of "experts" and another clearly defined group of "masses." Rather, science dissemination existed in many different forms and for many different audiences. This is further evidence that *kēpǔ* permeated the overall vision of the new society—many people considered themselves responsible for it. Top-level research scientists and professional writers both contributed to the body of *kēpǔ* literature;

17. The abbreviation *kēpǔ* is not used in Táiwān but is familiar throughout mainland China.

18. Xī'nán jūnzhèng wěiyuánhuì wénjiào bù, ed., *Kēxué pǔjí gōngzuò shǒucè* [Handbook for Science Dissemination Work], vol. 1 (N.P.: Xī'nán jūnzhèng wěiyuánhuì wénjiào bù, 1951), 14. See also Shěn Qíyì et al., *Zhōngguó kēxué jìshù xiéhuì* [The Chinese Association for Science and Technology] (Běijīng: Dāngdài Zhōngguó chūbǎnshè, 1994), 29.

19. Guō Mòruò, "Kēxué de pǔjí yǔ tígāo" [Science Dissemination and Raising Standards], *Kēxué pǔjí tōngxùn*, no. 3 (1950): 35.

20. Fāng Qiě, *Cóng yuán dào rén tòushì: Láodòng zěnyàng chuàngzào le rénlèi běnshēn hé shìjiè* [A Penetrating Look at from Ape to Human: How Labor Created Humanity Itself and the World] (Shànghǎi: Shànghǎi biānyì shè, 1950), 83; Wáng Shān, *Láodòng chuàngzào rénlèi* [Labor Created Humanity] (Shànghǎi: Shànghǎi qúnzhòng shūdiàn, 1951), preface (no page number).

research institutes, government agencies, professional associations, and museums all organized *kēpǔ* activities.[21]

However, the number of distinct individuals and organizations engaged in science dissemination does not imply quite the diversity it may seem. For one thing, even ostensibly nongovernmental organizations were closely associated with relevant official organs. Moreover, the system of "concurrent positions," in which China's limited number of scientists "play[ed] functionally related roles in more than one organization," meant that the same people were often in charge of science dissemination work in several different organizations.[22] Some of the leading members of the Institute of Vertebrate Paleontology and Paleoanthropology held important positions in other institutions involved in science dissemination.[23]

Such overlapping networks helped bring continuity to science dissemination materials on human evolution, which no doubt otherwise would have reflected more clearly the diverse backgrounds of the people creating them. Jiǎ Lánpō was fond of saying that Peking Man did not simply "fall from the sky," but must have had predecessors not yet dis-

21. Scientists at IVPP were among the most influential, although members of Fùdàn University's anthropology department were also significant participants. The observation of a member of the American Paleoanthropology Delegation in 1977 was equally true of the 1950s and 1960s: "Senior people, more than their junior colleagues, take on such assignments." Kwang-chih Chang, "Public Archaeology in China," in *Paleoanthropology in the People's Republic of China: A Trip Report of the American Paleoanthropology Delegation*, ed. W. W. Howells and Patricia Jones Tsuchitani (Washington, D.C.: National Academy of Sciences, 1977), 135. Some scientists—like Liú Hòuyī, the paleontologist, writer, and in the 1970s editor of IVPP's *Fossils* magazine—were at least as celebrated for their popular as for their professional work. Jiǎ Zǔzhāng, one of the founders of popular science writing in China, published a book entitled *From Ape to Human* in 1950. Yè Yǒngliè, *Zhōngguó kēxué xiǎopǐn xuǎn, 1949–1976* [Selected Chinese Short Writings on Science, 1949–1976] (Tiānjīn: Tiānjīn kēxué jìshù chūbǎnshè, 1985), 3. (Yè himself is China's most noted popular science writer today.)

22. Richard Suttmeier, *Research and Revolution: Science Policy and Societal Change in China* (Lexington, Mass.: Lexington Books, 1974), 67.

23. While IVPP (古脊椎動物與古人類研究所) received its current name only in 1960, its institutional history dates to 1929, when it was the Cenozoic Research Laboratory under the Chinese Geological Survey. After 1949, the laboratory was subordinate to the National Planning and Steering Commission for Geological Works until 1953, when it became an independent research laboratory within the Chinese Academy of Sciences. In 1957, it attained the status of an "institute" as the Institute of Vertebrate Paleontology. In 1960 "Paleoanthropology" was added to the name to reflect the important place of such research in the institute's historic and current work—after all, the Cenozoic Research Laboratory was created around the Peking Man research. (For the sake of simplicity, I refer to the institute as IVPP throughout.) With 103 people on staff in 1957, it was only half the size it would become in the 1970s; it continued to grow during the post-Máo era, reaching 250 staff members by 1994. Rèn Bǎoyì, ed. *Zhōngguó kēxuéyuàn gǔjǐzhuī dòngwù yǔ gǔrénlèi yánjiùsuǒ* [The Chinese Institute of Vertebrate Paleontology and Paleoanthropology] (Běijīng: Zhōngguó kēxuéyuàn gǔjǐzhuī dòngwù yǔ gǔrénlèi yánjiùsuǒ, 1994), 1. Statistical information for 1957 comes from Zhōngguó kēxuéyuàn niánbào (Chinese Academy of Sciences Annual Report) for that year, found in the IVPP archives. I have not since been able to find this periodical for years before 1977 in any American library or in catalogs of the Chinese National Library or the Chinese Academy of Sciences Library.

covered.[24] Similarly, the people who produced materials on human evolution for popular consumption during the 1950s did not appear suddenly on the scene when the new state was established in 1949. Rather, they were for the most part people who had been active in science, education, and publishing under the Nationalist regime and who were now, with remarkable rapidity, adjusting their ideas to accommodate the demands of the new political order. Some, like Péi Wénzhōng, had been involved in communist activities before the revolution. Others were more like Jiǎ Lánpō, who was so unfamiliar with the Communist Party that he failed to recognize the face or name of senior party member and former director of the Yán'ān Academy of Sciences, Xú Tèlì, when he visited the fossil exhibition hall where Jiǎ was working shortly after the revolution.[25]

Although books published in the early 1950s had often been started before the revolution, most authors seem to have taken care to incorporate Engels's theory that labor created humanity and other Marxist ideas into their texts before publication. As a whole, however, the books owed much to the authors' rich sets of experiences and connections to Western, nonsocialist scientists and scientific ideas. Sometimes this influence was explicit, but more often it silently shaped the texts while the names of the scientists, associated as they were with capitalism and imperialism, were left out.[26]

The influence of Soviet authors was far more transparent and unproblematic, beginning in the socialist camp before 1949 and lasting well after the Sino-Soviet split in 1960.[27] This influence further helped create a consistent feeling to Chinese popular science materials on human evolution. Mikhail Ilin's enormously popular series on human history, known to the English-reading world as *How Man Became a Giant*, was second only to Engels's *From Ape to Human* in its impact on Chinese popular books on human evolution.[28] The series began by integrating the basic

24. Jiǎ Lánpō, interview with Dennis Etler, 26 September 1999. In the early 1960s, Jiǎ Lánpō and Péi Wénzhōng debated the question of whether or not Peking Man represented the earliest humans.

25. Jiǎ Lánpō, interview with Dennis Etler, 25 September 1999.

26. In 1961, Francis Hsiu offered a different perspective based on his survey of 1950s Chinese journal articles. He found that anthropological scientists were "still Euro-American oriented more than they seem[ed] to be Russian oriented." He noted more references to English, French, and German ("especially English") than to Russian sources, and he saw the continued influence of "Euro-American" methodology in their research. The difference between our two findings probably stems from the different ways scientists presented their work in popular and professional contexts. Francis Hsiu, "Anthropological Sciences," in *Sciences in Communist China*, ed. Sidney Gould (Washington, D.C.: American Association for the Advancement of Science, 1961), 147.

27. M. A. Gremiatskii [Gélièmǐyàcíjī], *Rénlèi shì zěnyàng qǐyuán de* [How Humans Originated], trans. Liú Quán and Wú Xīnzhì (Běijīng: Kēxué chūbǎnshè, 1964).

28. M. Ilin [Yīlín] and E. Segal [Xièjiāěr], *Rén zěnyàng biànchéng jùrén* [How Humans Became

theory of human origins as put forward by Engels in 1876 with other ideas circulated and evidence found since that time. Many Chinese writers followed Ilin in presenting, for example, an anecdote about the 1925 Scopes trial in the United States in which a schoolteacher in Tennessee challenged a law against teaching evolution in the classroom and the local inhabitants protested that they were "not monkeys" and "would not be made monkeys of."[29] The story helped reinforce the political message that the supposedly advanced United States was actually held back by religious belief.

At least one other popular Soviet book on human origins was translated into Chinese and published before 1949, and many others quickly followed the revolution.[30] All of these translated books tended to dwell on very similar themes, covering not only current knowledge about the evidence for human evolution but also a history of the major discoveries and theoretical advances. The issues they addressed and examples they employed became the core issues and examples covered by the majority of books written on the subject by Chinese authors.

Museums offered yet another arena in which people learned about human evolution. They served dual functions as research institutions and places of dissemination, and they were usually staffed with university professors and scientists who also held positions in various institutes. Most museums fell under the responsibility of the Museum Department of the Culture Ministry, which Péi Wénzhōng led in the first few years.

In 1949, the new government took over the old Běipíng [Běijīng] History Museum at Āndìngmén. The core permanent exhibit was planned in March 1951 as Exhibit of the Comprehensive History of China (中國通史陳列). In order to "meet the needs of all the cadres studying the history of social development," they "began by planning a Chinese Primitive Society Exhibit" and invited members of IVPP and other institutions to assist in its construction. This became the model for museums all over the country that were in the process of creating their own exhibits "from the perspective of historical materialism." In 1951 alone, 251,400 people

Giants) (Běijīng: Kāimíng shūdiàn, 1951). Chinese translations were published in Shànghǎi (1947) and in Kāifēng (1948); it was reprinted many times after 1949.

29. M. Ilin and E. Segal, *How Man Became a Giant*, trans. Beatrice Kinkead (Philadelphia: J. B. Lippincott Company, 1942), 38; Fāng Qiè, *Cóng yuán dào rén*, 5–6; Yáng Yè, *Wǒmén de zǔxiān* [Our Ancestors] (Hànkǒu: Wǔhàn gōngrén chūbǎnshè, 1952), 7; Jiǎ Zǔzhāng, *Cóng yuán dào rén* [From Ape to Human] ([Shànghǎi?]: Kāimíng shūdiàn, 1950), 2–3.

30. In 1946, a press in Shànghǎi (then still under Nationalist rule) published G. A. Gurev's *How Humans Developed*, and it was republished several times after the revolution. Gurev [Gǔlièfú], *Rénlèi shì zěnyàng zhǎngchéng de* [How Humans Developed], trans. Chén Yìngxīn (Shànghǎi: Kāimíng shūdiàn, 1950 [1946]). See the bibliography for other titles.

visited the History Museum's exhibit on primitive society.[31] Printed museum guides dating from 1959 and the early 1960s document that the first thing visitors were meant to learn about history at the museum was that labor created humanity.[32]

While the History Museum began with Engels's theory, the Natural History Museum ended with it. The transition from ape to human was the culmination of the paleontology room that also displayed dinosaurs, early birds, and other fossil organisms.[33] Planning for the construction of the museum began in 1950, and Péi Wénzhōng and the IVPP director, Yáng Zhōngjiàn, played leading roles. The museum did not open until 1959; when it did, Yáng became the director, a post he held until his death in 1979.[34]

Also in the Běijīng area was the Peking Man Site Exhibition Hall at Zhōukǒudiàn. A makeshift exhibit was completed soon after 1949. In 1951, the deputy director of the Chinese Academy of Sciences, Zhú Kēzhēn, visited the site and recommended the construction of a formal exhibition hall. A small hall opened to the public in 1953, and in 1955 it expanded to accommodate the growing number of visitors. During these years, the state constructed what was by the standards of the time a major highway to link Běijīng and Zhōukǒudiàn; to this day it is called Jīng-Zhōu Road (京周公路) despite the decline in Zhōukǒudiàn's importance. Approximately 100,000 people, about 2,000 of them foreign, visited the site in 1956.[35] This exhibition was somewhat different from those at history and nature museums. While visitors were exposed to the theory that labor created humanity, the exhibit itself focused much more closely on the site, its fossils, and its excavation history.[36]

---

31. Běijīng bówùguǎn xuéhuì, *Běijīng bówùguǎn niánjiàn (1912–1987)* [Yearbook of Běijīng Museums (1912–1987)] (Běijīng: Běijīng Yànshān chūbǎnshè, 1989), 155–156, 172, 182.

32. Zhōngguó lìshǐ bówùguǎn, *Zhōngguó lìshǐ bówùguǎn yù zhǎn shuōmíng* [Chinese History Museum Exhibit Guide] (Běijīng: Wénwù chūbǎnshè, 1959), 2; Zhōngguó lìshǐ bówùguǎn, *Zhōngguó lìshǐ bówùguǎn tōngshǐ chénliè shuōmíng* [Guide to the Comprehensive History Exhibit of the Chinese History Museum] (Běijīng: Wénwù chūbǎnshè, 1964), 1; and others. By this time, the History Museum had moved to its new and impressive location on the east side of Tiānānmén Square. Museums in other cities followed a similar pattern. A 1958 guidebook to the Xiàmén University Anthropology Museum indicates that the first hall included an explanation of the theory that labor created humanity. Lín Huìxiáng, *Xiàmén dàxué rénlèi bówùguǎn chénlièpǐn shuōmíng shū* [Guidebook to the Exhibited Materials at the Xiàmén University Anthropology Museum] (Xiàmén: Xiàmén dàxué rénlèi bówùguǎn, 1958), 4.

33. Zhōngyāng zìrán bówùguǎn, *Gǔshēngwù, dòngwù, zhíwù chénliè jiǎnjiè* [Brief Introduction to the Paleontology, Zoology, and Botany Exhibits] (Běijīng: Zhōngyāng zìrán bówùguǎn, 1961), 1–5.

34. Běijīng bówùguǎn xuéhuì, *Běijīng bówùguǎn niánjiàn*, 493.

35. Jia Lanpo and Huang Weiwen, *The Story of Peking Man: From Archaeology to Mystery*, trans. Yi Zhiqi (Beijing: Foreign Languages Press, 1990), 214–215.

36. Visitors who signed the 1956 guest book often discussed "Labor created humanity," but the pamphlet published in 1956 made no mention of this or of any other aspect of materialist

Nonetheless, one visitor wrote in the exhibition hall's guest book that the "ironclad evidence" of fossils encountered at Zhōukǒudiàn swept away all doubts that "labor created humanity" the way cadre classes the visitor had attended in 1949 could not.[37]

The state further created several organizations whose specific mandate was to produce materials on human evolution and organize their dissemination through a variety of activities for cadres, workers, peasants, and soldiers. In 1950 and 1951, the Culture Ministry's Science Dissemination Bureau (文化部科學普及局) organized lectures, slide shows, exhibits, and other forms of dissemination. Regional and local chapters of the Science Dissemination Bureau made the story of "ape to human," and specifically the theory that labor created humanity, one of their main priorities. An exhibit in Shànghǎi received 1,800 people an hour for a total of 418,726 visitors.[38] Almost 100,000 people attended the Spring Festival Scientific Knowledge Exhibition in Běijīng, where "From Ape to Human" was one of three focal areas.[39] Similar exhibits were mounted in cities all over China.[40]

In October 1951, the Bureau of Science Dissemination merged with another related bureau in the Culture Ministry. At that time, it turned over most responsibility for science dissemination to another organization that was already engaged in similar activities, the All-China Association for the Dissemination of Scientific and Technological Knowledge (中華全國科學技術普及協會, hereafter Science Dissemination Association).[41] The Science Dissemination Association was founded in August 1950 alongside the professionally oriented All-China Federation of Scientific Societies. These two organizations merged in 1958 to form the Chinese Association for Science and Technology (CAST, 中國科學技術協會), which remains in operation today and in 2002 was named by law the principal organization in charge of science dissemination.[42] They were from the begin-

---

interpretations of human evolution. Many of the visitors had surely already encountered the theory in other educational contexts, and the guides at Zhōukòudiàn may also have emphasized it. See Zhōngguó kēxuéyuàn gǔjǐzhuī dòngwù yánjiūshì, *Zhōngguó yuánrén zhī jiā* [Peking Man's Home] (Běijīng: Rénmín měishù chūbǎnshè, 1956).

37. Zhōukǒudiàn Peking Man Site Exhibition Hall Guest Book, [22?] July 1956.

38. Huádōng wénhuà bù kēxué pǔjí chù, "Shànghǎi 'Cóng yuán dào rén' zhǎnlǎn huì zǒngjié" [Summary of Shànghǎi's "From Ape to Human" Exhibition], *Kēxué pǔjí tōngxùn*, no. 10 (1950): 216.

39. "Shǒudū chūnjié kēxué zhīshí zhǎnlǎn huì jīngyàn zǒngjié" [Summary of the Capital Spring Festival Scientific Knowledge Exhibition], *Kēxué pǔjí tōngxùn*, no. 2 (1950): 24.

40. "Yīnián lái de kēxué pǔjí yùndòng" [The Science Dissemination Movement after One Year"], *Kēxué pǔjí tōngxùn*, no. 10 (1950): 210.

41. Shěn Qíyì et al., *Zhōngguó kēxué*, 43.

42. Zhōnghuá rénmín gònghéguó kēxué jìshù pǔjí fǎ [People's Republic of China Science and Technology Dissemination Law], 2002, article 2, section 12.

ning considered separate from the government itself, much like the trade unions, the most influential of which represented workers, youth, and women, respectively. Nonetheless, they were funded by the government and not coincidentally shared the government's priorities. Like the governmental and party apparatuses, they had national-level organizations and provincial, county, and municipal chapters, each of which drew up its own constitution. The Běijīng chapter, for example, determined that members would include anyone who had adequate scientific or technological knowledge or relevant production experience and whom two current members had recommended.[43] Activities included lectures, demonstrations, slide shows, exhibits, movies, magazines, newspaper supplements, radio programs, community posters, and songs. In 1951, for example, the Science Dissemination Association reported that it reached a total of 589,063 people at lectures, 1,090,024 people at movies and slide shows, and 973,052 people at exhibits.[44]

Like the Science Dissemination Bureau, the association made knowledge of human origins a top priority. In this, they were also following their counterpart and model in the Soviet Union.[45] In 1951, the association's journal *Science Dissemination Work* published a lecture by Péi Wénzhōng (who also served as vice-director of the association's planning committee) entitled "Labor Created Humanity"; it was designed for science dissemination workers throughout China to emulate.[46] In the same issue, it also printed sample materials for a slide show entitled "From Ape to Human" created by the Shànghǎi Municipal History Museum and the Shànghǎi Science and Technology Slide Show Work Association.[47] Documents from 1954 and 1955 indicate that "labor created humanity" remained one of fourteen or fifteen topics on which they planned to educate workers, peasants, soldiers, and regular cadres.[48]

43. "Běijīng shì kēxué jìshù pǔjí xiéhuì chóubèi wěiyuánhuì zhāngchéng (cǎoàn)" [Draft Constitution by the Běijīng Science and Technology Dissemination Association Planning Committee], n.d. (probably February 1951), Běijīng Municipal Archives 10.1.22: 32.

44. "1951 nián yèwù gōngzuò gàikuàng biǎo" [Chart for Work (Accomplished in) 1951], Zhōngguó kēxué jìshù pǔjí xiéhuì archives.

45. China's Science Dissemination Association sent a delegation to Moscow in 1954 to learn from its Soviet counterpart. The materials they collected and published indicate that science dissemination in both countries was already organized very similarly and covered similar topics. See Zhōnghuá quánguó kēxué jìshù pǔjí xiéhuì, *Zhōnghuá quánguó kēxué jìshù pǔjí xiéhuì fǎng Sū dàibiǎo tuán zīliào huìbiān* [Compilation of Materials from the All-China Scientific and Technological Knowledge Dissemination Association's Delegation to the Soviet Union], vol. 1 (Běijīng, 1955).

46. Péi Wénzhōng, "Láodòng chuàngzào le rén (jiǎngyǎn cáiliào)" [Labor Created Humanity (Lecture Materials)], *Kēxué pǔjí gōngzuò*, no. 4 (1951): 93–100.

47. Shànghǎi shìlì lìshǐ bówùguǎn and Shànghǎi kējì huàndēng gōngzuò huì, "Cóng yuán dào rén" [From Ape to Human], *Kēxué pǔjí gōngzuò*, no. 4 (1951): 112–116.

48. "Kēpǔ 1954 nián gōngzuò jìhuà zǒngjié" [Summary Plan for 1954 Science Dissemination

Over the years, the Science Dissemination Association's scope and authority grew. By 1958, when it merged with the professional Federation of Scientific Societies to become CAST, it had established a Science Dissemination Press with national, provincial, and municipal branches and had taken over the operation of six of the country's most important science dissemination magazines, including *Popular Science* (科學大眾), *Popular Science Monthly* (科學畫報, literally *Science Pictorial*), and the Chinese version of the Soviet magazine *Knowledge Is Power* (知識就是力量). The first of these had been launched during the republican era and had remained under the direction of the Science Society of China in Shànghǎi until 1957.[49] The Science Dissemination Association also formed collaborative partnerships with trade unions and with relevant government organizations, such as cultural centers (文化館) and health and forestry agencies.[50] It worked particularly closely with the cultural centers, which took up science dissemination as one of four mandates.[51]

While multiple organizations and individuals were involved in science dissemination in China during the 1950s and 1960s, they were connected to one another in various institutional ways. This helped them cooperate in identifying priorities and crafting a coherent and consistent story of human origins. A certain diversity of opinion did prevail on some key scientific issues (for example, the geographic birthplace of humanity), but the core theme of the materials remained remarkably uniform: specifically, the use of "Labor created humanity" to instill a materialist worldview and thereby eliminate alternative "superstitious" or "idealist" perspectives.

Work], Zhōngguó kēxué jìshù xiéhuì archives; "1955 kēpǔ xiéhuì chángqī guīhuà" [1955 Long-Term Plan for the Science Dissemination Association], Zhōngguó kēxué jìshù xiéhuì archives. The others were the sun, moon, and stars; wind and rain; the human body; germs; plant growth; animal evolution; natural disasters; the great ancestral country; the flourishing of the great peace; social development; electricity; industry; scientific research applications; and atomic power and the future.

49. Shěn Qíyì et al., *Zhōngguó kēxué*, 54, 356–357.

50. Shěn Qíyì et al., *Zhōngguó kēxué*, 47–49; "Běijīng shì kēxué jìshù pǔjí xiéhuì chóubèi wěiyuánhuì èr nián lái gōngzuò bàogào" [Report on Two Years of Work by the Běijīng Science and Technology Dissemination Association Planning Committee], 21 January 1953, Běijīng Municipal Archives 10.1.26: 4–9. Cultural centers are organized under the Culture Bureau at the county or township level. Their responsibilities include safeguarding cultural relics, managing performing troupes, and producing educational exhibits.

51. "Wénhuà guǎn xuānchuán gànbù huìyì shàng duì kēpǔ gōngzuò de pīpíng hé jiànyì" [A Cultural Center Propaganda Cadre Makes Criticisms and Offers Opinions on Science Dissemination Work], 28 December 1957, Běijīng Municipal Archives 10.1.81: 6–7.

Darwin "Strikes a Blow" for Materialism

Many science dissemination materials published in the 1950s and 1960s sought to "squash superstition" (破除迷信) by presenting well-known stories about natural phenomena and then providing scientific explanations as replacements. "Science and superstition are opposites," asserted the author of a book entitled *Talking about Heaven and Earth, Squashing Superstition.*[52] The task of science, such materials often explained, was to "crack the mysteries" of natural phenomena and so provide a materialist understanding of the universe.[53]

Cuthbert O'Gara, a Catholic bishop who was in China in 1949, later recollected that in seminars organized immediately upon the communists' assumption of control, "The very first, the *fundamental*, lesson given was man's descent from the ape—Darwinism!"[54] Chinese books on human evolution, like their Soviet counterparts, used Darwin and Engels as a one-two materialist punch: Darwin to demonstrate the commonality of humans and animals, and Engels to assert the primacy of labor. The chief target in this program was religious idealism (the philosophical antithesis of materialism), and it was therefore religious creation stories that served as antagonists in these narratives. The title of the opening chapter of one Soviet book translated into Chinese asked bluntly, "Do You Believe in the Bible or in Science?" While focusing on a criticism of the Genesis story in the Bible, the author, G. A. Gurev, also included a discussion of other stories, for example, the Egyptian ram-headed god Khnum's creation of humans on a potter's wheel—an image of which appeared in many Soviet and Chinese books on human evolution in the 1950s and beyond.[55] Chinese writers largely adopted the anti-Christian crusade of the Soviet materials, and they often added references to Chinese creation stories—most commonly of the male Pángǔ, who created the world by splitting heaven and earth, and the female Nǚwā, who created humans out of mud.[56]

52. Sūn Wǔ, *Tán tiān shuō dì, pò míxìn* [Talking about Heaven and Earth, Squashing Superstition] (Tiānjīn: Tiānjīn rénmín chūbǎnshè, 1964), 3. On other efforts in this period to suppress superstition, see Steve A. Smith, "Local Cadres Confront the Supernatural: The Politics of Holy Water (*Shenshui*) in the PRC, 1949–1966," *The China Quarterly* 186 (2006): 999–1022.

53. Sūn Wǔ, *Tán tiān shuō dì*, 5–8; Zhāng Yìzhěn, Hán Fǔ, et al., *Jiěkāi kēxué zhī mí* [Cracking the Mysteries of Science] (Shànghǎi: Shàonián értóng chūbǎnshè, 1962).

54. Cuthbert M. O'Gara, *The Surrender to Secularism* (St. Louis: Cardinal Mindszenty Foundation, 1989 [1967]), 11.

55. [G. A.] Gurev [Gǔlièfú], *Rénlèi shì zěnyàng zhǎngchéng de* [How Humans Developed], trans. Chén Yìngxīn (Běijīng: Zhōngguó qīngnián chūbǎnshè, 1953), 5.

56. Wáng Shān, *Láodòng chuàngzào rénlèi*, 1; Guō Yìshí, *Rénlèi shì cóng nǎlǐ lái de* [Where Humans Came From] (Běijīng: Tōngsú dúwù chūbǎnshè, 1955), 1; Huáng Wànbō, Shěn Wénlóng, and Hú

Some authors offered a sympathetic, if patronizing, interpretation of the origins of religious explanations for natural phenomena. For Gurev, religion was created by "savages" who were "like little children because they did not understand very much" about the natural world.[57] Similarly, Jiǎ Lánpō in his picture book *Our Ancestors of 200,000 Years Ago* said that superstitions arose out of early humans' fear and misunderstanding of natural phenomena like thunder and lightning.[58] At least two Chinese books noted a similarity in several of the origin stories: the Christian, Egyptian, and Chinese stories all described humans as having been created out of dust, clay, or mud, and thus probably emerged when pottery became an important technology in these societies.[59] (No one, however, appears to have risked mixing science and superstition by suggesting that it was Nǚwā's labor that created humanity!)

Other materials took a harsher stand on religion by focusing on its later use by the ruling classes to deceive and exploit the workers. The Soviet writer M. S. Plisetskii dedicated an entire chapter of his *How Humans Originated and Developed* to the question of why capitalism supports religion, with substantial discussion also of the relationship between religion and imperialism.[60] A 1950 Chinese book proclaimed, "This kind of religious superstitious thought is very damaging to workers . . . making them think they have no power to conquer oppression."[61] The Chinese books sometimes further referred to religious accounts of human origins derisively as "ghost stories," in this way linking them to other forms of superstition slated for eradication under socialism.[62] Accounts of creation stories in these materials were in all cases set up to be torn down

---

Huìqīng,*Wǒmén de zǔxiān* [Our Ancestors] (Běijīng: Běijīng kēxué pǔjí chūbǎnshè, 1958), 1; Fāng Shàoqīng [pseud. for Fāng Zōngxī], *Gǔ yuán zěnyàng biànchéng rén* [How Ancient Apes Became Human] (Běijīng: Zhōngguó qīngnián chūbǎnshè, 1965), 4; and Wú Rǔkāng, *Rénlèi de qǐyuán hé fāzhǎn* [Human Origins and Development] (Běijīng: Kēxué pǔjí chūbǎnshè, 1965), 6. I have found only two sources that discuss Buddhist creation stories: Wáng Shān, *Láodòng chuàngzào rénlèi*, 1; and Běijīng shūdiàn, ed., *Rén cóng nǎlǐ lái* [Where Humans Come From] (Běijīng: Běijīng shūdiàn, 1951), preface (no page number).

57. Gurev, *Rénlèi shì zěnyàng zhǎngchéng de* (1953), 10.

58. Yáng Háonìng (art) and Jiǎ Lánpō (text), *Wǒmén de zǔxiān 2: Wǒmén èrshíwàn nián de zǔxiān* [Our Ancestors, vol. 2: Our Ancestors 200,000 Years Ago] ([Tiānjīn?]: Zhīshí shūdiàn, 1951), 47.

59. Huáng Wéiróng, *Zhōngguó yuánrén* [Peking Man] (Shànghǎi: Shàonián értóng chūbǎnshè, 1954), 1–2; and Wú Rǔkāng, *Rénlèi de qǐyuán* (1965), 6.

60. M. S. Plisetskii [Pǔlíxīcíjī], *Rénlèi zěnyàng shēngchǎn hé fāzhǎn de* [How Humans Were Produced and Developed], trans. Bì Lí (Shànghǎi: Zhōnghuá shūjú, 1951), 17–21.

61. Wáng Xiǎoshí, *Cóng yuán dào rén: Tōngsú jiǎnghuà* [From Ape to Human: A Simple Account] (Shànghǎi: Xīnyà shūdiàn, 1950), 61. See also Huáng Wéiróng, *Zhōngguó yuánrén*, 3; Guō Yìshí, *Rénlèi shì cóng*, 19–20; Péi Wénzhōng, *Rénlèi de qǐyuán hé fāzhǎn* [Human Origins and Development] (Běijīng: Zhōngguó qīngnián chūbǎnshè, 1956), 1; Wú Rǔkāng, *Rénlèi de qǐyuán* (1965), 8.

62. Yáng Háonìng and Jiǎ Lánpō, *Wǒmén de zǔxiān 2*, preface (no page number); Dǒng Shuǎngqiū, *Rén shì zěnyàng lái de* [How People Came to Be] (Chángshā: Húnán rénmín chūbǎnshè, 1957), 2.

by science—a pattern that began in the republican era but became much fiercer in the People's Republic. Darwin was usually given first honors: he was credited with "conquering religion" and acclaimed as a "great scientist" who showed that humans are a part of the natural world and were not created by God.[63]

The heroic status that Darwin enjoyed in the Soviet Union and socialist China has been somewhat obscured by the infamous politics surrounding Lysenkoism. Trofim Lysenko was a Soviet agronomist who conducted experiments during the 1930s that purported to show that organisms could be forced to adapt to environmental conditions, and that their descendants would inherit these adaptations. While scientists in other countries were building what today remains the powerful field of modern genetics, Lysenko was insisting on the validity of the inheritance of acquired characteristics. Lysenkoism's appeal came partly from its formidable potential utility (were it only sound) in improving strains of agricultural crops[64] and from its association with Ivan Vladimirovich Michurin, portrayed as "the earthy, grizzled farmer whose personal involvement with production is the source of endless creative innovation."[65] Lysenkoism first came to China in the late 1940s, and until 1956 it dominated Mendelian genetics, with proponents of the latter frequently finding themselves criticized as "idealists" or even as followers of "fascist eugenics." After 1956, Lysenkoism lost its monopoly, and both schools enjoyed some latitude in research.[66]

Yet even during the height of Lysenkoism, Darwin's contribution to science and to materialism was not in question. Nor was this inconsistent: Darwin himself had no accepted mechanism for the transmission of traits from one generation to the next, and he even allowed for the possibility of the inheritance of acquired characteristics in some cases.[67] It was the geneticists (especially Gregor Mendel, Thomas Morgan, and Charles

63. Fāng Qiĕ, *Cóng yuán dào rén*, 8; V. K. Nikol'skii [Níkēĕrsíjî], *Yuánshĭ shèhuì shĭ* [The History of Primitive Society], trans. Páng Lóng (Shànghăi: Zuòjiā shūjú, 1953), 48.

64. Loren Graham, *Science in Russia and the Soviet Union: A Short History* (Cambridge: Cambridge University Press, 1993), 124. Graham specifically notes that "to the end of his days [Lysenko] never applied his biological scheme to human beings."

65. Laurence Schneider, "Learning from Russia: Lysenkoism and the Fate of Genetics in China, 1950–1986," in *Science and Technology in Post-Mao China*, ed. Denis Fred Simon and Merle Goldman (Cambridge, Mass.: Council on East Asian Studies, Harvard University, 1989), 47. Note that in China Lysenko's theories were typically identified as Michurinism rather than Lysenkoism.

66. Schneider, "Learning from Russia."

67. Charles Darwin, *The Descent of Man, and Selection in Relation to Sex*, 2nd ed. (New York: Clarke, Given, and Hooper, 1874 [1870–1871]), 36–40. On page 40, Darwin concludes, "Consequently we may infer that when at a remote epoch the progenitors of man were in a transitional state, and were changing from quadrupeds into bipeds, Natural Selection would probably have been greatly aided by the inherited effects of the increased or diminished use of the different parts of the body."

Weissman) who bore the brunt of Lysenkoist attacks. Interestingly, one of the criticisms leveled against experimental genetics was that it took a religious-like view of the body as an impermanent vessel for genes that live on from one generation to the next and are immutable with respect to the environment.[68]

While such criticisms painted Mendelian geneticists as wholly un-redeemable, Darwin's many acknowledged shortcomings did not knock him off the pedestal on which Marx and Engels had placed him. Both philosophers made numerous laudatory references to Darwin's work. Marx wrote in an 1860 letter to Engels that Darwin's *The Origin of Species* was "the book which, in the field of natural history, provides the basis for our views."[69] Engels later wrote that, with respect to knowledge of historical evolution, "Darwin must be named before all others. He dealt the metaphysical conception of nature the heaviest blow by his proof that the organic world of today, plants, animals, and man himself, are the products of a process of evolution going on through millions of years."[70] Both Marx and Engels saw their own work as part of a larger intellectual movement characterized by a materialist and developmentalist understanding of the world, to which Darwin had made an enormous contribution.

Chinese and Soviet authors of books on human evolution did sometimes echo Marx's criticism that Darwin had suffered from idealist and capitalist influences. For example, a 1950 Chinese book noted that Darwin's idealist perspective caused him to see only the effects of competition and not of cooperation.[71] A Soviet writer further linked Darwin's emphasis on competition to the influence of Malthus's theories of population.[72] Most commonly, however, such writings portrayed Darwin's theory of evolution as merely incomplete, rather than philosophically flawed. As Gurev put it, Darwin "did not completely resolve the question of human origins . . . [because] he did not completely resolve how humans separated from the biological world."[73] Péi Wénzhōng and Jiǎ Lánpō's 1954 book, *Labor Created Humanity*, similarly explained that Darwinism had a "fundamental gap" because Darwin "had not yet realized the basic difference between humans and the animal world, and so

68. M. S. Plisetskii [Pǔlièxuécíjī], *Rénlèi qǐyuán de kēxué jiěshì yǔ zōngjiào chuánshuō* [Scientific Explanation and Religious Myths of Human Origins], trans. Huáng Dèngzhōng (Shànghǎi: Liányíng shūjú, 1950), 61.

69. Karl Marx, "Marx to Engels, 19 December 1860," in *Karl Marx, Frederick Engels: Collected Works*, ed. E. J. Hobsbawm et al. (Moscow: Progress Publishers, 1975–), 41:232.

70. Frederick Engels, *Socialism: Utopian and Scientific*, in *Karl Marx, Frederick Engels*, 24:301.

71. Fāng Qiě, *Cóng yuán dào rén*, 9.

72. Plisetskii, *Rénlèi shì zěnyàng*, 56.

73. Gurev, *Rénlèi shì zěnyàng zhǎngchéng de* (1953), 18.

was unable to point out the differences in the causes for human evolution and the causes for animal evolution."[74] Engels himself had filled this gap. As Wú Rǔkāng neatly summarized, "If we say that Darwin liberated humans from the hands of God and returned us to the animal world, then Engels separated us from the animal world, allowing people to see clearly that humans are laborers and the transformers of nature."[75] Engels's thesis that labor created humanity thus formed a necessary complement to Darwinism, which had failed to identify the key difference between humans and (other) animals: whereas other animals adapt to the natural world, humans transform themselves and the natural world through labor.

While some materials on human origins also addressed native beliefs—most commonly about Nǚwā, but also sometimes from the Buddhist tradition—the chief superstition targeted was Christian creationism. This was somewhat strange in a country with only a few million Christians out of a total population of more than 500 million.[76] Several scientists I spoke with in 2002, however, indicated they thought Christianity was a "problem" for China. This may be because some of their fieldwork has been in areas where Western missionaries had established churches. Another factor explaining the attention given to Christianity was, no doubt, the sheer power of the Soviet example. In the early years, it probably also resulted from the party's concern about the influence of republican-era missionary education on intellectuals and other urbanites. For example, biologists educated during the republican era had almost always attended a missionary school at some point.[77] As time went on and such reasons faded, continued bashing of Christian "idealism" was a way of reminding people of the superiority of socialism, especially as practiced in China.

In any case, science dissemination in China was premised on the notion that the people were hampered by superstition and needed help from above to acquire scientific knowledge. However, there are multiple other ways science dissemination could have been configured. Here a comparison with recent "public understanding of science" movements in the West is instructive. The dominant view among British academics

74. Péi Wénzhōng and Jiǎ Lánpō, *Láodòng chuàngzào le rén* [Labor Created Humanity] (Běijīng: Zhōnghuá quánguó kēxué jìshù pǔjí xiéhuì, 1954), 1–2.

75. Wú Rǔkāng, *Rénlèi de qǐyuán* (1965), 11.

76. There were about 3 million Catholics and fewer than 1 million Protestants in China in 1949. Richard Madsen, *China's Catholics: Tragedy and Hope in an Emerging Civil Society* (Berkeley: University of California Press, 1998), 137.

77. Laurence Schneider, *Biology and Revolution in Twentieth-Century China* (Lanham, Md.: Rowman and Littlefield, 2003), 280.

and policy makers seeking to improve public understanding of science through better dissemination has been critically characterized as a "deficit model," which "sees the public as blank slates or empty vessels—as minds in deficit that need scientific information in order to be replete."[78] Thus, the relevant category in Britain has more often been "ignorance" than "superstition." This model has often been assumed to apply as well to the public understanding of science movement in the United States, but Jane Lehr's recent reconsideration shows a significant difference.[79] Celebrated astronomer and science popularizer Carl Sagan and others have led a crusade against what they consider to be antiscientific forms of knowledge, such as belief in alien abductions, repressed memories, the power of crystals, or urine therapy.[80] Thus, somewhat like their Chinese counterparts, American advocates for science education have focused on modes of thought ("scientific" versus "superstitious") rather than content ("knowledge" versus "ignorance"). Still, there are critical differences between American and Chinese priorities, the most important being the American emphasis on connecting science literacy with responsible participation in a democratic society.[81] In contrast to all these formulations, some Western writers on science and the public have sought to "replace the notion of 'public ignorance' with a much richer pattern of social relations and personal understandings."[82] Such accounts have acknowledged that laypeople often have their own interpretations of the phenomena in question, and that they are agents in reformulating disseminated scientific knowledge and in arguing for their positions based on their own senses of social identity.

Popular knowledge of science can thus be characterized in multiple ways, and so the specific way adopted in socialist China is significant.

78. Jane Gregory and Steve Miller, *Science in Public: Communication, Culture, and Credibility* (New York: Plenum Trade, 1998), 17. The term "deficit model" is often credited to Brian Wynne. See his "Knowledges in Context," *Science, Technology, and Human Values* 16, no. 1 (1991): 113.

79. Jane Lehr, "Social Justice Pedagogies and Scientific Knowledge: Remaking Citizenship in the Non-science Classroom" (Ph.D. diss.: Virginia Tech University, 2006), especially 62. Lehr notes that the focus on process rather than content is recognized as the "second phase" in the public understanding of science movement in the United Kingdom, whereas it has been central to the U.S. model since the early 1980s.

80. Carl Sagan, *The Demon-Haunted World: Science as a Candle in the Dark* (New York: Ballantine Books, 1997); Martin Gardner, *Did Adam and Eve Have Navels? Discourses on Reflexology, Numerology, Urine Therapy, and Other Dubious Subjects* (New York: W. W. Norton, 2000). Sagan's *The Demon-Haunted World* has been translated into Chinese and in 2002 was sold at Běijīng book fairs side by side with tantalizing tales of the paranormal. Sagan would be spinning in his grave, if his commitments to rationality allowed it.

81. See Sagan's *The Demon-Haunted World* and chapter 2 of Lehr's "Social Justice Pedagogies."

82. Alan Irwin and Brian Wynne, eds., *Misunderstanding Science? The Public Reconstruction of Science and Technology* (New York: Cambridge University Press, 2003 [1996]), 9.

On one hand, the term "superstition" acknowledges that people are thinking something, even if it is not what one wants them to be thinking, while "ignorance" suggests there are holes in people's heads that you simply fill with science. In other words, science disseminators in China have engaged directly with the issue of religious belief, whereas the "deficit model" dodges the issue by discussing only "ignorance" and thus avoids explicitly requiring the eradication of existing knowledge forms. A "superstition model," however, does not allow for their survival: science replaces them. Thus, adopting such a model of science dissemination made possible in socialist China a conception of scientific knowledge strongly hierarchical even by standards of science dissemination, and one that legitimated, or even required, an attack on popular culture.

Scientists Feel the Heat

Attacking superstition was a state priority to which scientists were uniquely able to contribute. For many scientists, a commitment to scientific methods and principles predisposed them to be hostile to religious or otherwise superstitious views. The replacement of such perspectives with a scientific worldview was thus a value that political and scientific elites shared. The role of scientists in this endeavor was even codified in Máo's seminal article "On New Democracy," which in the earliest years of the People's Republic still defined the political environment. Scientists, however, were not the only ones to entertain hopes in the first few years regarding their fate under the new regime. In participating in attacks that helped to crush this hope for people of faith, scientists should perhaps have reflected on their own shaky position as purveyors of truths only sometimes in harmony with official ideology.

Máo's "On New Democracy" had provided for the possibility that "even religious people" could join the proletariat in a united front against imperialism and feudalism although the party would "never approve of their . . . religious doctrines."[83] In 1950 and early 1951, Protestants and Catholics had productive meetings with Premier Zhōu Ēnlái in which Zhōu went so far as to give provisional assurance that ties between Chinese Catholic churches and the Vatican could be maintained.[84] In January 1951, a Christian critic of evolutionary science was even able to

---

83. Mao Tse-tung, *Selected Works*, vol. 2, 381.
84. See John Tong, "The Church from 1949 to 1990," in *The Catholic Church in Modern China: Perspectives*, ed. Edmond Tang and Jean-Paul Wiest (Maryknoll, N.Y.: Orbis Books, 1993), 8–9.

publish a book espousing a creationist interpretation of human origins. The book—entitled *The Question of Human Origins* and published by the Catholic Education Joint Committee—attempted through persuasive argument to deny the validity (or at least the certainty) of scientific claims about human evolution. In a manner similar to that employed by creationists today, the author cited famous scientists including Péi Wénzhōng and Henry Fairfield Osborn to reveal that scientists themselves recognized the spottiness of the fossil record and the difficulty interpreting what existed. He then noted, "Usually when we read about evolutionary theory, we come across the sentence [referring to Darwin's contribution]: 'And so the ghost story that God created humans was toppled.' What we have seen above is that the question of human evolution is an unproven theory."[85] After raising a few "problems" with evolutionary theory, he concluded that human evolution required a designer (計劃者).[86] Had the author waited a few months, his book would never have been published. By summer 1951, the Resist America–Aid Korea Movement (the domestic front of Chinese participation in the Korean War) began to target Protestant and Catholic churches in China: religious leaders were persecuted, and government oversight of all church-related activities tightened dramatically.[87]

Scientists watched the crackdown on religion with little apparent concern for their own activities. Rather, they seem to have seen themselves safely on the side of the state in the construction of a new society that valued science, and many intellectuals continued to believe in the early People's Republic that their contributions to social and national reconstruction would be appreciated. This hope was to be squashed repeatedly in the political movements that followed, but many intellectuals continued to cling to it nonetheless. This story of optimism and disappointment is vividly seen in the experiences of Liú Xián, an influential but politically beleaguered anthropologist who had studied in London with the notable Sir Arthur Keith, had founded China's first anthropology department at Jǐ'nán University in 1947, and subsequently helped found China's most important post-1949 anthropology department at Fùdàn University in 1952.[88]

As a colleague optimistically wrote in the preface to Liú Xián's 1950

85. Lǐ Qīngbō, *Rénlèi qǐyuán wèntí* [The Question of Human Origins] (Shànghǎi: Xīnshēng chūbǎnshè, 1951), 23.

86. For example, he asked why, when the forests disappeared, some apes remained apes and did not evolve into humans. Lǐ Qīngbō, *Rénlèi qǐyuán wèntí*, 24, 30.

87. Tong, "The Church," 10–11.

88. Gregory Guldin, *The Saga of Anthropology in China: From Malinowski to Mao* (Armonk, N.Y.: M. E. Sharpe, 1994), 66, 68, and 100.

book entitled *The History of Development from Ape to Human*, humans had been developing at an astonishing rate since the Stone Age of Peking Man, and "having entered the era of 'New Democracy,' we must all use our hands and our brains in order to create."[89] Eight months later, however, Liú Xián's book fell under attack in a review published in the *People's Daily* by a reader named Hán Wénlǐ (he is not otherwise identified).[90] According to Liú's own recollections, bookstores quickly responded by removing the book from their shelves.[91] Hán's main line of criticism centered on Liú's "idealist" (as opposed to "materialist") portrayal of our ape ancestors as having succeeded in making the transition from the trees to the ground through their bravery in accumulating new experiences and their ability to use these experiences to think of a plan of action. Indeed, Liú's narrative is marked by a strongly voluntarist interpretation of evolution reminiscent of some of the early twentieth-century thinkers who wrote on the subject. The apes in his account seemed to map out their own evolutionary trajectory. This is not surprising, since Liú must have been influenced by his teacher, Arthur Keith, who focused heavily on the role of the brain in human evolution and saw the development of the human brain as the result of a latent tendency in this direction.[92]

Yet it was not precisely voluntarism that Hán opposed, but rather the existence of such voluntarism *before* the necessary first step of engaging in labor. Voluntarism was, after all, a core element of Chinese Marxism, beginning with its founder, Lǐ Dàzhāo, and continuing with Máo Zédōng.[93] After quoting a relevant passage of Liú's book, Hán demanded, "Is this not equivalent to saying that ancient anthropoid apes who had not yet begun to labor were already able to think?" Several paragraphs later, he concluded, "What we must realize is that the ancient apes'

89. Lú Yúdào in Liú Xián, *Cóng yuán dào rén*, preface, 2.
90. Hán Wénlǐ, "Jiǎn píng Liú Xián zhù *Cóng yuán dào rén fāzhǎn shǐ*" [A Brief Criticism of Liú Xián's *A History of the Development from Ape to Human*], *Rénmín rìbào*, 17 June 1951, 6.
91. "Shànghǎi zhīshí jiè gāi guànchè 'bǎijiā zhēngmíng' wèntí" [Shànghǎi Intellectual Circles Take a Turn at the Question of "100 Schools Contending"], *Guāngmíng rìbào*, 1 May 1957, 2. Liú also noted his attempt to publish a reply in the *People's Daily*, but the editors returned his letter, saying that his book was "Mr. Darwin's viewpoint" and that he should study "Mr. Michurin's viewpoint" (i.e., Lysenkoism) more fully.
92. Peter Bowler, *Theories of Human Evolution: A Century of Debate, 1844–1944* (Baltimore: Johns Hopkins University Press, 1986), 168, 207–208. See also Arthur Keith, *Concerning Man's Origin* (New York: G. P. Putnam's Sons, 1928), 25–29. Keith was steeped in a scientific tradition that understood evolution as driven by internal trends (orthogenesis). Liú's narrative strongly reflects this view. Hán's critique of Liú was in this respect actually closer to the emergent consensus among scientists internationally that evolution occurred through adaptation to changing environmental conditions. The role of the environment was thus important both to the neo-Darwinian critique of teleology and to Lysenkoism.
93. Maurice Meisner, *Li Ta-chao and the Origins of Chinese Marxism* (Cambridge, Mass.: Harvard University Press, 1967).

transformation into humans did not depend . . . on 'bravery' or 'deter-mination,' but rather on the long and slow [process of] learning how to labor."[94]

Liú, however, was no rebel. His ambition was not to challenge Engels.[95] Nor is it likely that he intended to write a book that changed the conventional understanding of labor's role in human evolution. Rather, the most probable explanation is that Liú simply failed in his attempt to incorporate Engels's theory that labor created humanity into his pre-existing understanding of human origins. In 1950, there was still enough uncertainty about the political winds and enough belief in the principles of the New Democracy to allow the publication of books on human evolution that did not even mention Engels or his theory that labor created humanity.[96] In January 1951, it was even possible to publish a creationist account of human origins. Criticism of Liú in May 1951—along with at least one other, similar attack on another popular book about human origins in June[97]—signaled the hardening of the orthodox narrative of human evolution. One could now be attacked as Liú Xián was for failing to interpret Engels "correctly," despite apparently earnest efforts in this direction.

Hán's critique foreshadowed years of formal criticism and interference in Liú Xián's research and teaching activities. In 1955, he was again targeted during a Chinese Academy of Sciences campaign against idealism (during which Yáng Zhōngjiàn, Péi Wénzhōng, Jiǎ Lánpō, and other members of IVPP gathered to criticize his book), and he was further caught up in a new offensive waged by supporters of Michurin biology at Fùdàn University.[98] Though the editors of Science Bulletin solicited his

94. Hán Wénlǐ, "Jiǎn píng," 6. Hán had several other objections to Liú's portrayal of human evolution, including Liú's understanding that climate change happened very suddenly. However, it is "anthropomorphism" that one of his students remembers as his chief political label in later years. Interview data, 2002.

95. My reading of Liú's book differs significantly from Schneider's treatment of it in the hardcover edition of Biology and Revolution, 152–153. Schneider sees Liú providing a conscious alternative to Engels's theory that labor created humanity, while I see him attempting unsuccessfully to integrate his own ideas with Engels's. But note that in the paperback edition (2005), Schneider has modified his interpretation to fall somewhere between these two positions.

96. Zhū Xǐ, Wǒmén de zǔxiān [Our Ancestors] (Shànghǎi: Wénhuà shēnghuó chūbǎnshè, 1950 [1940]).

97. See Zhōu Lóng, "Píng Láodòng chuàngzào rén" [Criticizing Labor Created Humanity], Wénhuì bào fùkān, 19 June 1951. As in Liú Xián's case, the review criticized the book for, among other faults, attributing the fruits of labor to apes. The book in question was Tú Jīngzōng and Chén Guāngyì, Láodòng chuàngzào rén [Labor Created Humanity] (Shànghǎi: Shànghǎi wénhuà shūdiàn, [1951?]).

98. Schneider, Biology and Revolution, 153. Sūn Shǒudào, "Píng Liú Xián zhù 'Cóng yuán dào rén fāzhǎn shǐ' shū zhōng de wéixīnzhǔyì hé qítā fāngmiàn de cuòwù" [Criticizing Idealist and Other Types of Errors in Liú Xián's book The History of Development from Ape to Human], Kēxué tōngbào [Science Bulletin], no. 4 (1955): 74–81; "Kēxué gōngzuòzhě jǔxíng zuòtánhuì tǎolùn 'Cóng yuán

opinion as part of a forum the journal was to print, they deemed his contribution insufficiently self-critical and declined to publish it. Liú recalls that in 1956 he was made to "wear many hats" (the expression for political labeling), including "idealism" and "anthropomorphism" (擬人論), and that his book was denounced for promoting "reactionary political thought" and "harming the study movement."[99] When Máo Zédōng launched his Hundred Flowers Movement in 1956, Liú Xián was among many intellectuals who took the opportunity to voice their frustrations, and he joined several others in writing about his experiences in a short article in the *Guāngmíng Daily*—the source of the recollections cited here. However, the "blooming and contending" of the Hundred Flowers Movement revealed more dissatisfaction than Máo had been prepared to face, and the Anti-Rightist Movement quickly followed. Students then in the anthropology department of Fùdàn University recall that Liú was banished to library work because of his "rightist" label, although he was at some point (probably during the early 1960s) allowed to teach a few classes on the use of classical texts in the study of anthropology.[100]

The attack on Liú Xián was but one of a great many trials that scientists suffered when critics deemed their work tainted with "idealism" or other politically undesirable influences. Nonetheless, few if any scientists or others involved in science dissemination seem to have changed their tune on superstition. Whatever their feelings about the state's heavy-handed blows on idealist scientific writings, they appear to have remained committed to their roles as squashers of superstition. At the very least, they continued to play these roles with distinction.

---

dào rén fāzhǎn shǐ' yī shū de cuòwù" [Science Workers Hold Meeting to Discuss the Errors in the Book *The History of Development from Ape to Human*], *Shēngwùxué tōngbào* [Biology Bulletin], no. 11 (1955): 5. Note, however, that neither the *People's Daily* review nor articles published in *Science Bulletin* and *Biology Bulletin* made any reference to Michurin. The closest the *Science Bulletin* article came was in criticizing an implicitly anti-Lamarckist passage of Liú's book (p. 102), where he emphasized that while the development of the brain was the key human distinction, it was incorrect to say that the increased use of the brain to create tools was the cause of the brain's development. This, too, was consistent with his mentor's perspective: Keith subscribed to a teleological view of human evolution but not a Lamarckian "use-inheritance" theory (Bowler, *Theories of Human Evolution*, 207). Lysenkoism/Michurinism did not play an important role in Chinese discourse on human evolution. The tendency not to apply Lysenkoism to human evolution may have been related to the strong condemnation of eugenics in both the Soviet Union and China. On Lysenko's explicit denunciations of eugenics, see Graham, *Science in Russia*, 123–124.

99. "Shànghǎi zhīshí jiè gāi guànchè 'bǎijiā zhēngmíng' wèntí," 2.

100. Interview data, 2002.

The Pursuit of Monsters

Attacks on superstition had another consequence that most who participated in them probably did not forsee. In repeatedly conjuring up demons to destroy, science disseminators helped to keep them alive and in circulation. They even made them into legitimate subjects of scientific inquiry.[101]

In addition to stories regarding the supernatural origins of humanity, ghosts, dragons, and other supernatural beings and mysterious animals were frequent targets of science dissemination materials.[102] A 1957 article entitled "Strange Creatures of Legend" addressed supposed sightings of legendary animals and suggested alternative explanations more in line with scientific knowledge. For example, the *xīngxing* (猩猩), which sat on the ape-human boundary, were probably just monkeys or bears that people mistook for mythical creatures.[103] A book for children on human origins similarly informed its readers that legends grandmothers told about half-ape, half-human "*yěrén*" in fact described regular apes.[104] Dragons, too, faced extinction in such materials. For example, a 1956 book entitled *Do Spirits and Ghosts Really Exist?* noted, "Chinese emperors used dragons to trick people in order to consolidate their own ruling position." While some people thought dragon bones proved the existence of dragons, the book explained that these were just the remains of ancient reptiles called "dinosaurs."[105] The authors neglected to mention that dragon bones also included the fossilized bones of other animals—and that knowledge of them had led to important scientific discoveries like Peking Man.

101. Dahpon Ho, drawing on Joseph Levenson, has made a similar point about Cultural Revolution–era attacks on the "four olds": "Instead of slaying the old world, the Cultural Revolution hoisted cultural relics out of limbo and into the melee of modernity. Cultural relics suddenly assumed a living significance to both destroyers and would-be protectors." Dahpon Ho, "To Protect and Preserve: Resisting the 'Destroy the Four Olds' Campaign, 1966–1967," in *The Chinese Cultural Revolution as History*, ed. Joseph Esherick, Paul Pickowicz, and Andrew Walder (Stanford, Calif.: Stanford University Press, 2006), 93.

102. For an example of a book that included supernatural origins of humanity on a long list of other superstitions, see Āndōng shì kēxué jìshù xiéhuì, *Kēxué jīchǔ zhīshí* [Basic Scientific Knowledge] (Shěnyáng: Liáoníng rénmín chūbǎnshè, 1959), 147–148.

103. Tán Bāngjié, "Chuánshuō zhōng de yìshòu" [Strange Creatures of Legend], in *Kēpǔ xuānchuán shǒucè* [Science Dissemination Handbook], vol. 1, ed. Shànghǎi kēxué pǔjí chūbǎnshè (Shànghǎi: Shànghǎi kēxué pǔjí chūbǎnshè, 1959 [1957]), 24–25. This article originally appeared in the 22 May 1957 issue of the newspaper *Dàgōng bào*. Note that the name *xīngxing* is also used to translate "orangutan," and sometimes refers more generally to "ape"—gorillas are "large *xīngxing*" and chimpanzees are "black *xīngxing*."

104. Wáng Xiǎoshí, *Cóng yuán dào rén*, 65–66.

105. Jílínshěng kēxué jìshù pǔjí xiéhuì, *Zhēn yǒu shénguǐ ma?* [Do Spirits and Ghosts Really Exist?] (Chángchūn: Jílín rénmín chūbǎnshè, 1956), 4–5.

Science disseminators further kept monsters alive by highlighting cases of atavism as evidence for evolution in their books and articles. Such examples were prominent in some Chinese anthropological writings published before 1949 but became much more so after the revolution. The most common examples presented were the occasional births of people with tails, extra nipples, or thick hair covering their entire bodies.[106] Darwin himself had included such examples in his 1871 *The Descent of Man*.[107] The last of these conditions was related to the coat of downy hair known as lanugo that covers a human fetus during the sixth month. The existence of lanugo suggested that human fetuses passed through an apelike stage in the womb, and the occasional retention of that hair into adulthood further indicated that, with respect to body hair, some individuals remained in that stage. Examples of atavism provided vivid evidence for humans' animal past: they were anomalies that apparently confirmed rather than disrupted the linear, developmentalist conception of the evolution of humans from animals. However, images of hairy babies and children with tails simultaneously kept alive a tacit possibility that monsters might reemerge to confuse this boundary.

When reports of such monsters did arise, paleoanthropologists were called upon to investigate. In 1958 the Soviet Yeti Research Commission organized an expedition to investigate yeti, inspiring Chinese investigators—including IVPP scientists—to incorporate a yeti component in their 1959 survey of the Mount Everest region. Péi Wénzhōng collected reports of similar creatures from Xīnjiāng, Inner Mongolia, and other parts of China and kept up a correspondence with the Soviet researchers.[108] In the early 1960s, the Chinese Academy of Sciences sent a team—led by IVPP's noted paleoanthropologist Wú Rǔkāng, who was then engaged in research on the fossil ape *Gigantopithecus*—to Xīshuāngbǎnnà in Yúnnán Province to investigate reports of *yěrén*-like creatures. The investigation was brief, and the participating scientists suggested that the animal in

106. Gurev, *Rénlèi shì zěnyàng zhǎngchéng de* (1953), 25; Dǒng Shuǎngqiū, *Rénlèi shì zěnyàng*, 6–7; Wú Rǔkāng, *Rénlèi de qǐyuán* (1965), 14–16.

107. Darwin, *The Descent of Man*, 21–22, for lanugo and body hair; 41–49, on other forms of "reversion."

108. Liú Mínzhuàng, *Jiěkāi 'yěrén' zhī mí* [Solving the "*Yěrén*" Mystery] (Nánchāng: Jiāngxī rénmín chūbǎnshè, 1988), 58–59; Shàng Yùchāng, "Cóng Zhū Fēng dìqū "xuěrén" kǎochá tánqǐ" [Talking about Yeti Investigations from the Mount Everest Area], *Huàshí*, no. 3 (1979): 6–8. Péi Wénzhōng's personal collection, now stored in IVPP's library, contains a box marked "yeti" (雪人). It includes correspondence with Soviet researchers and Chinese officials in Inner Mongolia and Xīnjiāng and Soviet reports in Russian. In letters to Soviet colleagues, Péi Wénzhōng repeatedly expressed his doubts about the existence of yeti and *yěrén*.

question might be a gibbon, but that their results were not enough to verify or deny the existence of *yěrén*.[109]

The scientists involved generally seem to have been very skeptical about the scientific potential of *yěrén* stories, although they typically took a "wait-and-see" approach and supported the investigation of such claims. For example, a 1958 article jointly written by the IVPP scientists Péi Wénzhōng, Wú Rǔkāng, and Zhōu Míngzhèn noted that the question was important with respect to Darwinism and Engels's theory that labor created humanity, and so they encouraged further research on the subject. For the time being, however, they considered stories of yeti and *yěrén* to be exactly what one would expect from "eras prior to the development of science or minority cultures in mountainous regions."[110] The very next year, another time-tested explanation for anomalous creatures got some new exercise. Recalling the "white-haired girl" of Yán'ān, a woman reportedly came out of hiding after living an "ape-like existence" in the Sìchuān wild for twelve years to escape "local despots."[111]

Conclusion

There is no doubt that scientists and other science disseminators sincerely desired to instill in their fellow Chinese people a scientific worldview and believed that this would benefit the nation as a whole. Scientific and political elites had a common interest: to liberate the population from the shackles of superstition and create a strong country founded on science. The project of science dissemination was, however, framed in an antagonistic way: engagement in a *battle* against superstition incurred certain consequences. In attacking religious forms of knowledge, scientists helped strengthen the authoritarian character of the state ideology, thus indirectly contributing to their own persecution. At the same time, efforts to expose and annihilate monsters ironically increased their visibility and even their legitimacy as subjects of inquiry in official scientific discourse.

Pursuing a strongly hierarchical agenda of science dissemination based on an understanding of the general population as superstitious also played into the hands of those who sought to emphasize the popu-

109. Lǐ Jiàn, *Yěrén zhī mí* [The *Yěrén* Mystery] (Wǔhàn: Zhōngguó dìzhì dàxué chūbǎnshè, 1990), 22; Liú Mínzhuàng, *Jiēkāi 'yěrén' zhī mí*, 91.
110. Péi Wénzhōng, Wú Rǔkāng, and Zhōu Míngzhèn, "'Xuěrén' zhī mí" [The Yeti Mystery], *Guāngmíng rìbào*, 27 February 1958, 3.
111. Walter J. Meserve and Ruth I. Meserve, "'The White-Haired Girl': A Model for Continuing Revolution," *Theatre Quarterly* 24 (Winter 1976–1977): 30.

list side of the two-way flow of knowledge Máo set out in his "Talks at the Yán'ān Forum." The implicit elitism of discourse on superstition lent weight to the perception that intellectuals—including scientists—represented a class divorced from the "broad masses of the people." Resentment to intellectuals' superior attitudes in turn undermined the inclusive ideals of the New Democracy and "united front," fueling instead the kind of bottom-up, mass science correctives that characterized the more radical periods of the People's Republic.

THREE

"The Concept of Human": In Search of Human Identity, 1940–1971

The Question of a Universal Human Nature

In 1942, in his concluding speech at the Yán'ān Forum on Literature and Art, Máo Zédōng set out his definitive statement on human nature: "Is there such a thing as human nature? Of course there is. But there is only human nature in the concrete, no human nature in the abstract. In a class society there is only human nature that bears the stamp of a class; human nature that transcends classes does not exist."[1] His inspiration was a passage from Marx's *Theses on Feuerbach*: "The human essence is no abstraction inherent in each single individual. In its reality it is the ensemble of social relations."[2] Máo's authoritative interpretation set limits on what could be said about "humanity": in the strictest possible terms, there was no universal human nature. This, then, was the flip side of the New Democracy. Progressive natural scientists and members of the Chinese bourgeoisie had a role to play in the revolution, but class differences would not be forgotten.

1. Mao Tse-tung, *Selected Works of Mao Tse-tung* (Peking: Foreign Languages Press, 1967–1971), 3:90. Quoted also in Donald J. Munro, *The Concept of Man in Contemporary China* (Ann Arbor: University of Michigan Press, 1977), 21.
2. Karl Marx, "Theses on Feuerbach," in *Ludwig Feuerbach and the End of Classical German Philosophy*, Frederick Engels (Beijing: Foreign Languages Press, 1976), 63.

The consequences for the cultural sphere were profound. Beginning in 1958 with the emphasis on proletarian literature accompanying the populist Great Leap Forward, China's chief official literary newspaper published numerous articles rejecting "humanism" (人道主義).[3] Of particular import was an article by Yáo Wényuán (later to become one of the leaders of the Cultural Revolution). Yáo took a cue from Máo's pronouncement at the Yán'ān forum, when in 1960 he led a criticism of "the theory of human nature" (人性論) in literature and art. The specific target of his attack was the influential intellectual and state official Wáng Rènshū, who wrote under the pen name Bā Rén. Wáng had called for less "dogma" and more "human feeling" (人情) in literature and art and had posited certain "universal" human needs (such as romantic love) and likes (such as fragrant flowers and bird song). Yáo countered that "human feeling that transcends class cannot exist in a class society," and he accused Wáng of promoting the capitalist variety of human feeling.[4] In similar fashion, the novelist Oūyáng Shān suffered an attack five years later for expressing the idea of love as a human emotion that transcended class.[5]

During the early 1980s, some intellectuals sought to use the political thaw following Máo's death to explore humanist ideas, including the notion of "human rights" (人權). They drew inspiration from neo-Marxists in the West who focused on the themes of humanism and alienation in Marx's early writings. The intellectual Wáng Ruòshuǐ (then a leading official at the *People's Daily*) in particular made waves by saying that the human was "the starting point of Marxism" and by suggesting that socialism itself had produced conditions in China oppressive to the human as a human.[6] Wáng and others involved in the discussion of Marxist humanism came under attack in 1983 during a political campaign against "spiritual pollution."[7] The issue reemerged in different form in the 1990s, when intellectuals began calling for a revival of a "humanistic spirit" (人文精神) to counter the commercialism created by the economic reforms. Interestingly, it was not party officials but

3. D. W. Fokkema, "Chinese Criticism of Humanism: Campaigns against the Intellectuals, 1964–1965," *China Quarterly* 26 (1966): 68–81. The newspaper was *Wényì bào*.

4. Yáo Wényuán, "Pīpàn Bā Rén de 'rénxìng lùn'" [Criticizing Bā Rén's Theory of Human Nature], *Wényì bào*, no. 2 (1960): 32. A reprinting of Wáng's original article begins on page 41 of the same issue. See also Munro, *The Concept of Man in Contemporary China*, 58–61.

5. Fokkema, "Chinese Criticism of Humanism," 80.

6. Shiping Hua, *Scientism and Humanism: Two Cultures in Post-Mao China, 1978–1989* (Albany, N.Y.: State University of New York Press, 1995), 98, 100–101.

7. Yan Sun, *The Chinese Reassessment of Socialism, 1976–1992* (Princeton, N.J.: Princeton University Press, 1995), 131–133.

postmodern scholars who raised the class issue this time, pointing to the elitism of the humanists and defending consumerism as a populist force.[8]

This history of hostility to humanism is so familiar to China scholars that few would be surprised by a statement Máo made on the eve of the Cultural Revolution. In an address on education, Máo used "human" as a quintessential example of a "concept" (概念) that lacked a reality in the "concrete" (具體) world. "The concept of human leaves out too many things. We do not see males and females, adults and children, Chinese people and foreign people, revolutionaries and counterrevolutionaries. All we have left are the characteristics separating humans from other animals. Who has ever seen a 'human'? We can see only [concrete examples like] a Mr. Zhāng or a Mr. Lǐ [張三李四]. Similarly, we cannot see the concept of a 'house,' but only concrete 'houses' like the Western-style buildings of Tiānjīn or the courtyard houses of Běijīng."[9] In the context of this study, Máo's statement *is* surprising. After decades of intensive dissemination of Engels's theory that labor created humanity, it is strange that Máo should have dismissed the characteristics defining humanity as insufficiently "concrete."[10]

The evolutionary theory that labor created humanity supported a kind of humanism—an understanding of human identity—that remained influential despite Máo's denial of the very "concept of human." Propagating Engels's brand of humanism, like squashing superstition, was an area of overlap for scientific and political elites, and thus a way for scientists to secure a role in the new state apparatus. Both missions existed in contradiction to the more radical side of Maoist ideology, where elite knowledge was suspect and class struggle dominated. Scientists and other intellectuals involved in science dissemination could not escape class struggle, but they could accomplish much in areas where they were of use to the state.

When scholars in the humanities could not explore what it meant to be human, their counterparts in the natural sciences thus had considerably more freedom. Science was perhaps the last place ideologues expected humanist values to breed. It was also safer to look for human

---

8. Wang Hui, *China's New Order: Society, Politics, and Economy in Transition*, ed. Theodore Huters (Cambridge, Mass.: Harvard University Press, 2003), 93.

9. Zīliào shì, "Máo Zédōng duì wéngé zhīshì huìbiān," *Zǔguó* 66 (vol. 1969, no. 9): 42. My translation differs somewhat from that in Munro, *The Concept of Man in Contemporary China*, 22.

10. He was certainly very familiar with Engels's text. He referred to it obliquely in a speech on 6 June 1950, where he said that while some "idealists" held that God created humans, "we say, From Ape to Human." Mao Tse-tung, *Selected Works*, 5:34.

identity in the ancient past, before the emergence of class society. Máo had said at the Yán'ān forum, "As for the so-called love of humanity, there has been no such all-inclusive love since humanity was divided into classes."[11] The implication, though somewhat muted, was that there had been a time when a universal human love, and so a universal human identity, had flourished.

What, then, was the nature of the human identity science dissemination on human evolution constructed? "Universal human nature" is itself far from a universal concept. Given a notion that all people share a common root identity, there are still endless ways to describe what makes up such an identity. In socialist China, labor was the core. From this starting point came a distinct vision of the lives that the earliest humans lived, summed up in the term "primitive communism." Scientific and political elites did more than just propagate theory; they created a vivid picture of primitive human life with endearing characterizations of the early ancestors whose labor had created humanity.

Labor as the Core of Human Identity

In theorizing on the emergence of humans from apes, Engels pinpointed tools as the critical issue: "The tool implies activities that are specifically human."[12] But Engels and his socialist followers were hardly alone in this focus. Jane Goodall's discovery that chimpanzees make tools—albeit very simple ones and certainly not "even the crudest of stone knives," as Engels had said—elicited widespread excitement and prompted the famous paleoanthropologist Louis Leakey to write, "Now we must redefine tool, redefine Man, or accept chimpanzees as humans."[13] In like spirit, Kenneth P. Oakley titled his 1956 book *Man the Toolmaker*, a title adopted by at least a dozen later books, films, and other popular and educational materials.[14] His book opened: "Man is a social animal, distinguished by 'culture': by the ability to make tools and communicate

11. Mao Tse-tung, *Selected Works*, 3:91.

12. Frederick Engels [Ēngésī], *Cóng yuán dào rén* [From Ape to Human], trans. Cáo Bǎohuá and Yú Guāngyuǎn (Shíjiāzhuāng: Jiěfàng shè, 1950 [1948]), 22. This quotation comes from an excerpt of Engels's *The Dialectics of Nature* included in all editions of this volume.

13. Jane Goodall, "Essays on Science and Society: Learning from the Chimpanzees: A Message Humans Can Understand," *Science* 282, no. 5397 (1998): 2184. Goodall made her famous discovery in 1960.

14. Kenneth P. Oakley, *Man the Toolmaker* (Chicago: University of Chicago Press, 1959 [1956]). I thank Lín Shènglóng of the Institute of Vertebrate Paleontology and Paleoanthropology for directing me to this book. It was very likely available to Chinese scientists during the 1950s: the image on page 107 of a cave painting representing a person collecting honey from a beehive is found also in several Chinese books of this period.

ideas."[15] Tool manufacture and language thus constituted key elements in the definition of humanity for both Oakley and Engels, despite their distance in time, space, discipline, and political perspective.

Yet there was an important difference between Oakley and Engels not to be overlooked. For Oakley and many other authors in Western, non-socialist countries, the overarching concept linking tool manufacture and language was culture, while for Engels and his followers in socialist countries it was labor. Moreover, according to Oakley, the use and production of tools were activities "*in the first place* dependent on adequate powers of mental and bodily coordination, but which in turn *perhaps* increased those powers" (emphasis added). He went on, citing Sir Wilfrid Le Gros Clark, to say that "the real difference between what we choose to call an ape and what we call man is one of mental capacity."[16] Oakley could thus emphasize tool manufacture and even make it essential to the definition of humanity while still considering the development of the brain to have preceded tool use.

This was very much in opposition to Engels's analysis. For Engels, the development of the hand made possible more complex forms of labor, which encouraged "mutual assistance and joint cooperation," which in turn required language. It was only under these conditions that the early human brain grew in size and "perfection."[17] This is important: although Engels theorized a type of positive feedback loop in which labor and social stimulus encouraged brain development and a larger brain encouraged more complex forms of labor and social interaction, there was no question for him that labor and not brain growth initiated the loop. For Engels, this followed directly from a materialist—as opposed to an idealist—understanding of natural phenomena. While his contemporaries did not necessarily share his particular philosophical perspective, they did actively debate whether the development of the brain or erect posture emerged first. Indeed, this question remained at the center of debates about human evolution in the nonsocialist West at least until the mid-twentieth-century theoretical revolution in Darwinism.[18] The consensus then settled on the position Engels had taken, although apparently none of the participating scientists interpreted this as labor's having created humanity. And even after this reversal, the significance of "mental capacity" remained central in defining humanity

---

15. Oakley, *Man the Toolmaker*, 1.
16. Oakley, *Man the Toolmaker*, 2–3.
17. Engels, *Cóng yuán dào rén*, 5–7.
18. Peter Bowler, *Theories of Human Evolution: A Century of Debate, 1844–1944* (Baltimore: Johns Hopkins University Press, 1986), 149–185.

for many Western authors. "Tool manufacture" thus held very different meanings in different political contexts.

The focus on labor was the most significant way the socialist state reconstructed human identity. The idea that the experience of labor lay at the root of a common humanity dovetailed neatly with Máo-era policies in which intellectuals underwent reeducation through labor in the countryside. Some might argue that what had been created was not so much an inclusive human identity as a narrower class identity that claimed humanity only for those who worked with their hands. But the early People's Republic was marked by a profound optimism about the possibility for transformation: thought reform and labor reform were premised on the notion that people could shed their bourgeois (or otherwise problematic) pasts and become proper socialists.[19] This optimism existed in contradiction to another strain in Chinese communist thought that emphasized the importance of "class background." Yet even at its zenith in the Cultural Revolution, the notion that people were permanently fixed in class categories did not go unchallenged: the youth Yù Luókè wrote a famous essay which drew on quotations from Máo to argue that the concept of "class background" was a "new type of racism." Yù insisted instead that actions determined identity.[20] Yù was executed for his efforts, but his ideas were consistent with a powerful theme in Maoist thought. In this context, the refrain "Labor created humanity" was both a threat to intellectuals' privileged status and their greatest hope to remain included in the new society. If they were willing to roll up their sleeves, they could reclaim their human identity.

The use of labor as the primary explanatory variable in human evolution had further consequences. Treating the transition from ape to human as the first chapter of the history of social development effectively privileged sociology over biology. This can be seen in statements made by both Soviet and Chinese writers in the 1950s. The Soviet author of a book entitled *The History of Primitive Society* criticized "capitalist" historians for talking about the "prehistoric," since the period to which they referred was also history and analyzable as such, that is, with respect to tools of production and productive power.[21] A Chinese academic writing in 1951 similarly stated that Darwin's theory was incomplete because it

---

19. See Donald Munro's discussion of the "malleability of man" in *The Concept of Man in Contemporary China.*

20. Lu Xiuyuan, "A Step toward Understanding Popular Violence in China's Cultural Revolution," *Pacific Affairs* 67, no. 4 (1994–1995): 559.

21. V. K. Nikol'skii [Níkēěrsījī], *Yuánshǐ shèhuì shǐ* [The History of Primitive Society], trans. Páng Lóng (Shànghǎi: Zuòjiā shūjú, 1953), 1.

was impossible to explain human evolution solely through biological, as opposed to historical, mechanisms.[22] Soviet and socialist Chinese materials on human evolution contained a great deal of biological analysis of the commonalities between humans and other animals, but it was society that made humans special. This fit nicely with official discourse more generally. As an article printed in a 1957 issue of *Liberation Daily* put it, "That which determines why a person is a human being does not lie in his natural essence but lies in his social essence, in his social nature."[23]

An interesting parallel can be drawn to the controversies surrounding sociobiology today. While in socialist China it was natural science that faced rewriting as an extension of social science, many critics in the West today worry about the incursions of biology into sociology—for example, in sociobiological explanations of warfare, slavery, and rape. Western critiques of sociobiology are numerous.[24] The political context of socialist China made a similar, though more explicit and highly charged, criticism almost inevitable. In 1979, the Chinese paleoanthropologist and party member Wú Rǔkāng went on record against E. O. Wilson's landmark *Sociobiology: The New Synthesis*. He noted its similarity to Spencerian social Darwinism and cautioned readers about its potential social and political consequences. He suggested that sociobiology had emerged in a capitalist context as a response to the revolutions in the third world, the rise of the American working class, and the civil rights and feminist movements.[25] More generally, those skeptical of sociobiological approaches have decried the reduction of human behavior to biological principles. China offers the reverse case: only in the late 1980s were scientists able to challenge the dominance of social over biological elements in popular narratives of human evolution.

If the widespread application of Engels's theory effectively privileged the social over the biological with respect to analytical frameworks of human evolution, it also encouraged a more material manifestation of this privilege. According to Engels, labor made possible the "final, essential distinction" between animals and humans—the ability to "master" nature, that is, to transform the natural world to serve human goals

22. Jiǎn Bózàn in Lín Yàohuá, *Cóng yuán dào rén de yánjiū* [Research on from Ape to Human] (Běijīng: Gēngyún chūbǎnshè, 1951), preface, 4.

23. Quoted in Munro, *The Concept of Man in Contemporary China*, 16.

24. Steven Rose, Leon J. Kamin, and R. C. Lewontin, *Not in Our Genes: Biology, Ideology and Human Nature* (New York: Pantheon Books, 1985); Anne Fausto-Sterling, *Myths of Gender: Biological Theories about Women and Men* (New York: Basic Books, 1992 [1985]).

25. Wú Rǔkāng, "Píng Wēiěrxùn *Shèhuìshēngwùxué—xīn de zōnghé*" [Review of Wilson's *Sociobiology: The New Synthesis*], *Gǔjǐzhuī dòngwù yǔ gǔrénlèi* 17. no. 1 (1979): 89–90. I thank Lín Shènglóng for calling my attention to this review.

rather than simply adapting to nature, as the other animals do.[26] Celebrating human mastery over nature worked to undergird the Chinese socialist state's disastrous environmental policies. The famine after the failed Great Leap Forward can be traced to such an attitude: at that time, economic radicals declared that a voluntaristic spirit could compel the land to produce many times the amounts it had in the past. Socialist educational materials, beginning well before the 1949 revolution, supported this perspective by painting human history, beginning with Paleolithic cave dwellers, as a series of successful "struggles with nature."[27] Science disseminators thus not only defined humans as social rather than biological creatures; they further identified an antagonistic relationship with nature as essential to what it meant to be human.

Primitive Communism

Engels's *From Ape to Human* provided anthropologists with little help in interpreting how early humans lived and related to one another. The task of representing Peking Man's society required reference to other theoretical works, most important among them Engels's *The Origin of the Family, Private Property, and the State* and Stalin's elaboration and codification of primitive communism as the first stage in Marxist history. Engels wrote *The Origin of the Family* in 1884 based on Lewis Henry Morgan's 1877 *Ancient Society* and the late Marx's notes on Morgan and other anthropologists.[28] *The Origin of the Family*, more than any previous work by Marx or Engels, explicitly discussed social relations in preliterate societies. Stalin then cemented the "five stages" of human history (primitive communism or primitive society, slave society, feudalism, capitalism, and socialism) and made them universal for all peoples. This orthodox interpretation was imposed in 1937 on top of "lively debates" among Chinese Marxists on the character of ancient society and the appropriate way to study it.[29]

While Peking Man and other early humans left no written traces,

---

26. Frederick Engels, *The Part Played by Labor in the Transition from Ape to Man* (New York: International Publishers, 1950), 18.

27. Zhōu Róngwéi, *Rén hé zìrán dòuzhēng* [Humans' Struggle with Nature] (Nánjīng: Mín fēng yìnshūguǎn, 1953). For a pre-1949 socialist example, see "Yíge kēxué gōngzuòzhě de xìnyǎng" [The Beliefs of a Science Worker], *Kēxué shìjì* 1, no. 4 (1936): 3–4.

28. Frederick Engels, *The Origin of the Family, Private Property, and the State*, in *Karl Marx, Frederick Engels: Collected Works*, vol. 26, ed. E. J. Hobsbawm et al. (Moscow: Progress Publishers, 1975–), 26:139.

29. Gregory Guldin, *The Saga of Anthropology in China: From Malinowski to Mao* (Armonk, N.Y.: M. E. Sharpe, 1994), 120. See also Albert Feuerwerker, "China's History in Marxian Dress," in *History in Communist China*, ed. Albert Feuerwerker (Cambridge, Mass.: MIT Press, 1968), 29.

scientists and other science disseminators worked with the conviction that knowledge of the principles of production and social organization, combined with fossil evidence and a vivid imagination, could produce a picture of how they lived. Books and other materials disseminated in the early People's Republic of China employed a variety of techniques to assist readers in imagining life 500,000 years ago. The 1959 movie *Peking Man* used animated clay figures set against a painted backdrop; museum exhibits included dioramas and murals; many books were dominated by illustrations accompanied by interpretive prose or verse. A few authors employed literary devices to bridge the gap in time and allow the author greater room for conjecture: for example, one asked readers to imagine they were watching a movie, while another invited readers to board a boat that would sail back along the river of time in order to "meet our ancestors."[30]

Whatever the medium or technique, the portrayals shared a view of early human society as peaceful and harmonious, and of life as basic, crude, difficult, and dangerous. Peking Man lived in an era of primitive communism, forming small communities characterized by harmony within and danger without. Toolmaking, foraging, hunting, and other activities required communal labor, for only by working together could the early humans scrape together a meager existence and protect themselves from ferocious predators. In the picture book *Our Ancestors 500,000 Years Ago*, the artist depicted a group working together to make tools, next to which Jiǎ Lánpō wrote, "The first person from the left represents searching for stone tool materials from the riverbank; the second represents making flakes; the third represents refining a stone flake tool; the fourth represents learning from their construction experience; through this kind of cooperative labor, they are exchanging collective experience."[31] In *Human Origins and Development*, Péi Wénzhōng criticized "warmongers" (好戰分子, a common pejorative aimed at American imperialists) for believing that people were naturally competitive and combative, when in fact early human society was actually characterized

30. See Huáng Wéiróng, *Zhōngguó yuánrén* [Peking Man] (Shànghǎi: Shàonián értóng chūbǎnshè, 1954), 22; Chéng Wànfú and Qín Juéshí, eds., *Zhōngguó yuánrén: Wǒmén wǔshíwàn nián de zǔxiān* [Peking Man: Our Ancestors 500,000 Years Ago] (Nánjīng: Mínfēng yìnshūguǎn, 1953), 18. Martin Rudwick has pointed out the use of such a literary device in Pierre Boitard's 1861 *Paris before Men*. Boitard invoked a magical creature capable of time travel to help readers make the mental leap necessary to imagine an ancient world. Martin J. S. Rudwick, *Scenes from Deep Time: Early Pictorial Representations of the Prehistoric World* (Chicago: University of Chicago Press, 1992), 166–172.

31. Yáng Háonìng (art) and Jiǎ Lánpō (text), *Wǒmén de zǔxiān 1: Wǒmén wǔshíwàn nián de zǔxiān* [Our Ancestors, vol. 1: Our Ancestors 500,000 Years Ago] ([Tiānjīn?]: Zhīshí shūdiàn), 17.

by cultural exchange and peaceful interaction.[32] (Indeed, the outspoken anticommunist American popular writer on human evolution Robert Ardrey had pronounced an interpretation of human evolution in striking contrast with that of Engels: "Weapons had produced man, not man weapons.")[33] Moreover, no one was exempt from labor; no class divisions existed, and all members of society contributed to production.[34] Behind such statements could be heard the voice of Stalin saying, "Stone tools, and, later, the bow and arrow, precluded the possibility of men individually combating the forces of nature and beasts of prey. In order to gather the fruits of the forest, to catch fish, to build some sort of habitation, men were obliged to work in common if they did not want to die of starvation, or fall victim to beasts of prey or to neighbouring societies. Labour in common led to the common ownership of the means of production, as well as of the fruits of production. Here the conception of private ownership of the means of production did not yet exist. . . . Here there was no exploitation, no classes."[35] As attractive as such a society might have seemed, one Chinese author warned that it was no use wishing to return to an earlier historical time. But no matter, he continued, for there was the "even more beautiful, fortunate, free, and happy society" of communism itself.[36] In China and other communist countries, then, the human identity created through popular paleoanthropology made communism itself feel more possible—it was in human nature. Such a construction was of particular significance for the Chinese state under Máo, who despite the disapproval of Soviet onlookers sought during the Great Leap Forward to accelerate the revolution and propel the country into a fully communist society.

In descriptions of early society, while class (although absent) received considerable attention, gender (although present) received very little explicit analysis. The most notable exception was Yáng Yè's 1952 *Our*

---

32. Péi Wénzhōng, *Rénlèi de qǐyuán hé fāzhǎn* [Human Origins and Development] (Běijīng: Zhōngguó qīngnián chūbǎnshè, 1956), 11.

33. Robert Ardrey, "A Slight (Archaic) Case of Murder," *Reporter*, 5 May 1955, 34–36. Quoted in Nadine Weidman, "Popularizing the Ancestry of Man: Ardrey, Dart, and the Killer Instinct" (paper presented at annual meeting of the the History of Science Society, 4 November 2006).

34. Jiǎ Lánpō and Liú Xiàntíng, *Cóng yú dào rén* [From Fish to Human] (Tiānjīn: Zhīshí shūdiàn, 1951), preface (no page number); Yáng Yè, *Wǒmén de zǔxiān* [Our Ancestors] (Hànkǒu: Wǔhàn gōngrén chūbǎnshè, 1952), 35–37.

35. Joseph Stalin, "Dialectical and Historical Materialism," in *Problems of Leninism* (Běijīng: Foreign Languages Press, 1976), 862–863. The essay was first published in September 1938. Note that Stalin's reference to the dangers of "neighboring societies" does suggest a potential for aggression; still, he uses this point to emphasize the need for harmonious cooperation.

36. Yáng Yè, *Wǒmén de zǔxiān*, 35–36.

Ancestors, which summarized in highly simplified form Morgan's analysis of the development of mating and marriage practices among what Morgan had called the "savages," the lowest stage of human history. In the earliest years of this stage, humans mated at random and formed groups based on age similarity. During the middle period, they began to avoid mating with immediate family members and formed groups based on sex in which females gathered and men hunted. In the last years (before the transition to what Morgan called "barbarism"), these early humans evolved even more restrictive mating habits and formed groups according to the matriline. Yáng Yè further noted that primitive societies had existed without notions of "male superiority and female inferiority."[37]

The more detailed depictions of life in Peking Man's time generally followed the notion that "men hunt and women gather."[38] In the 1954 book *Peking Man*, for example, the author asked readers to imagine they were watching a movie and then described the following scene: "Three or four males have taken big clubs, stone axes, and other weapons and are leaving the cave to go hunting; some of the females are foraging for food to eat and sticks and leaves to burn, while other females are remaining in the cave to watch over the fire in preparation for roasting wild game."[39] Readers unfamiliar with anthropological maxims like "men hunt and women gather" would probably have found such portrayals reasonable nonetheless, since they resonated with the gendered divisions of labor typically found in rural China.

One critique of these materials from a feminist standpoint could certainly be their naturalization of contemporary gender roles. Still more significant, however, is their near total failure to take the opportunity, as Yáng Yè did, to highlight issues of gender equality. Nor have I found any evidence that authors were ever criticized on the basis of this failure, though criticism on other points was not uncommon. One could even imagine a feminist analysis based on Máo's 1965 critique of "the concept of human": Human identity as presented in many depictions of primitive human life "left out many things." That gender was virtually absent as an analytical question in popular materials on human origins,

---

37. Yáng Yè, *Wǒmén de zǔxiān*, 27–28.

38. Chéng Míngshì and Fāng Shīmíng, *Cóng yuán dào rén tōngsú huà shǐ* [A Simple Pictorial History from Ape to Human] (Shànghài: Rén shì jiān chūbǎnshè, 1951), 148; Fāng Qiě, *Cóng yuán dào rén tòushì: Láodòng zěnyàng chuàngzào le rénlèi běnshēn hé shìjiè* [A Penetrating Look at from Ape to Human: How Labor Created Humanity Itself and the World] (Shànghài: Shànghǎi biānyì shè, 1950), 49.

39. Huáng Wéiróng, *Zhōngguó yuánrén*, 23.

while another social issue—namely, labor—was primary, supports the argument of many China scholars that women's equality was a promise from the revolutionary period, celebrated in Máo's famous quotation that "women hold up half the sky," but left largely unfulfilled in the early People's Republic of China.[40]

Peking Man as a National Ancestor

Further complicating the question of human identity is the significance of Peking Man as a particularly *Chinese* human ancestor. Peking Man was not widely embraced as a human ancestor during the republican era in China or abroad. Even in 1950, few popular or professional Chinese writings on human evolution unambiguously traced the modern human lineage back to Peking Man. Many continued to consider her an offshoot or insisted that the evidence was not yet sufficient to determine conclusively whether Peking Man was ancestral to modern humans. In 1950 Péi Wénzhōng wrote, "We still have no way of knowing which of the ape-humans (猿人) is a human ancestor. Perhaps none of them is a human ancestor, and he [our ancestor] is still waiting for us to discover him."[41] One important work reprinted in 1950 even suggested that Peking Man in some ways resembled modern Europeans more than modern Asians.[42] But by 1952, things had changed. Peking Man had become indisputably a direct ancestor of modern humans, and thus an incarnation of earliest human identity. In the process, however, Peking Man also took on a decidedly nationalist feel, becoming a potent symbol of the Chinese people and creating tension between broadly inclusive and narrowly nationalist representations of humanity.

Beginning in the second decade of the twentieth century and lasting through the Second World War, the tendency among scientists was to notice the differences among fossil hominids rather than the similarities, and so to place them not only in separate species, but in separate genera.[43] Thus, *Pithecanthropus erectus* (Java Man), *Sinanthropus pekinensis* (Peking Man), and the Neanderthals, among others, were all understood

---

40. Kay Ann Johnson, *Women, the Family, and Peasant Revolution in China* (Chicago: University of Chicago Press, 1983); Margery Wolf, *Revolution Postponed: Women in Contemporary China* (Stanford, Calif.: Stanford University Press, 1985); Judith Stacey, *Patriarchy and Socialist Revolution* (Berkeley: University of California Press, 1983).

41. Péi Wénzhōng, *Zìrán fāzhǎn jiǎnshǐ* [A Short History of Natural Development] (Běijīng: Liányíng shūdiàn, 1950), 58.

42. Zhū Xǐ, *Wǒmén de zǔxiān* [Our Ancestors] (Shànghǎi: Wénhuà shēnghuó chūbǎnshè, 1950 [1940]), 163.

43. Bowler, *Theories of Human Evolution*, 75.

to have been side branches on the human tree. They had become extinct, replaced by our direct ancestors, who were typically missing from the fossil record.

Franz Weidenreich was an exception to this pattern. Weidenreich had taken over as director of research at the Peking Man site of Zhōukǒudiàn after Davidson Black's death in 1934. He held to a linear model of human evolution in which human populations in any single evolutionary period represented different races of only one species. There were thus no "replacements" and no "dead ends," but only continuous evolution from one form into another. Weidenreich identified key apparent morphological similarities between Peking Man and modern Mongoloid peoples, and between other fossils and the modern peoples among whom they were found. Each of these populations retained specific racial characteristics while following the basic, shared "trends" of human evolution—an orthogenetic model somewhat similar to that of Arthur Keith. For Weidenreich, Peking Man was not only a human ancestor, but specifically an ancestor of the Chinese people.[44]

In 1950, scientists at a series of symposia at Cold Spring Harbor in New York brought the "modern synthesis" in evolutionary biology to bear on paleoanthropology. For several decades, this revolution in the life sciences had been working to integrate the previously disparate work of paleontologists, geneticists, and others studying evolutionary processes. At Cold Spring Harbor, participating scientists radically simplified the human family tree, placing Peking Man within the species *Homo erectus*, held to be directly ancestral to *Homo sapiens*.[45] While Weidenreich's theory of racial continuity was sidestepped, the "single-species" thesis to which he was committed became the theoretical foundation for paleoanthropology for years to come.[46]

The transition to this linear view of human evolution, in which Peking Man and other fossils lay on the main trunk of the family tree, occurred in China during the first few years of socialist rule. There is no direct evidence that Chinese scientists changed their positions based on

44. Franz Weidenreich, *Apes, Giants, and Man* (Chicago: University of Chicago Press, 1946); Franz Weidenreich, "Some Problems Dealing with Ancient Man," *American Anthropologist* 42, no. 3 (1940): 380–383.

45. Ernst Mayr, "Taxonomic Categories in Fossil Hominids," *Cold Spring Harbor Symposia on Quantitative Biology* 15 (1950): 109–118.

46. Robert N. Proctor, "Three Roots of Human Recency: Molecular Anthropology, the Refigured Acheulean, and the UNESCO Response to Auschwitz," *Current Anthropology* 44, no. 2 (2003): 213–239. Significantly, the modern synthesis also cemented the opposition to orthogenetic theories of Keith and Weidenreich. Unlike Hán Wénlǐ in his critique of Liú Xián, however, scientists at Cold Spring Harbor were not interested in the evolutionary role of labor; natural selection was their focus.

the new international consensus. The coincidence, however, is striking. Since Chinese scientists were still in contact with their foreign colleagues during these years, it would be difficult to argue that the transformations in China bore no relation to those occurring abroad.[47]

The uncertainty over Peking Man's ancestral status disappeared from published accounts around 1952. After this point, Chinese authors increasingly referred specifically and repeatedly to Peking Man as "our ancestor." In one book with particularly flowery language, the phrase "our ancestor" appeared twenty-four times in fewer than one hundred sentences, and ten of these cases specified "our ancestor Peking Man" (我們的祖先北京人). [48] At least two books from the 1950s further pointed to the apparent similarities in bone structure between Peking Man and her purported modern progeny (in one case identified as "Mongoloids" and in the other as "northern Chinese people").[49]

While the transformation of the human family tree made it much easier to call Peking Man an ancestor in scientific terms, other factors were responsible for Peking Man's becoming celebrated as such in materials produced for general audiences. Part of what was at stake in the question of Peking Man's ancestral status was the antiquity of Chinese people's roots in China. One author perceived a threat to national pride in earlier theories about Chinese people originally coming from Babylon, central Asia, Japan, and other places.[50] With Peking Man, "the Chinese people finally found their own ancestors, proving that Chinese people and Chinese culture have been born and raised in Chinese soil and are not imported goods."[51] Once accepted as an ancestor, Peking Man extended China's claim to be China back half a million years, and thus supported Chinese national identity.

The story of human evolution became still more "Chinese" through the interweaving of paleoanthropological evidence and allusions to the Chinese classics. Ancient texts traced human society back to early figures like Nest Dweller, Flint Maker, and the Divine Husbandman, all of whom contributed important inventions to the creation of civilization. History textbooks from 1950 and 1951 presented the Peking Man fossils together with these legendary figures.[52] The legends served to put flesh on the

47. Yáng Zhōngjiàn's personal library, stored at IVPP, contains materials he received from the American Museum of Natural History in the 1950s.

48. Chéng Wànfú and Qín Juéshí, *Zhōngguó yuánrén*, 14–28.

49. Fāng Shàoqīng [pseud. for Fāng Zōngxī], *Gǔ yuán zěnyàng biànchéng rén* [How Ancient Apes Became Human] (Běijīng: Zhōngguó qīngnián chūbǎnshè, 1958), 48; Yáng Yè, *Wǒmén de zǔxiān*, 10.

50. On the "western origins theory" in the republican era, see chapter 1, notes 32–34.

51. Yáng Yè, *Wǒmén de zǔxiān*, 6.

52. Zhōngguó lìshǐ yánjiū huì, *Gāojí zhōngxué Zhōngguó lìshǐ* [Upper Secondary School Chinese

bones; they offered an idea of what life was like for "our ancestors." Such texts also emphasized that the sequence of legendary figures conformed substantially with modern knowledge of the past. "According to general laws of evolution, humans first lived in the trees, then invented fire, then invented fishing and hunting, then animal husbandry, and finally agriculture. This generally corresponds to the sequence of Nest Dweller, Flint Maker, Páoxī [also known as Fúxī, inventor of snares for trapping fish and game], and the Divine Husbandman."[53]

Unlike the history textbooks, popular science materials on human evolution produced for the general population very rarely discussed Flint Maker or other legendary figures. One cause of this omission may have been the difficulties involved in attacking "superstitious" accounts of human origins while simultaneously making positive use of other legends. Fúxī—the inventor of snares named above—commonly appeared in the ancient stories as Nǔwā's mate: to laud one and denounce the other in the same text presented obvious problems. Nonetheless, a few books did refer to the legends. Most notably, Yáng Yè used Nest Dweller, Flint Maker, and Fúxī to illustrate changes in early human social organization based on the work of Lewis Henry Morgan and Frederick Engels. Nest Dweller lived an egalitarian life in which all members shared the fruits of their labor with all the others. In Flint Maker's time, males hunted and females gathered, and each group shared with the other. With Fúxī came matrilineal society and more complex production relationships, but still no private property.[54]

Some authors tapped into the classics in different ways. Several books credited the ancient Chinese philosopher Zhuāngzǐ with describing the origins of language or developing a theory of evolution (albeit a primitive one) "thousands of years before Darwin."[55] One author devoted considerable space to the description in the Book of Rites (禮記) of the

---

History], vol. 1 (Běijīng: Xīnhuá chūbǎnshè, 1950), 4–5. Qián Huáběi rénmín zhèngfǔ jiàoyù bù, ed., *Gāojí xiǎoxué lìshǐ kèběn* [Upper Elementary School History Textbook], vol. 1 (Běijīng: Rénmín jiàoyù chūbǎnshè, 1951 [1950]), 2.

53. Zhōngguó lìshǐ yánjiū huì, *Gāojí zhōngxué*, 5. A similar statement can be found in a book first published in Shànghǎi in July 1949 and reprinted in November: Dàowěi wénhuà chūbǎnshè, ed., *Zhōngguó lìshǐ* [Chinese History] (Shànghǎi: Dàowěi wénhuà chūbǎnshè, 1949), 2. Such statements followed the approach set out by Xú Xùshēng, *Zhōngguó gǔshǐ de chuánshuō shídài* [The Legendary Era of Chinese Ancient History], rev. and enlarged ed. (Běijīng kēxué chūbǎnshè, 1960 [1943]). In painstaking detail, Xú analyzed legends from classical Chinese texts and associated them with historical times, places, and peoples.

54. Yáng Yè, *Wǒmén de zǔxiān*, 25–28. See also Jiǎ Lánpō, *Zhōngguó yuánrén* [Peking Man] (Běijīng: Zhōnghuá shūjú, 1962), 24.

55. Fāng Shàoqīng, *Gǔ yuán zěnyàng* (1958), 4; Fāng Qiě, *Cóng yuán dào rén*, 52; Jiǎn Bózàn in Lín Yàohuá, *Cóng yuán dào rén*, preface, 2–3.

7 Depiction of human ancestors living in nests in the trees before climate change forced them
to begin a terrestrial lifestyle. Such images resonated with classical Chinese descriptions of
the earliest people (the Nest Dwellers) frequently recirculated in history textbooks of the
republican and socialist eras. Reproduced from Zhāng Mín, Shèng Liángxián, and Shěn Tiězhēng, *Láodòng chuàngzào le rén* [Labor Created Humanity] (N.p.: Huádōng rénmín chūbǎnshè, 1954 [1952]), 17.

earliest people, who lived in nests and ate uncooked grasses and meats, and others told the story "from ape to human" in a manner that resonated with the older tales, either with respect to specific activities like nest building or through general descriptions of the difficult and crude lives "our ancestors" lived (figure 7).[56]

The variety of ways stories from the classics appeared or failed to appear in the new writings on human origins demonstrates an ambivalence about the value of Chinese tradition. While the new state condemned much of Chinese religion, philosophy, and social customs as "superstitious" or "feudal," some elements of the old tradition served instead to inspire nationalist spirit. Writers of popular science materials

56. Fāng Qiě, *Cóng yuán dào rén*, 70–71. Official handbooks for history teachers specifically prescribed such use of the Book of Rites to illustrate the transition from primitive communism to slave society. Joseph Levenson, "The Place of Confucius in Communist China," in Feuerwerker, *History in Communist China*, 60.

and school textbooks drew on the classics in order to create a *Chinese* history for China, and to infuse human identity with a Chinese flavor.

Nothing did more to make human evolution a nationalist concern than the excitement and frustration over the Peking Man fossils' whereabouts. The fossils had gone missing in 1941 while en route to the United States for safekeeping. In a book published in 1950, one author wrote of the fossils' disappearance as "a great loss for our nation's most precious 'people's property' [最寶貴的人民財產]," and suggested that Americans had concocted "nefarious plots" and harbored "wild ambitions" to plunder this treasure.[57] Newspaper articles printed in 1950 and 1951 brought more specific charges. *People's Daily*, the Hong Kong paper *Dàgōng bào*, the *New York Times*, and other newspapers around the world detailed allegations, attributed to Péi Wénzhōng, that the missing Peking Man fossils were being secretly held at the American Museum of Natural History in New York. These claims found ready believers in China. The Chinese public was already angry about illegal American acquisition of many valuable cultural relics.[58] The Korean War provided a political context in which such sentiments could reach sustained fury.

Years after the Korean War, charges against Americans continued to find widespread appeal in the People's Republic. In 1956, visitors to the Peking Man Site Exhibition Hall at Zhōukǒudiàn often highlighted this issue in the comments they wrote in the guest book. One member of the Institute of Architectural Science fumed, "We saw many fossils, but none of them are the original, real fossils. The Japanese and American imperialists used despicable methods to steal all the original, real fossils. These (fossils) all belong to Chinese people; we should demand payment of this debt from the imperialists."[59] Another, from the People's Liberation Army, wrote, "During my visit, one thing that I regretted was that in many cases we can see only models and cannot see the real items because they were plundered by the Americans. This gives me an even deeper understanding of American imperialism. It's true, we should denounce their criminal theft of cultural relics in front of the entire world."[60] A screenplay published in 1961 that took the "stolen" fossils as its theme bore the title "Nation's Vengefulness, Family's Enmity."[61]

---

57. Fāng Qiě, *Cóng yuán dào rén*, 79.

58. See Cheng Te-k'un, "Archaeology in Communist China," in Feuerwerker, *History in Communist China*, 53.

59. Zhōukǒudiàn Peking Man Site Exhibition Hall Guest Book, 19 April 1956.

60. Zhōukǒudiàn Peking Man Site Exhibition Hall Guest Book, 29 June 1956.

61. Wāng Suìhán, "Guó hèn, jiā chóu" [Nation's Vengefulness, Family's Enmity]," *Diànyǐng wénxué*, no 5 (1961): 18–33.

Scientists around the world sorely felt the loss of the Peking Man fossils. Yet it was not nearly as damaging to science as it was bolstering of Chinese nationalism. The excellent casts made at Peking Union Medical College in the 1930s and the meticulous descriptions by Weidenreich have largely sufficed for research purposes, as a Chinese science magazine predicted in 1946.[62] One Chinese paleoanthropologist explains today, "The loss of the Peking Man fossils is a spiritual loss, a loss of the heart," rather than a scientific loss, "because they had already been studied very carefully and are not likely to give any new, very important information."[63] The enormous attention paid to the loss of the fossils, particularly their loss at another nation's hands, arose from a nationalist feeling of attachment inspired by the notion that they represented the ancestors of the Chinese people.

All the World Is One Human Family

People who produced popular materials on human evolution in the early People's Republic juggled many conflicting influences and priorities. Not least of these were the competing paradigms of nationalism and internationalism. A century of struggle against imperialism, culminating in the Korean War, had created a widespread conviction that nationalism and national identity were necessary for the preservation of the country and people. On the other hand, China was now, by virtue of its communist leadership and relationship to the Soviet Union, a member of a larger community that extolled internationalism, specifically in the form of international socialism. In discourse on international socialism the Chinese nation as such took a back seat to the global struggle between the capitalist countries bent on oppressing the "colored races," on one hand, and the socialist countries engaged in spreading the liberating force of international socialism on the other.

Materials on human evolution contributed to this discourse by emphasizing the shared humanity of all peoples—by constructing, as it were, a universal human identity. It was not that Soviet and Chinese scientists dismissed the relevance of racial categories themselves. Rather, science dissemination materials (particularly those for more educated audiences) described in detail the different physical characteristics that defined the white, yellow, black, and brown (Australian) races. Such

---

62. Jì Zhī, "'Běijīng rén' tóugǔ xiàluò bù míng" [The Whereabouts of Peking Man's Skull Are Unclear], *Kēxué zhīshí* 1, no. 2 (1946): 42.

63. Interview with the author, 8 October 2002. Moreover, postwar discoveries have provided scientists with access to original fossil material.

writings sometimes also suggested that racial differences mapped onto differences in levels of cultural advancement (figure 8).[64] Authors were often careful, however, to distinguish their interest in race from that of "capitalists," whose notions of racial superiority and inferiority they emphatically disputed.[65]

Trumpeting this theme thus offered an excellent opportunity to paint capitalist countries in a negative light. Some writings on human origins produced in the nonsocialist West explicitly justified imperialism on the grounds that the races were inherently separate and unequal. Moreover, Marx and especially Lenin had written extensively and persuasively on the connections between capitalism and imperialism. With such easy targets and ready ammunition, Soviet and Chinese writers had little trouble foregrounding the struggle against imperialism in their narratives of human origins.

In a 1954 popular book, Péi Wénzhōng and Jiǎ Lánpō attacked polygenic theories. The "European and American capitalist-class pseudo-scholars . . . insist that human evolution progressed separately in several different regions." What is more, they continued, the capitalists "nakedly and irresponsibly assert that white people have white ancestors, black people have black ancestors, and yellow people have yellow ancestors." The authors then accused the "pseudo-scholars" of "despising humanity" and using "the preposterous theory that humans have many species origins in order to justify the phenomenon of racial inequality in the capitalist countries." The Soviets, they concluded, "had created an accurate theory of human origins," based on monogenism, or single-origins theory.[66] This kind of polemic was not by any means rare. Rather, it was almost the rule that the more in-depth books on human origins emphatically denounced the racism found in "capitalist" anthropology and genetics.[67] "Capitalist scholars in imperialist countries . . . have a ghost in their minds that bewitches their powers of reasoning," said

64. M. F. Nesturkh [Nisētúěrhè], *Rén de zǔxiān* [Human Ancestors], trans. Wú Xíngjiàn (Shànghǎi: Kēxué jìshù chūbǎnshè, 1956), 32–39; Lín Yàohuá, *Cóng yuán dào rén*, 149–157. Simpler books merely stated the existence of racial groups and provided pictures or diagrams, without offering analysis. Běijīng shūdiàn, ed., *Rén cóng nǎlǐ lái* [Where Humans Come From] (Běijīng: Běijīng shūdiàn, 1951), 28. Guides at the Peking Man site exhibition also discussed race. One visitor challenged a guide who mistakenly categorized Indians (from India) as members of the black, rather than the white, race (Zhōukǒudiàn Peking Man Site Exhibition Hall Guest Book, [15?] August 1956).

65. Nesturkh, *Rén de zǔxiān*, 36. Nesturkh explicitly contrasted his own racial discourse with that of "capitalists."

66. Péi Wénzhōng and Jiǎ Lánpō, *Láodòng chuàngzào le rén* [Labor Created Humanity] (Běijīng: Zhōnghuá quánguó kēxué jìshù pǔjí xiéhuì, 1954), 16–17.

67. For treatment of specific eugenicists like Francis Galton as well as modern geneticists (Mendel, Morgan, and Weissman), see Péi Wénzhōng, *Rénlèi de qǐyuán*, 3.

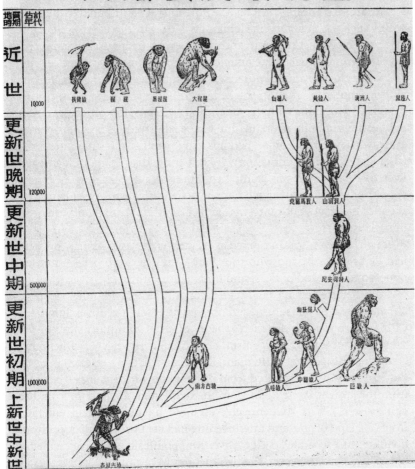

人類發展系統圖

8　Phylogenetic tree showing evolutionary relationships of humans and apes. The four human groups at the upper right are (*left to right*) the "white race," "yellow race," "Australians," and "black race." Note that the white and yellow races are represented by workers, while the Australians and black race are represented by tribal people. The Australian is also stooped over, which probably reflects the idea (present in Morgan's *Ancient Society*) that Australian aborigines are the most primitive of all the world's peoples. The two human figures on the second rung down represent Cro-Magnon Man (ancestral to Europeans) and Upper Cave Man (ancestral to Asians). The stooped figure below them is Neanderthal Man, below which is the head of Heidelberg Man, followed by Peking Man (note the similarity to the female image in figure 1) and then Java Man. The large figure to their right is *Gigantopithecus*, and the offshoot to their left is *Australopithecus*.

　　Reproduced from Lín Yàohuá, *Cóng yuán dào rén de yánjiū* (Běijīng: Gēngyún chūbǎnshè, 1951), 26.

Fāng Shàoqīng (the pen name of the important biologist Fāng Zōngxī) in 1958. "What ghost? . . . The ghost of the feeling of racial superiority."[68]

The anti-American politics of the Korean War contributed to the force with which early writers on human evolution in the People's Republic attacked imperialist strains in anthropology. Materials on human evolution sometimes singled out American anthropological writings as especially egregious examples of racism and imperialism. For example, one author pointed in 1950 to the American book *The Negro a Beast*, published in 1900, along with the American Earnest A. Hooton's more recent publications on human origins and race, as illustrations of anthropology at its worst.[69] It was ironic—perhaps even paradoxical—that one of the chief targets of these criticisms was none other than Franz Weidenreich. Even as Peking Man was gaining the status of ancestor of the Chinese people for which Weidenreich had so vociferously argued, Weidenreich's theories of racial continuity were attacked as emblematic of American imperialism.

It was further ironic inasmuch as Weidenreich was a German Jew who had fled the Nazis, and who thus had every reason to believe himself opposed to the kind of racist and imperialist anthropology of which he was accused. He had lived in America for only a short time after escaping Germany, and he had returned to America after war had driven him from China. Moreover, Weidenreich's theory of racial continuity was not the same as the notorious polygenism. He had explicitly rejected earlier polygenic theories like that of Hermann Klaatsch, who had suggested that the modern races had originated from different anthropoid apes ("Negroes from the gorilla, Mongolians from the orangutan, and whites from the chimpanzee").[70] In Weidenreich's account, all humans had emerged from one kind of ape, and interbreeding had kept the human races from dividing into separate species. Given the politically sensitive character of the race issue, however, the two theories were easy to conflate.

Weidenreich had had a strong mentoring relationship with Jiǎ Lánpō. Jiǎ's 1950 *Peking Man* argued that Peking Man was a direct ancestor of modern humans, and he specifically cited Weidenreich's evidence to prove it. Nonetheless, he took issue with Weidenreich's theories on racial continuity. He noted Weidenreich's identification of features shared by Peking Man and modern "Mongoloids" (the major racial group that includes peoples of East Asia) and his conclusion of racial continuity

---

68. Fāng Shàoqīng, *Gǔ yuán zěnyàng* (1958), 63.

69. Fāng Qiě, *Cóng yuán dào rén*, 7–8. Charles Carroll, *The Negro a Beast* (St. Louis: American Book and Bible House, 1900).

70. Weidenreich, *Apes, Giants, and Man*, 24.

between these two forms. Jiǎ then expressed disagreement with this formulation, saying that the modern races were too similar to have such ancient roots, and that the split must have occurred after the Neanderthal period.[71] Even so, Péi Wénzhōng wrote a preface to Jiǎ's book in which he professed to be concerned that Jiǎ had assimilated ideas from Weidenreich and had thus been tainted with idealism. Péi suggested tactfully that Jiǎ should read more Marxist philosophy before revising the book for a second edition.[72] By 1952, Jiǎ had come to understand the importance of explicitly distancing himself from his foreign teacher. In an article entitled "A Look at American Imperialist Methods of Aggression from the Viewpoint of Anthropology," Jiǎ blasted Weidenreich as a representative of the idea that the modern races have different evolutionary origins.[73]

Fāng Qiě, the author of the 1950 book *From Ape to Human*, tied criticism of Weidenreich to the domestic issue of Hàn chauvinism in a 1951 book entitled *Sons and Daughters of the Chinese Nationality*. According to Fāng, Weidenreich's claim that Peking Man was "the direct ancestor of the Chinese people" bore striking resemblance to an idea popular among Chinese people in the past that they were the "descendants of the Yellow Emperor." Fāng Qiě charged that the appeal of this notion lay in the Yellow Emperor's reputed conquering of the Miáo-nationality leader Chīyóu, thus glorifying the majority Hàn people. In actuality, Fāng went on, the notion that the Chinese people were descendants of the Yellow Emperor was "a narrow-minded, nationalistic type of defensiveness, a mistaken, high-and-mighty stele [erected to glorify] Hàn chauvinists." He then cited Máo Zédōng on the long history of interaction among the many ethnic groups (民族) currently living in China and quoted Stalin's statement that "nationalities are not communities made up of a single race [種族]."[74] This vehement insistence on incorporating China's many "minority nationalities" into a larger definition of the "Chinese people" (中華民族) reflected the state's effort to attain the loyalty of those nationalities and thus maintain as closely as possible the boundaries of the former Chinese empire. Interestingly, this new Chinese nationalist (one might even say imperialist) project gained strength from assaults on the "chauvinistic" Hàn nationalism of old.

71. Jiǎ Lánpō, *Zhōngguó yuánrén* [Peking Man] (Shànghǎi: Lóngmén liánhé shūjú, 1950), 135–138.

72. Péi Wénzhōng in Jiǎ Lánpō, *Zhōngguó yuánrén* (1950), preface, 2–3.

73. Jiǎ Lánpō, "Yóu rénlèixué de guāndiǎn kàn Měi dìguózhǔyì de qīnlüè shǒuduàn" [A Look at American Imperialist Methods of Aggression from the Viewpoint of Anthropology], *Kēxué dàzhòng*, no. 5 (1952): 110.

74. Fāng Qiě, *Zhōnghuá mínzú de ěrnǚ* [Sons and Daughters of the Chinese Nationality] (Shànghǎi: Shànghǎi biānyì chūbǎnshè, 1951), 3–6.

The criticism of Weidenreich and of racist and imperialist trends in paleoanthropology as a whole served as a foundation for what was to become one of the defining features of science dissemination materials on human evolution in socialist China. Chinese authors did not confine themselves to exposing racism in the works of their politically unsavory colleagues overseas; they also had a positive message. The chapters in which such attacks occurred were usually the concluding chapters and were optimistically entitled "The Modern Races Are All One Family," "All the World's Humans Are One Family," "All the World Is One Human Family," or some other variation on this theme.[75] A movie produced in 1959 to accompany the Peking Man site exhibit at Zhōukǒudiàn concluded with a charming painting of a group of young people of different races all happily and resolutely holding hands (figure 9). The narrator proclaimed, "All the world's humans—no matter the color of their skin, no matter their race—went through the same process of development."[76] A 1954 book written for youth approached this concept from another direction. While the movie highlighted the sameness of all peoples' past development, this book concluded with a prediction about their present and future development. Before the October Revolution, the author explained, the Soviet Union's many minority nationalities had suffered from oppression, which kept their lives simple and their cultures backward. Under socialism they had developed quickly and were now on the same level as the most advanced nationalities. This was the promise, he implied, that international socialism held for the rest of the world.[77] In later years, the identification of anthropological theories of racial equality with revolutionary politics only became stronger and more militant. Figure 10 shows a Cultural Revolution take on the same theme: here, however, representatives of the races cease to hold hands and instead use their fists to crush imperialism.

What kinds of existing racist ideas were to be found among the people, and to what extent the antiracist propaganda sunk in, are troubling questions. African exchange students in China during the 1960s and 1970s regularly encountered ignorance, disdain, fear, and at times even violence.[78] According to the account of Emmanuel Hevi, a Ghanaian

75. See, for example, Lín Yàohuá, *Cóng yuán dào rén*, 149; Huáng Wéiróng, *Zhōngguó yuánrén*, 33; Péi Wénzhōng and Jiǎ Lánpō, *Láodòng chuàngzào le rén*, 16; Péi Wénzhōng, *Rénlèi de qǐyuán*, 11; Wú Rǔkāng, *Rénlèi de qǐyuán hé fāzhǎn* [Human Origins and Development] (Běijīng: Kēxué pǔjí chūbǎnshè, 1965), 73.

76. Shànghǎi kēxué jiàoyù diànyǐng zhìpiànchǎng, *Zhōngguó yuánrén* [Peking Man], 1959.

77. Huáng Wéiróng, *Zhōngguó yuánrén*, 35.

78. Philip Snow, *The Star Raft: China's Encounter with Africa* (Ithaca, N.Y.: Cornell University Press, 1988).

9   Still photo from the 1959 movie *Zhōngguó yuánrén* (Peking Man), according to which "modern humans" of all races share the same history of evolutionary development. This image suggests that they also share a present and future characterized by cooperation and solidarity under the banner of international socialism, with Chinese people set to play a central role. From Shànghǎi kēxué jiàoyù diànyǐng zhìpiànchǎng, *Zhōngguó yuánrén*, 1959.

10   Men of different races coming together to overthrow imperialism. Reproduced from Shànghǎi zìrán bówùguǎn, *Cóng yuán dào rén* (Shànghǎi: Shànghǎi rénmín chūbǎnshè, 1973), 110.

student who lived in China from 1960 to 1962, even medical doctors could not be relied upon to refrain from asking "why your skin is so black if you ever wash." Believing that doctors by virtue of their profession would understand the nature of skin pigmentation, Hevi concluded that the question arose from "malice, pure and simple."[79] Hevi believed that the racism he experienced was not "of the kind that springs spontaneously from the people," but rather "a deliberate attempt by the Communist Party to assert and to make the African accept once and for all the idea of the superiority of Yellow over Black."[80] As further proof of the ultimate responsibility of the state for entrenched racism, Hevi noted that given its control of the press, the "government need just publish one article on the subject in the *People's Daily* and that will be the end."[81]

But in actuality, intellectual and political elites working on behalf of the state published a great number of articles and books proclaiming the races equal and equating racist ideology with fascism and imperialism. Much more convincing than Hevi's analysis are the later reflections of a Chinese man who had reached adolescence during the 1960s in China. In a 1989 letter to the *New York Times*, Gao Yuan responded to the anti-African violence then erupting on Chinese college campuses: "Although racism is adamantly opposed by the Chinese Government, it is undeniably part of the psychological makeup of China's citizens, even the educated elite." Gao went on to contrast the recent racist behavior of the post-Máo reform era with his recollections of the Máo era, when students "held rallies to express solidarity with oppressed peoples, including America's blacks." He concluded, "Whatever one might have thought about Africans or Arabs at Chinese universities, nobody would have expressed open hostility to students from fraternal third world countries."[82] The picture we get is a state working very hard to promote an official policy of racial equality and solidarity—and supported greatly in this endeavor by international socialist discourse on human origins—in the face of deep-seated racist ideas that permeated all classes of society. The failure of the state's propaganda lay foremost in its unwillingness to acknowledge and confront the problem within Chinese society. As long as racism officially remained a characteristic only of recognized capitalist, imperialist societies like the United States, Chinese people could

---

79. Emmanuel John Hevi, *An African Student in China* (New York: Praeger, 1962), 187.
80. Hevi, *An African Student*, 134–135.
81. Hevi, *An African Student*, 186–187.
82. Gao Yuan, "In China, Black Isn't Beautiful," *New York Times*, 25 January 1989.

actively participate in solidarity rallies without ever questioning their own racist assumptions.[83]

Conclusion

To borrow Máo's words, the human identity constructed in popular paleoanthropology at times very consciously "left out many things," including the differences between "Chinese people and foreign people." More often than not, it also left out discussion of "males and females." And, most important, it left out class. In short, through dissemination of knowledge about human origins and evolution, the socialist state promoted a universal human identity.

This is not to say that the particular shape human identity took was uninformed by commitments to such social categories as class and nation. Rather, what it meant to be human was inseparable from a kind of class consciousness that glorified manual labor, and it slipped in and out of an identity that was at least as specifically Chinese as it was broadly human. If we were to take the same approach to Chinese paleoanthropology as Donna Haraway, for example, has taken to Western primatology, we would emphasize such constraints on a universal human identity. Máo in his own way would no doubt have approved of such a take. While a great deal separates these two thinkers, Máo shared Haraway's concern for exposing the way liberal humanists assume that their concept of the "human" is a universal truth. As he put it at the Yán'ān forum, "We uphold the human nature of the proletariat and of the masses of the people, while the landlord and bourgeois classes uphold the human nature of their own classes, only they do not say so but make it out to be the only human nature in existence."[84]

Our task, however, is different from Haraway's and Máo's. Whereas they highlighted the dangers in embracing humanism without recognizing one's own commitments to specific social identities, in Máo-era China the converse position was dominant. What is needed, then, is to reveal the political significance of a human identity that existed in addition to—and despite—the emphasis on class and nation. In its humanism it was universal—not in the sense that it was the only possible way of understanding what it meant to be human, but in the sense that it presented certain essential features said to be held in common by all people of the world.

83. Philip Snow makes a similar point on pages 203–204 of *The Star Raft.*
84. Mao Tse-tung, *Selected Works,* 3:90.

Human identity was in actuality instrumental in the construction of the socialist state and society. Defining humans in terms of labor not only glorified the laboring classes but also allowed for the possibility that, through labor, intellectuals, capitalists, and members of other suspect classes could reform. Portraying human evolution as a series of "struggles with nature" also helped to promote a voluntaristic spirit—a notion that human will could conquer the natural world. The slogan "all the world is one human family" was undeniably an expression of a universal human identity. But unlike its counterpart in the West (as in the famous Family of Man exhibition of photographs in New York in 1955), the international socialist version of humanism was meant to inspire revolutionary struggle against imperialism.

Finally, and most important, the very notion of communism itself in some sense depended on the portrait of primitive society painted in materials on human evolution. Far more than was the case in the Soviet Union, Máo hoped and expected that communism would be achieved in the near future. The promise of that better world seemed more plausible in light of primitive communism as a natural state that existed before the emergence of class differences.

Humanism, like the political vision of the New Democracy, was an ideal precious to many intellectuals in socialist China. Both exalted commonality and cooperation while downplaying social differences and class struggle. While scholars of literature and art were repeatedly stymied in their explorations of humanism, scientists and other science disseminators were actually charged with creating a human identity based on scientific knowledge of evolution. In this sense, constructing a human identity was an area of overlap between intellectual and state interests similar to the project to disseminate science and squash superstition. Such state-supported efforts could not, however, wipe away the equally state-supported cultivation of class struggle. The skepticism Máo and others expressed regarding elite claims to knowledge about human nature extended also to elite forms of knowledge generally. While Engels made it possible for—or even required—science writers to talk about humans, their privileged position as purveyors of scientific truth was far from secure. Again and again during the Máo era, one would split into two: the concept of human would fragment into discrete social identities, and the presumed stability of scientific knowledge would be rent by class struggle.

FOUR

"Labor Created Science": The Class Politics of Scientific Knowledge, 1940–1971

Top-Down and Bottom-Up Approaches to Popularizing Science

One of the main goals of science dissemination, or *kēpǔ*, in socialist China was to smash widely held "superstitious" beliefs and provide the masses with scientific explanations for the phenomena in question. In the case of human origins, science disseminators sought to replace ideas that God or the Chinese legendary figure Nǚwā created humanity with Engels's theory that labor created humanity. Yet the characterization of the masses as superstitious and in need of correction from above rubbed up uncomfortably against Maoist calls to "learn from the masses."

Once again, the story begins at Yán'ān, where Máo called on intellectuals to "draw nourishment from the masses to replenish and enrich themselves so that their specialities [did] not become 'ivory towers,' detached from the masses and from reality and devoid of content or life."[1] This apparently harmonious relationship was belied by the coercive methods through which Máo enforced this program of

---

1. Mao Tse-tung, *Selected Works of Mao Tse-tung* (Peking: Foreign Languages Press, 1967–1971), 3:84.

"enrichment," which included sending intellectuals accused of "individualism" to labor alongside the peasants. After the revolution, ideas about the necessary relationship between knowledge and labor periodically offered serious challenges to the scientific authority on which *kēpǔ* was based. During these times in particular, mobilization programs sought to create "mass science," a truly proletarian enterprise based on the contributions of the laboring classes.

Richard Suttmeier, writing in 1974, recognized the "two rather different premises underlying" science dissemination and mass science. He noted that the first was fundamentally a "'top-down' philosophy of knowledge" that saw expertise residing in scientific institutions, whereas the second was "bottom-up" and considered workers to be sources of "untapped expertise" based on their intimate understanding of production processes.[2] I will extend this astute observation much further. Rather than harmonious interdependence, the top-down and bottom-up approaches to popularizing science lay in profound contradiction. Uncertainties about authority and knowledge, embedded in larger "class struggles" cultivated by party officials, gave rise to decades of contest and change in the arenas of both art and science. Moreover, this conflict was inherent in Máo's own writings: factional politics and divergent interests, though undoubtedly present, were not the only cause of the confusion.

Science dissemination crystalized into a clear position that was well articulated, given a specific name (and even a well-known abbreviation), and provided with specific organizations to carry out its mission. Even within this program, however, there was room for class-based criticism: scientists and others involved were frequently cited for failure to anticipate and fulfill the needs of the "broad masses." Science dissemination was, moreover, problematic to its core in that it fostered an intellectually elitist attitude toward knowledge.

Mass science, on the other hand, was never so clearly conceptualized; it had no single easily recognizable name and abbreviation.[3] Nor were there organizations comparable to the Science Dissemination Bureau or Science Dissemination Association to oversee it. Mass science was, however, consistent with the commitment to overturning social hierarchies

---

2. Richard Suttmeier, *Research and Revolution: Science Policy and Societal Change in China* (Lexington, Mass.: Lexington Books, 1974), 126.

3. It was generally referred to as "mass scientific research," "science of mass character," or "masses doing science," among other names (群衆科學研究, 群衆性科學, 群衆辦科學). Richard Suttmeier has discussed it as "technological mobilization" in his *Research and Revolution*. I have chosen the term "mass science," since this more closely matches the Chinese terms.

that was so fundamental to the vision of the revolution and the new society. The chief problem was that it was often difficult to articulate exactly what the masses offered science that did not fall into the categories of superstition and "remnants of feudal society." For this reason mass science during less radical periods frequently took a far less extreme form, in which professionals were supposed to guide the masses and provide structure to their unsystematic ideas and experiences.[4] During the more radical periods, however, the privileged position of intellectual elites was questioned to the point that scientists would have to "follow the masses"—or at least pretend to—even in their own research.

Not only did experiences vary across time, but scientific discipline also mattered greatly. There was a big difference between encouraging peasants to breed new strains of rice and inviting them to contribute to the study of human evolution. Until the Cultural Revolution no one suggested that peasants or workers be asked for their opinions on the origins of humans, much less that such opinions be incorporated into science dissemination materials. Being a field science, however, paleoanthropology offered an obvious place for the laboring classes. Peasants and mining workers frequently discovered fossils, and scientists relied on local labor to help in excavation work. Nonetheless, even in such cases the familiar worry about superstition arose. Class politics did not exist simply as rhetoric; rather, struggles over the production and dissemination of knowledge were often rooted in social differences and in conflicting interpretations over proper divisions of labor in a revolutionary society.

Science Dissemination for Whom, by Whom?

In his "Talks at the Yán'ān Forum," Máo asked the question, "literature and art for whom?" The answer was: "first for the workers, the class that leads the revolution." Second came the peasants, third the "armed workers and peasants" (i.e., the soldiers), and only fourth "the urban petty bourgeoisie and . . . the petty-bourgeois intellectuals."[5] Máo further insisted that dissemination was supposed to come about through a process that involved "learning from the masses" about their own needs, with needs defined both in the negative and the positive—in terms of both their degrees of ignorance and their goals. For the scientists and

4. Laurence Schneider, *Biology and Revolution in Twentieth-Century China* (Lanham, Md.: Rowman and Littlefield, 2003), 13.
5. Mao Tse-tung, *Selected Works*, 3:75, 77.

science writers who modeled their dissemination efforts on Máo's "Talks at the Yán'ān Forum," however, all this was easier said than done. They faced the pressure of serving multiple audiences: cadres, political leaders, and other scientists, in addition to the "great masses of the people." It was not always easy to achieve the "correct" balance, given competing political priorities and the value placed on "serving the people."

In the preface to his 1950 book *Peking Man*, Jiǎ Lánpō claimed to have "spent an entire night" deciding "for whom to write this book." In the end, he determined to "choose a method that both the elite [雅] and the popular [俗] can appreciate." By including academic discussion, he hoped to help advance scientific debate. By discussing the results of research in a simple style, he intended that "the broad masses [would] first recognize what Peking Man is, and then on the basis of this general [普遍] understanding, they [would] demand to raise their standards [提高] a level."[6] Here, Jiǎ echoed the "Talks at the Yán'ān Forum," where Máo asserted, "The people demand dissemination [普及] and, following that, higher standards [提高]."[7] In actuality, it is hard to imagine many people of limited education reading Jiǎ's book. It abounds with specialized terminology and theoretical concepts, and most of the drawings are very technical. That Jiǎ felt the need to pretend that the book was written for "the broad masses" indicates the pressures on scientists to popularize. That his specialist book was published nonetheless, however, points to a widespread understanding that professional needs must also be met.

In a report on its work during the first eight months of 1953, the Běijīng Science Dissemination Association documented similar trouble. Of all the people who had heard their lectures on science so far that year, only 17,166 (or 9.4%) had been workers. The numbers were considerably better for peasants (51,850, or 28.4%), but by far the greatest number of recipients had been cadres (82,076, or 44.9%).[8] The authors suggested several reasons for this discrepancy. Factory workers were often too busy to stop work to listen to lectures. Moreover, they were interested in learning concrete solutions to specific technological problems, while association members considered this beyond their own capabilities and

6. Jiǎ Lánpō, *Zhōngguó yuánrén* [Peking Man] (Shànghǎi: Lóngmén liánhé shūjú, 1950), author's preface, 2.

7. Mao Tse-tung, *Selected Works*, 3:83. For consistency, I translate *pǔjí* as "dissemination" rather than "popularization."

8. The source provided only numbers for workers, cadres, and peasants and listed a percentage only for workers. I derived the total and then the percentages for cadres and peasants. The three groups combined equal only 82.7%. The remaining 17.3% may have been soldiers (the fourth major social category), but it is somewhat strange that the authors did not mention them.

were more inclined to disseminate general scientific knowledge. Aside from these difficulties, however, lay the "subjective [problem] of a lack of effort," the report said in the familiar style of self-criticism. "We are so busy giving lectures to cadres that we do not have enough energy left over for factory work." The solution, they said, was to redouble their efforts and to pay closer attention to the specific needs of factories.[9]

The Peking Man Site Exhibition Hall at Zhōukǒudiàn faced even greater challenges in serving multiple audiences. The two thousand foreign visitors documented in the 1956 guest book received more than their share of attention, not only because of their special status but also because domestic visitors often came in larger groups with their classes or work units.[10] Many of the Chinese who signed the guest book, moreover, were college students and science professionals. The content of the exhibit reflected these demographics, leaving less well educated visitors at something of a loss. The guest book contains many requests to improve explanations and visual displays, particularly to assist "the masses" in understanding the significance of the fossils and artifacts they encountered. One visitor wrote, "While I was here, I saw that some of the visiting comrades thought the fish fossils were very strange. Upon seeing the fossilized dog [sic, hyena] feces, they thought that dogs pooped out rocks. They are young peasants, and they don't really understand the rationale behind this place." The commentator thought this indicated a need for explanations better tailored to diverse audiences.[11] Another visitor put it more bluntly: "I wish there were a complete and systematic introduction so that the exhibition hall would be suitable not only for scientific research but also for regular people."[12]

Beginning in 1956, the Science Dissemination Association began cooperative efforts with the All-China Federation of Trade Unions to help adapt science dissemination to the needs of workers.[13] A 1959 document from the Běijīng branch indicated how far they had moved in that direction. "Workers, peasants, and soldiers," the association asserted, "will never again be considered merely 'disseminatees' [被普及者], but will also be 'disseminators' [普及者]." The association had established

9. Yīn Zǔyīng et al., "Běijīng shì kēxué jìshù pǔjí xiéhuì: 1953 nián yī zhì bā yuè yèwù bàogào" [Běijīng Municipal Science and Technology Dissemination Association: Work Report for January–August 1953], 18 October 1953, Běijīng Municipal Archives, 10.1.31: 15–27, 23.

10. For example, a group of more than one hundred students went together on 13 April 1956.

11. Zhōukǒudiàn Peking Man Site Exhibition Hall Guest Book, 29 June 1956. Because hyenas eat large amounts of bone, their feces fossilize very well. Note that the word for hyena in Chinese literally means "fierce dog" (獵狗). For other examples, see [21?] July 1956 and 9 September 1956.

12. Zhōukǒudiàn Peking Man Site Exhibition Hall Guest Book, 16 March 1956.

13. Suttmeier, Research and Revolution, 126.

programs in which "advanced producers" and "technological innovators" from among the masses gave science dissemination lectures according to their own experience-based knowledge.[14]

Such activities seem to have done justice to the vision Máo had laid
out in his "Talks at the Yán'ān Forum." But it was one thing to give the
masses a voice in dissemination on agricultural and industrial subjects;
"basic science" subjects like human origins were another story. There the
elimination of superstition continued to be central, the knowledge flow
thus tending to remain almost exclusively one way and top down.

Ivory Towers and Cow Sheds

Even as they were engaged in a top-down endeavor that valued their
knowledge over that of the masses, scientists faced repeated challenges
to their authority, not only in the realm of science dissemination but
also in scientific research itself. Their expertise was both a professional
requirement and a political liability. Depending on the shifting political winds, scientists received special privileges or special punishments
designed to undermine those privileges.

Until the Rectification Campaign of 1942–1944, party policy at Yán'ān
favored the Academy of Sciences director Xú Tèlì and others who focused
on building a foundation for a strong future science profession through
education in basic science. With the Rectification Campaign came not
only attacks on intellectuals in the world of art and literature, but also
a victory for the populist scientist Luò Tiānyǔ. Luò would later become
the foremost proponent of Lysenkoism in China. In Yán'ān his cause was
more general: he taught agricultural biology by requiring his students
to go among the peasants and gather specimens in the mountains and
fields. He ridiculed intellectuals like Xú Tèlì for teaching about esoteric
scientific subjects using foreign books. Such practices served to alienate
science cadres from the masses and prevented them from addressing
concrete problems posed in the here and now.[15]

The victories in the resistance and civil wars released the communists
from these conditions of exceedingly scarce resources and extremely
pressing needs. The early 1950s saw a return to the ideal of "new democ-

---

14. "Běijīng shì kělián, kěpǔ bā nián lái de gōngzuò hé jiànlì shì kēxué jìshù xiéhuì de bàogào
(chūgǎo)" [Report on Eight Years of Work in Běijīng's Federation of Science and Science Dissemination Association, and Establishing the Běijīng Association of Science and Technology (First Draft)],
29 September 1959, Běijīng Municipal Archives, 10.1.104: 1–8, 5–6.

15. James Reardon-Anderson, *The Study of Change: Chemistry in China, 1840–1949* (Cambridge:
Cambridge University Press, 1991), 352–359.

racy" that Máo had defined in 1940 at Yán'ān. Many intellectuals were optimistic about their futures under the new state, and party leaders in the early years "elevated the value of expertise to a point which almost seemed to be higher than the value of 'red.'"[16] Those who fought too hard for a radical approach to science could be criticized, as Luò Tiānyǔ was for "ingratiating himself with the backward masses" and failing to help scientists in their efforts to reform.[17]

There were early signs, however, that intellectuals would have to be careful about what they wrote and how they acted in order to preserve their status as friends, rather than enemies, of the people. Even when they tried to play by the rules, scientists like Liú Xián sometimes found themselves held up as negative examples. Such criticisms did not merely serve to establish an orthodoxy with respect to human origins and other subjects. Even more important, they served to undermine the authority scientists held by virtue of their education, to show that they were not immune from challenges by laypeople.

Debates over Lysenkoism and Morganism played a central role in the class politics of scientific knowledge. Promoters of Lysenkoism branded Morganist genetics "bourgeois" science conceived in the "ivory tower," in contrast with their own theories, which were rooted in the earthy experience of the heroic farmer figure Michurin. The Hundred Flowers Movement of 1956, however, briefly turned the tide, as Máo attempted to distinguish Chinese socialism from the recently revealed atrocities of Stalin's regime. Máo called for a respect for diversity of opinion with the slogan "Let one hundred flowers bloom and one hundred schools of thought contend." Overturning the notion that sciences like genetics were inherently bourgeois, the Hundred Flowers Movement argued instead for the position that science had no "class character."[18]

This situation was not to last. While Máo's insistence that cadres be "both red and expert" (又紅又專) appeared to give almost equal value to political and intellectual contributions to society,[19] during the Anti-Rightist Movement of 1957 scientists and intellectuals accused of being "expert" but not "red" faced harsh criticism and labor reform. The Běijīng Science Dissemination Association, for example, declared that

16. Franz Schurmann, *Ideology and Organization in Communist China* (Berkeley: University of California Press, 1966), 51.

17. Laurence Schneider, "Learning from Russia: Lysenkoism and the Fate of Genetics in China, 1950–1986," in *Science and Technology in Post-Mao China*, ed. Denis Fred Simon and Merle Goldman (Cambridge, Mass.: Council on East Asian Studies, Harvard University, 1989), 50.

18. Schneider, *Biology and Revolution*, 167, 177, 192, 226.

19. Mao Tse-tung, "Be Activists in Promoting the Revolution" (7 October 1957), in *Selected Works*, 5:489.

"the problem" of rightism was "relatively severe" among members of the association, and "even more severe" within the association's standing committee. The association found some of these people to have been engaged in "counterrevolutionary activities" and to have been "practicing science dissemination with ulterior motives."[20] While the fates of these individuals were not related in this document, it was not uncommon for those attacked in the Anti-Rightist Movement to spend months or even years in the countryside "learning from the masses" and being punished for their transgressions, real or invented.[21]

The Great Leap Forward, in 1958–1960, was an ambiguous time for science. On one hand, science was trumpeted as the means through which the masses, organized in communes, would conquer nature, allowing China to make a "great leap" into a socially and technologically advanced future. On the other hand, an extreme optimism enforced from above compelled local officials and scientists alike to throw empirical considerations to the wind in the name of "confidence" (信心) as they continually ratcheted up their production estimates. There were a few important successes in this period, notably the kelp farms developed by marine biologist Zēng Chéngkuí (C. K. Tseng) and run by hundreds of fishers in coastal provinces.[22] However, the twenty to thirty million deaths in the famine that accompanied the Great Leap testified to the overall dismal failure of the program.[23]

During this period, the war against superstition took a fascinating turn. While salient as ever, the term increasingly referred to beliefs held by intellectual and political elites in addition to those held by "the masses." Moreover, the newly identified superstitions were explicitly related to class politics. One book from 1958, edited by the Science Dissemination Press, called for the "eradication of unrealistic practices and ideas that make science out to be mysterious, rule bound, instititional, and one sided, and that claim one must have specialists and modern

20. Běijīng shì kēpǔ xiéhuì, "1957 nián gōngzuò bàogào (cǎogǎo)" [1957 Work Report (Draft)], 15 January 1958, Běijīng Municipal Archives 10.1.69, 11.

21. IVPP appears to have suffered relatively little during the Anti-Rightist Movement. Guldin concludes from interview data that this was true for most natural scientists compared with their counterparts in the social sciences. Gregory Guldin, *The Saga of Anthropology in China: From Malinowski to Moscow to Mao* (Armonk, N.Y.: M. E. Sharpe, 1994), 157. Yao, however, sees the movement to have been seriously detrimental to the lives of scientists and the progress of science. Shuping Yao, "Chinese Intellectuals and Science: A History of the Chinese Academy of Sciences (CAS)," *Science in Context* 3, no. 2 (1989): 455.

22. Peter Neushul and Zuoyue Wang, "Between the Devil and the Deep Sea: C. K. Tseng, Mariculture, and the Politics of Science in Modern China," *Isis* 91, no. 1 (2000): 79.

23. Judith Banister estimates thirty million excess deaths during the Great Leap Forward. *China's Changing Population* (Stanford, Calif.: Stanford University Press, 1987), 85.

equipment to develop science."[24] The same year, the Science Dissemination Association went to a rural county to observe science dissemination work there. One participant enthused, "We see very clearly that our superstitious ways of thinking about science are more severe than those of the peasants. . . . In the past we always thought that we could call something science only if it were written in a book or talked about by people with academic status; neither the masses' production experience nor their inventions were considered science, and it was thought China's rich scientific heritage was not worth talking about."[25] Superstition was thus radically reconfigured during the Great Leap Forward era to include the notion, held by scientists and other educated people, that the masses were incapable of practicing science. Also evident here is a common juxtaposition of national- and class-based critiques of science characteristic of the late 1950s through the early Cultural Revolution.[26]

The disaster following the Great Leap Forward sent the radicals to the margins for several years. Along with the implementation of more moderate economic and educational policies, scientific elites were once again allowed greater authority. The press made room for academic debates—for example, in a series of articles in which Jiǎ Lánpō and Péi Wénzhōng argued back and forth over Peking Man's status as the "earliest human."[27] The Institute of Vertebrate Paleontology and Paleoanthropology scientist Wú Rǔkāng traveled to Tanzania's Olduvai Gorge to study recently unearthed hominid fossils there. Beginning in 1960, scientists met in symposia to discuss issues of science and society in relative freedom. The party called these "meetings of immortals," because of their flexible agenda and tolerance for divergent opinions.[28] At these meetings, scientists busily set about restating their position in society, resulting in a key document called "Fourteen Articles on Scientific Work." This document resolved the problem of "red and expert" in a new way: scientists

24. Kēxué pǔjí chūbǎnshè, ed., *Qúnzhòng kēxué yánjiū wénjí* [Collection of Mass Science Research] (Běijīng: Kēxué pǔjí chūbǎnshè, 1958), 1.

25. Kēxué pǔjí chūbǎnshè, ed., *Kēxué pǔjí hé yánjiū gōngzuò wèi shēngchǎn dàyuèjìn fúwù* [Science Dissemination and Research Serves the Great Leap Forward in Production] (Běijīng: Kēxué pǔjí chūbǎnshè, 1958), 117.

26. Rensselaer W. Lee III, "Ideology and Technical Innovation in Chinese Industry, 1949–1971," *Asian Survey* 12, no. 8 (1972): 648.

27. Articles appeared in 1961 issues of *People's Daily* and 1961 and 1962 issues of *New Construction* (新建設). See also Jiǎ Lánpō and Huáng Wèiwén, *Zhōukǒudiàn fājué jì* [The Excavation of Zhōukǒudiàn] (Tiānjīn: Tiānjīn kēxué jìshù chūbǎnshè, 1984), 182–187.

28. Yao, "Chinese Intellectuals and Science," 458; Nie Rongzhen, *Inside the Red Star: The Memoirs of Nie Rongzhen*, trans. Zhong Renyi (Beijing: New World Press, 1988), 714. China watchers at the time suggested that such meetings were thought reform sessions in a milder form. Merle Goldman, "The Fall of Chou Yang," *China Quarterly* 27 (1966): 142; Dennis Doolin, "The Revival of the 'Hundred Flowers' Campaign: 1961," *China Quarterly* 8 (1961): 37–40.

served socialist China through their expertise, and for this reason they were also "red." The "Fourteen Articles" further declared science to be without "class character."[29] A 1961 article in *Red Flag* took up these pronouncements and went so far as to argue for a separation of scientific and political questions.[30]

The early 1960s also saw arguments that science was a productive force rather than part of the superstructure.[31] The significance of this for scientists lay in their ability to be considered genuine "workers" rather than suspect intellectuals. At a national conference on science and technology held in Guǎngzhōu in 1962, Foreign Minister Chén Yì pronounced the slogan "Take off the hat and put on the crown"—in other words, remove the political label of "bourgeois intellectuals" and celebrate scientists as "intellectuals of the working people."[32]

Such relatively moderate policies were short lived. As early as 1964, events in the cultural sphere pointed to a return to radicalism, and in 1966 "big character posters" criticizing bourgeois education appeared at Běijīng University, students all over the country formed revolutionary groups known as "red guards," and the media announced the beginning of the Cultural Revolution. Merle Goldman notes that scientists did not become the targets of Cultural Revolution persecution until the purge of Liú Shàoqí in the winter of 1966–1967. The reasons, she argues, were that science was supposedly less ideological than other intellectual endeavors, that Máo and others "did not claim to have the knowledge of science that they did of history, philosophy, and literature," and that the leadership recognized the value of science to the state. She further suggests that Máo personally "wanted to shield [scientists] from the kind of violent attacks that hit nonscientific intellectuals."[33] Attacks on the natural sciences were certainly less thorough and consistent than attacks on other forms of intellectual work. Nonetheless, the question of whether science was a productive force or part of the superstructure was settled during the Cultural Revolution in favor of the superstructure. Moreover, as "the masses" moved to "occupy the domain of the superstructure" (工農兵佔領上層建築領域), scientists, having been first labeled intellectuals, were then pushed out of their territory.[34] Literally, they were told to

---

29. Nie Rongzhen, *Inside the Red Star*, 715–719. Yao, "Chinese Intellectuals and Science," 459.

30. James H. Williams, "Fang Lizhi's Big Bang: Science and Politics in Mao's China" (Ph.D. diss.: University of California, Berkeley, 1994), 1:293.

31. Suttmeier, *Research and Revolution*, 96–97.

32. Nie Rongzhen, *Inside the Red Star*, 723.

33. Merle Goldman, *China's Intellectuals: Advise and Dissent* (Cambridge, Mass.: Harvard University Press, 1981), 135–138.

34. Yao, "Chinese Intellectuals and Science," 470. James Williams makes a strong claim that

"stand aside" (靠邊站) while the "working class led everything" (工人階級領導一切).

As radical as the Great Leap Forward's attack on specialist knowledge had been, the Cultural Revolution went much further. During the Great Leap Forward it was still necessary to be both red and expert. In 1958 a common slogan proclaimed, "Technology is a treasure that cannot be mastered without education" (技術是寶, 沒有文化學不了).[35] The Cultural Revolution, on the other hand, emphasized such slogans as "the lowliest are the smartest and the most elite are the most foolish" (卑賤者最聰明, 高貴者最愚蠢).[36] As one foreign scholar puts it, "Where the Leap model had emphasized both professional and mass science—'walking on two legs'—in the [Cultural Revolution] the professional leg was essentially bitten off."[37]

In 1967, "worker propaganda teams" began running educational and research institutions around the country. Scientists of the IVPP considered themselves fortunate to have been taken over by the Number One National Cotton Factory, whose workers engaged in relatively little violence at the institute compared with workers in many other places.[38] As elsewhere, the new authorities at IVPP forced a group of scientists to stay in a "cow shed" for about eight months. In some places in the rural areas, these were literal cow sheds. At IVPP, it was a room at the institute converted into a detention area. Jiǎ Lánpō recalls that the people assigned to watch him and the others usually had a good relationship with the scientists and would even run errands to purchase tobacco for them. "We played cards and chess (we drew the board on a piece of paper)—we could do anything, except we weren't allowed to read books." Once out of the "cow shed," Jiǎ was placed under house arrest

---

scientists were intellectuals in Máo-era China. "The Cultural Revolution especially victimized scientists in ways little different from other intellectuals, and produced similar responses, including political alienation." Williams, "Fang Lizhi's Big Bang," 2:679.

35. Kēxué pǔjí chūbǎnshè, Qúnzhòng kēxué, 34.

36. Máo coined this phrase in response to a report on worker innovations during the Great Leap Forward. Máo Zédōng, Jiànguó yǐlái Máo Zédōng wéngǎo [Mao Zedong's Manuscripts since the Founding of the PRC] (Běijīng: Zhōngyāng wénxiàn chūbǎnshè, 1998), 7:236. The slogan really took off, however, in 1968. A full-text search of People's Daily for 卑賤者最聰明 yielded 12 results for 1958–1961, 3 for 1966, and 250 for 1968–1976, with far fewer references after that time.

37. Williams, "Fang Lizhi's Big Bang," 2:364. Members of the radical group Science for the People wrote in their 1974 report on Cultural Revolution science that "walking on two legs" did not mean "cutting the stronger one off in favor of the weaker, as some western observers have implied." Science for the People, China: Science Walks on Two Legs (New York: Avon Books, 1974), 6.

38. The state's nationalist interest in paleoanthropology may also have contributed to its relative security. A few other fields—notably national defense, astrophysics, and bioscience—also continued to receive political support during the Cultural Revolution. Yao, "Chinese Intellectuals and Science," 466–467.

but surreptitiously resumed reading and writing, hiding his materials behind a newspaper or a volume of Chairman Máo's writings whenever someone came along.[39]

Some had a more difficult time. Yáng Zhōngjiàn, director of IVPP, recalls being forced to stand bent ninety degrees at the waist for long periods of time, to wear a heavy plaque around his neck with his political label written on it, and to wait for everyone else to finish eating before he could eat.[40] In the early 1970s, other IVPP scientists were sent down to the countryside to May Seventh cadre schools to "learn from the peasants" by engaging in manual labor alongside them.

Some scientists, however, had it easier. Yóu Yùzhù, for example, was able to keep working on various field assignments throughout the Cultural Revolution. He was even able to obtain permission from the worker propaganda team for Péi Wénzhōng to leave the institute and join a group doing fieldwork on *Gigantopithecus* in Húběi.

Scientists were not simply washed this way and that by the shifting currents of these years. Rather, they actively strove to protect themselves and advance their own positions. Paleoanthropologists and other scientists did not entirely let go of their special status. It was still important for them as professionals to demonstrate expertise publicly. Jiǎ Lánpō, for example, did not emphasize his lack of postsecondary education the way one might expect, given the political cachet of learning by doing.[41] Moreover, when they were not being beaten and compelled to engage in manual labor, intellectual elites were often afforded privileges far above those of ordinary people. For example, Jiǎ recalls that in 1964 while in the countryside carrying out the Four Cleanups, he received "special treatment" in the form of fifty extra *jīn* (about seventeen pounds) of flour a month, specifically because he was a professor.[42]

Nonetheless, throughout the socialist period, it simultaneously behooved paleoanthropologists to emphasize their affinity with laborers. They accomplished this in numerous ways. Péi Wénzhōng was known for dressing in shabby clothes and otherwise downplaying his class sta-

39. Jiǎ Lánpō, interview with Dennis Etler, 7 October 1999. Jiǎ did not mention house arrest in this interview, but see Guldin, *The Saga of Anthropology*, 195.

40. Lǐ Èróng, *Yáng Zhōngjiàn huíyì lù* [The Memoirs of Yáng Zhōngjiàn] (Běijīng: Dìzhì chūbǎnshè, 1983), 206–207.

41. I have found only one place that highlights this. In his preface to Jiǎ's 1950 book *Peking Man*, Péi Wénzhōng said that a book like this would be expected for someone with foreign education but was truly extraordinary for a "self-taught person" like Jiǎ. Jiǎ Lánpō, *Zhōngguó yuánrén* (1950), preface, 2. Since 1950 was hardly a very radical period for class struggle, this comment served the Paris-educated Péi at least as well as it did Jiǎ.

42. Jiǎ Lánpō, interview with Dennis Etler, 7 October 1999. The Four Cleanups sent urban cadres to the countryside to investigate corruption and other problems.

tus: "He didn't consider himself a scientist."[43] Jiǎ Lánpō similarly said he "was very uncomfortable being taken for a special person." He preferred to ride in a jeep rather than a fancy car that would "make people think I was an official." By riding in a jeep he "looked like an ordinary person" and "could make friends with the common people."[44] The work that paleoanthropologists did in the field was particularly easy to depict as labor. Jiǎ Lánpō described in detail the arduousness of excavation work in a 1962 popular book. "For each object [excavated]," he said, "science workers have expended a lot of sweat."[45] Throughout the Máo era, the term "science workers" referred collectively to scientists, technicians, and laborers engaged in scientific work; it further enabled scientists to downplay the intellectual aspects of their work and gain recognition as members of the laboring classes.

Cultivating a rugged persona is common to field scientists in socialist and capitalist countries alike.[46] The historian of science Martin Rudwick has uncovered a delightful anecdote from Victorian-era geology: "One eminent English geologist (Sedgwick) relished the story of how, when hammering in the field one day, a passing lady gave him a shilling; that evening, transformed in clothing and unrecognized, he met her at dinner in the gentleman's house where he was staying, and he amused the other guests—if not the lady—by displaying the shilling and telling of its charitable donation!"[47] While they are not immune to the elitism that afflicts so many of us who do intellectual work, it is nonetheless ironic that paleoanthropologists should have been sent down to the countryside to learn hard labor from the peasants. They could just as easily have learned it on a dig. Yet there was a method to the madness: targeted criticisms and mass campaigns served to undermine the very real privileges that scientists enjoyed and the elitism that characterized their relationship to the laboring classes.

Mass Science

Challenges to the top-down model of scientific knowledge went far beyond criticizing scientists and undermining their authority. There

43. Péi Shēn and Yóu Yùzhù, interview with the author, 29 October 2002.

44. Jiǎ Lánpō, interview with Dennis Etler, 7 October 1999. Jiǎ was talking about a field trip he made in 1976, but it reflected his attitude in earlier years as well.

45. Jiǎ Lánpō, Zhōngguó yuánrén [Peking Man] (Běijīng: Zhōnghuá shūjú, 1962), 11–12.

46. Naomi Oreskes, "'Objectivity or Heroism?' On the Invisibility of Women in Science," Osiris 11 (1996): 87–113; Bruce Hevly, "The Heroic Science of Glacier Motion," Osiris 11 (1996): 66–86.

47. Martin J. S. Rudwick, The Great Devonian Controversy: The Shaping of Scientific Knowledge among Gentlemanly Specialists (Chicago: University of Chicago Press, 1985), 38.

were also positive efforts to promote popular participation in science. Mass science was a political priority that arose first in Yán'ān—for example, with Luò Tiānyǔ's efforts to bring scientists to the fields to learn from peasants—and then repeatedly thereafter to challenge the notion that science was the province of elites. It involved both an acknowledgment that laborers contributed to science and active programs to increase such contributions.

Zhōu Jiànrén—the accomplished popular science writer, influential political figure, and brother of the famous writer Lǔ Xùn—wrote in 1962: "Knowledge is the product of society, not of individuals. The masses hold it in their hands."[48] That science is socially produced is arguably the most important finding of the academic field of science studies. While the class character of such claims is not nearly as pronounced in most Western writings as it was in socialist China, it has been important in the work of some scholars.[49] In a more direct and politically charged way, socialist Chinese writers on science emphasized the roles of manual labor and the working classes in the production of scientific knowledge.

As the title of one Great Leap–era book summarized it, "Labor created science." "In the very beginning," the authors suggested, "our human ancestors knew how to hunt with only very simple wooden clubs and stone tools. . . . Later, as the result of the unceasing practice of labor, their labor experience gradually accumulated, their knowledge slowly increased . . . and human scientific culture . . . reached greater and greater heights."[50] The authors made clear throughout that by "labor" they meant the manual labor of workers and not the mental labor of elites. The book's argument took as its basis Máo's influential 1937 essay "On Practice," in which he articulated the theory that knowledge both arises from and feeds back into practice. This theory produced such slogans as "practice is the mother of science," which was "an epistemological stress designed to negate the predictive or guiding role of theory, and the consequent hegemony of 'experts' over the process of innovation and change."[51]

48. Zhōu Jiànrén, *Kēxué zátán* [On Science] (Hángzhōu: Zhéjiāng rénmín chūbǎnshè, 1962), 77.

49. Steven Shapin, *A Social History of Truth: Civility and Science in Seventeenth-Century England* (Chicago: University of Chicago Press, 1994); Anne Secord, "Science in the Pub: Artisan Botanists in Early Nineteenth-Century Lancashire," *History of Science* 32, no. 3 (1994): 269–315; Alison Winter, "Mesmerism and Popular Culture in Early Victorian England," *History of Science* 32, no. 3 (1994): 317–343.

50. Zhōng gòng Nánchāng shì xuānchuán bù, *Láodòng chuàngzào kēxué* [Labor Created Science] (Nánchāng: Jiāngxīrénmín chūbǎnshè, 1959 [1958]), 14–16.

51. Lee, "Ideology and Technical Innovation," 647.

Unsurprisingly, the fields of agriculture and industry saw the most serious attention to principles of mass science. Already during the late 1940s, the Communist Party endeavored to foster workers' technological innovation in the "liberated areas." After the revolution, programs to encourage innovation continued: the party held innovators up as models, and the press celebrated their achievements, aided in part by the Science Dissemination Association and other organizations. In 1954, the All-China Federation of Trade Unions launched a formal campaign to "uncover the hidden potential of existing enterprises." The campaign, however, soon came under criticism for encouraging many "innovations" with little relevance to production issues. Nonetheless, the program for technological innovation resurfaced and became far more radical during the Great Leap Forward. Decentralizing production and establishing many small-scale enterprises (such as the notorious "backyard furnaces" for forging steel) were moves intended to maximize flexibility and thus cultivate workers' innovation. After the full scale of the Great Leap disaster became known, experts were brought in again to "listen" to the workers and then make their own authoritative decisions on design. But by late 1964, Máo was already calling for another mass movement in technological innovation, evidence that he was planning a return to radical politics. Again, model workers who had made contributions to improving agricultural or industrial technology appeared in the press as models for emulation.[52]

While the experience of peasants and workers was not as directly relevant to paleoanthropology as to farming and industry, the dependence of paleoanthropologists on local people to find and report fossils made it a field in many ways ideally suited to a celebration of workers' contributions to science. Jiǎ Lánpō began his 1952 book *Fossil Excavation and Preparation* by saying, "We in our laboratories have no way of knowing where the fossils lie. All information comes from the people."[53] In his popular 1962 book, *Peking Man*, he brought this point home with a revisionist history of the discovery of Peking Man. "Who first discovered and opened this 'treasure trove'?" he asked. "In the past," he answered, "some people said foreigners discovered it. But actually this is not at all the case. If it had not been for local workers quarrying and burning limestone who found the animal bones inside the cave, nothing would

52. Lee, "Ideology and Technical Innovation," 649–650, 655–657. On similar programs in the Soviet Union, see Kendall E. Bailes, *Technology and Society under Lenin and Stalin: Origins of the Soviet Intellectual Intelligentsia, 1917–1941* (Princeton, N.J.: Princeton University Press, 1978), 360–366.

53. Jiǎ Lánpō, *Huàshí de fājué hé xiūlǐ* [Fossil Excavation and Preparation] (Shànghǎi: Shāngwù yìnshūguǎn, 1957 [1952]), 1.

have aroused the attention of the scientific world and caused certain for-eigners to come. In the past, some other foreigners intentionally spread a skewed story saying that they discovered the cave. This kind of story conceals the local workers' accomplishments. Later some people even went so far as to give the credit for the discovery of the Peking Man skull to a foreigner. But this is even less true. Actually, the person who was responsible for the excavation was a Chinese science worker."[54] The "Chinese science worker" was Péi Wénzhōng, China's most celebrated paleoanthropologist. Again we see national and class identities merged in a critique of common portrayals of scientific research. The blanket term "science worker," moreover, masked the role of scientific elites. Of course, the conventional story equally masks the role of the workers who participated in the excavation. Jiǎ, on the other hand, told the story from the perspective of "an old worker," giving it a populist, if anony-mous, feel.[55]

This story has a Western counterpart in attempts to gain acknowl-edgment for the contributions to paleontology of Mary Anning, a poor woman of nineteenth-century Lyme who found and sold fossils for a living. (Anning is popularly cited as the "she" of the famous tongue twister who "sold seashells by the seashore.") Yet Anning received more credit and had more opportunities in the capitalist West than did most if not all lay Chinese fossil finders in their socialist society. She was able to work directly with geologists and to become an acknowledged author-ity in her own right, and admiring scientists named several species for Anning in the nineteenth and early twentieth centuries.[56] In contrast, Chinese peasants and workers who routinely found and handed over fossils have been all but invisible in the scientific reports that followed, warranting at most a clause indicating that workers from a certain com-mune made the discovery. Moreover, such participation did not, at least in the 1950s and 1960s, appear to have ever led any of them to the level of knowledge and engagement that Anning enjoyed.

54. Jiǎ Lánpō, *Zhōngguó yuánrén* (1962), 12. Burning limestone produces lime, which has mul-tiple uses in agriculture, construction, and the chemical industry.

55. Jiǎ Lánpō, *Zhōngguó yuánrén* (1962), 13–14. The "old worker" in question was actually the technician Wáng Cúnyì. See Jiǎ Lánpō and Huáng Wèiwén, *Zhōukǒudiàn fājué jì*, 46.

56. Hugh Torrens, "Mary Anning (1799–1847) of Lyme: 'The Greatest Fossilist the World Ever Knew,'" *British Journal for the History of Science* 28 (1995), 281–282.

Paleoanthropology and Popular Culture

Two memoirs by local people involved in important 1950s paleoanthropological excavations provide a more detailed picture of popular participation in paleoanthropology. They offer evidence of the working relationships among scientists, officials, and local peasants and workers. Of particular interest is the light they shed on the opportunities and frustrations that local knowledge of dragon bones posed for scientists in the field. Contradictions between the populist agenda and concerns about superstition are at their most obvious here.

A memoir published in 1989 on the excavations of Chángyáng Man in Húběi Province in 1956 discusses the ambiguous role of local people in paleoanthropology. The author, Gōng Fādá, was a member of the cultural administration section in Chángyáng and a participant in the excavation process. He reported that in 1943 people had come to Zhōngjiāwān in Chángyáng County to dig in a certain cave for dragon bones, which they then sold as medicine. A local villager, however, had claimed they were his family's "earth pulse and dragon spirit" (地脈龍神, in other words, powerful geomantic forces), and he had blocked up the cave with a boulder and forbade anyone to excavate there any longer. In 1956, after the revolution had eliminated such property rights, local people looking for sideline economic opportunities opened the cave and began excavating dragon bones again. News soon reached a biology professor in the county who brought his students to collect fossil specimens. Of the thousands of pounds of fossils the local cooperative had for sale, he chose enough to fill one crate. After removing a human upper jaw bone and two teeth, he sent the rest of the fossils he had purchased to the provinicial capital of Wǔhàn.

When the provincial cultural authorities discovered this, they sent a telegram to Gōng Fādá stating, "Chángyáng is destroying paleontological fossils; you must immediately investigate the matter and prohibit [any further destruction]." With help from someone sent by the provincial museum, and after some argument, Gōng managed to retrieve the human fossils from the biology professor. A subsequent investigation revealed that the original fossil had been a complete skull rather than just the jaw bone. The three people who had uncovered it failed to recognize its importance; they mistook it for a "dragon head" and carelessly dropped and destroyed it.[57]

57. Gōng Fādá, "Guóbǎo 'Chángyáng rén' de fāxiàn yǔ fājué" [The Discovery and Excavation of the National Treasure, Chángyáng Man], in *Chángyáng wénshǐ zīliào, dì wǔ jí: Mínzú wénhuà zhuān jí*

In 1957, Jiǎ Lánpō and others from IVPP and the provincial museum began excavating in Chángyáng. Gōng Fādá served as their logistics coordinator and hired about a dozen local people to work as laborers (民工) for twenty-one days on the dig. Gōng recalls, "While excavating, we laypeople [外行的人] frequently [but mistakenly] cried out in amazement and without restraint, 'Look, I found a human tooth' or 'Oh, this one is definitely a human bone!' After that, everyone's spirits were dampened and nobody believed each other anymore." On the morning of the last day, however, Gōng found what Jiǎ determined indeed to be a human tooth. The workers got the afternoon off and an extra portion at dinner, and Jiǎ dubbed the find "Chángyáng Man."[58]

The incident of the destroyed fossils led the county government to release a report calling for the protection of ancient artifacts and fossils, and the prohibition of the excavation and sale of dragon bones. Gōng considers the lesson to have been learned, and says that since that time whenever fossils were found they were duly reported to the proper authorities and left unharmed. However, he mourns the losses incurred before the report was released. "Because the 'Chángyáng Man' fossils were first discovered accidentally by ordinary people, they suffered a lot of damage. We never found any production tools belonging to 'Chángyáng Man' or any traces of their lives."[59]

Another important human fossil excavation for which we have a memoir from the local perspective is that of Dīngcūn Man in Shānxī Province. In 1953, Zhào Xiàngrú was assigned to organize workers to excavate and sell sand for the purpose of building an airport. He made the arduous work more enjoyable by inviting theater groups to put on performances some evenings and by presenting awards to model workers. One man of seventy years who voluntarily held a light for the workers during the night shift received ten pounds of rice.[60] Workers frequently came across fossils, and in the beginning many sold them to apothecaries. One worker received thirty yuán for a particularly large bovine horn fossil. When an official found out about this he called for a stop to this practice. They circulated the message to every worker, and, according to

---

(zhī yī) , ed. Zhōngguó rénmín zhèngzhì xiéshāng huìyì Húběi shěng Chángyáng Tǔjiā zǔ zìzhì xiàn wěiyuánhuì wénshǐ zīliào wěiyuánhuì (n.p., 1989), 10–11.

58. Gōng Fādá, "Guóbǎo 'Chángyáng rén,'" 12.

59. Gōng Fādá, "Guóbǎo 'Chángyáng rén,'" 12–13.

60. Zhào Xiàngrú, "Dīngcūn wénhuà yízhǐ fājué qiánhòu de wā shā gōngrén" [The Sand Excavation Workers during the Dīngcūn Culture Site Excavations], in Xiāngfén wénshǐ zīliào, dì bā jí, ed. Zhèngxié Xiāngfén xiàn wěiyuánhuì wénshǐ zīliào yánjiū wěiyuánhuì (n.p.: Zhèngxié Xiāngfén xiàn wěiyuánhuì wénshǐ zīliào yánjiū wěiyuánhuì, 1995), 161–162. From its tone, this account seems to originate in the Máo era.

Zhào, after that they "never again discovered anyone digging and selling [fossils] on their own."[61]

In 1954, Jiǎ Lánpō, Péi Wénzhōng, and more than thirty other "experts" arrived in Dīngcūn to conduct research. They worked out an arrangement with the group excavating sand. The workers would dig carefully and according to the scientists' instructions. If for this reason they did not produce enough sand, the State Council would chip in enough money for the workers to cover the difference. Zhào assigned twenty-five of the workers to assist the experts directly. The workers reportedly had great respect for the experts' ability to know where fossils were likely to be just by looking at the stratigraphy. They likened them to the "eight immortals crossing the sea" (八仙過海), an allusion that expresses admiration for talented individuals. The "experts" were almost as courteous. One reportedly told the workers, "We are archaeological experts; you are earth removal experts." When the archaeologists departed, they left the site under the protection of the workers. Zhào's pride in the local people's participation far exceeds that of Gōng from Chángyáng. "The original sand-digging workers," he said, "are not only the site's discoverers, but also partners in its excavation; not only the site's caretakers, but also its defenders."[62]

These two memoirs present both similarities and significant differences and thus help to indicate both the common aspects of 1950s popular paleoanthropology and the diversity of experience.[63] Labor relations appear quite differently in the two accounts. In Dīngcūn, we hear from a work team organizer about an efficient and valued workforce that shifted smoothly between industrial and scientific excavation projects. Scientists and workers demonstrated strong mutual respect, cooperated effectively, and even grew to share common interests in protecting the site. While Zhào Xiàngrú certainly had an interest in portraying his team in such a positive light, partial confirmation is available from the American Paleoanthropological Delegation, which visited Dīngcūn in 1975. The noted archaeologist Kwang-chih Chang was duly impressed by the seven members of the local production brigade whose responsibilities included preservation of the site. As he said, "Nobody could come here

---

61. Zhào Xiàngrú, "Dīngcūn wénhuà yízhǐ," 163.

62. Zhào Xiàngrú, "Dīngcūn wénhuà yízhǐ, 163–164.

63. There are many problems with using such memoirs as historical evidence. First, they are collected by the state and serve a variety of political purposes, especially to glorify selected aspects of local or national history. Second, they are typically written decades after the events occur and thus do not always accurately reflect the ideas of the time. Sometimes, however, they may reflect better than primary sources ideas that were held privately but for political reasons not written down.

and so much as pick up a pinch of dirt without the knowledge and consent of Mr. Ting and his six team-mates."[64] In contrast, the report from Chángyáng comes from an employee of the government agency responsible for the protection of local cultural artifacts and fossils. Far more clear in this case are the conflicts of interest among peasants seeking to enrich their local cooperative, provincial scientists exploiting this phenomenon for their own research, and government authorities bound to protect artifacts for national-level science. Moreover, Gōng Fādá highlighted the incompetence of the peasants who had destroyed precious scientific evidence, and he portrayed the lay laborers (himself included) mainly as enthusiastic bumblers whose efforts rarely panned out.

Common to both memoirs, and supported by other evidence, is the fundamental conflict between local practices of excavating and selling dragon bones and the scientific project of collecting fossils for research. Scientists knew that the dragon bone trade was an invaluable source of information about fossil locations. Moreover, they seem to have developed a genuine fondness for dragon bones. The Peking Man site was named Dragon Bone Hill, and scientists have often discussed dragon bones at length in their memoirs. The adoption of dragon bones as a kind of professional mascot had an important cultural significance: dragon bones made paleoanthropology more "Chinese."

Yet scientists were also well aware of the destruction caused by the dragon bone trade. Each time scientists took advantage of local people's knowledge to locate fossils, they were faced with the reality that many potentially valuable fossils had already been crushed for use as medicine. They also knew that apothecaries offered more for dragon bones than scientists could pay for fossils. Complicating matters, the dragon bone trade was a private "sideline industry" existing outside the socialist economy. Tolerated in the more moderate periods, such activities were officially discouraged, especially during more radical times. Of even greater importance, a 1950 law protecting cultural artifacts included a ban on fossil collection.[65] In a 1952 article, Jiǎ Lánpō tied these issues together with a nationalist interest in preventing fossils from leaving the country. He noted that, while the sale of dragon bones for medicine had a distinguished history, the practice should be discontinued because

64. Kwang-chih Chang, "Public Archaeology in China," in *Paleoanthropology in the People's Republic of China: A Trip Report of the American Paleoanthropology Delegation*, ed. W. W. Howells and Patricia Jones Tsuchitani (Washington, D.C.: National Academy of Sciences, 1977), 133.

65. Guójiā wénwù shìyè guǎnlǐ jú, ed., *Xīn Zhōngguó wénwù fǎguī xuǎnbiān* [Selected Laws on Cultural Property in New China] (Běijīng: Wénwù chūbǎnshè, 1987).

of its damaging consequences for cultural construction.[66] The memoirs make clear that in the 1950s scientists and officials were discouraging the dragon bone trade and encouraging rural people to report fossil finds through their local administrative hierarchies.

It is difficult to know what scientists and others thought of dragon bones as medicine. On one hand, they were established in the pharmacopoeia of traditional Chinese medicine. Máo was a strong supporter of strengthening Western medicine by infusing it with Chinese medicine and in 1958 made the widely circulated statement "China's medicine and pharmacology is a great treasure house."[67] Jiǎ Lánpō expressed skepticism about the efficacy and saftey of dragon bones in 1952, but this was two years before the state began to promote the pharmaceutical branch of Chinese medicine.[68] Pharmacopoeias published during the late 1950s and 1960s included dragon bones, and several scientists I spoke with in 2002 expressed a belief in their medical efficacy.[69] On the other hand, the very term "dragon bones" represented an ignorance of the "true" nature and significance of the objects, which implied the need for a replacement of the whole concept of dragon bones with that of fossils. When mentioned in books about fossils or evolution, they were almost invariably enclosed by quotation marks that indicated the author considered them not dragon bones at all, but fossils.

The ambiguous status of dragon bones in science was partly a function of their association with traditional Chinese medicine, whose proponents had struggled since the early twentieth century either to justify it in modern scientific terms or to preserve a "separate but equal" authority for it.[70] However, unlike most other items in the Chinese pharmacopoeia, dragon bones had a separate and incommensurable identity in modern science, and the difference inevitably threatened to put dragon bones in the category of superstition. The operating philosophy of science was at its core antirelativist.[71] As Máo had said in his 1940 "On New

66. Jiǎ Lánpō, "Yóu jué 'lónggǔ' zuò fùyè shēngchǎn tánqǐ" [On the Excavation of Dragon Bones as a Sideline Industry], Kēxué dàzhòng, no. 12 (1952): 388–389.

67. Kim Taylor, Chinese Medicine in Early Communist China, 1945–63: A Medicine of Revolution (London: Routledge Curzon, 2005), 120–121.

68. Jiǎ Lánpō, "Yóu jué 'lónggǔ.'" Taylor, Chinese Medicine, 77.

69. Nánjīng Zhōngyī xuéyuàn and Jiāngsū shěng Zhōngyī yánjiūsuǒ, Zhōngyàoxué [Chinese Pharmacology] (Běijīng: Rénmín wèishēng chūbǎnshè, 1959), 694–695; Zhōnghuá rénmín gònghéguó wèishēng bù yàodiǎn wěiyuánhuì, Zhōnghuá rénmín gònghéguó yàodiǎn 1963 nián bǎn yī bù [Pharmacopoeia of the People's Republic of China, 1963 Edition] (Běijīng: Rénmín wèishēng chūbǎnshè, 1964), 73.

70. The term "separate but equal" comes from Nathan Sivin, Traditional Medicine in Contemporary China (Ann Arbor: Center for Chinese Studies, University of Michigan, 1987), 19.

71. I say "at its core" to recognize the considerable room for ambiguity on this subject. Laurence

Democracy," "There is but one truth, and the question of whether or not one has arrived at it depends not on subjective boasting but on objective practice."[72] Thus, strictly speaking, "fossils" and "dragon bones" could not enjoy separate but equal status.

Dragon bones encapsulated the tensions between science and popular culture in socialist China. First, as *dragon* bones, they resonated powerfully with long-standing cultural ideas and practices. Rural communities in early twentieth-century northern China worshipped the Dragon God responsible for rivers and rain; the rituals they practiced shored up local hierarchies in irrigation associations.[73] As Gōng Fādá noted, local people in Chángyáng understood dragon bones to represent a powerful geomantic force. Second, dragon bones had a cash value important to local people looking to supplement their incomes. Neither the cultural nor the commercial significance of dragon bones meshed with the socialist state's vision of science and society.

How, then, were scientists supposed to balance the opposing prescriptions of replacing superstition with science and following the masses, when the masses showed every evidence of being wedded to "remnants of the feudal society"? For Máo and his followers, this was not an insurmountable contradiction, because they considered superstitions to have been imposed on the masses by elites over the centuries. In other words, eliminating superstition was an eradication not of real popular culture, but rather of false beliefs preventing the masses from making genuine contributions to science and society in general. This position is unacceptable to me and to most people who study culture in the West today. While religion has indisputably helped to maintain status quo, it has also often been an agent of change, for example, in sectarian rebellions. Moreover, people ("the masses") embrace religion as their own culture, and vigorously defend it whenever possible.[74] Thus, whether or not party

---

Schneider has argued that the Chinese Communist Party's acceptance of Lysenkoism in the 1950s stemmed from a belief that natural science was superstructural and thus not universal. See Schneider, "Learning from Russia," 51. The placement of science within the superstructure during the Cultural Revolution certainly suggests a relativist or constructivist epistemology. However, this view did not consistently characterize earlier periods and appears even during the Cultural Revolution not to have been fully articulated as a relativist position. Máo said in 1957, "Natural science is not [part of the] superstructure; but [because] it still requires people to do it, [it] can still be tainted with some people's class consciousness." Roderick MacFarquhar, Timothy Cheek, and Eugene Wu, eds., *The Secret Speeches of Chairman Mao: From the Hundred Flowers to the Great Leap Forward* (Cambridge, Mass.: Harvard University Press, 1989), 197.

72. Mao Tse-tung, *Selected Works*, 2:339.

73. Prasenjit Duara, *Culture, Power, and the State: Rural North China, 1900–1942* (Stanford, Calif.: Stanford University Press, 1988), 31–35.

74. Edward Friedman, Paul G. Pickowicz, and Mark Selden, *Chinese Village, Socialist State* (New Haven, Conn.: Yale University Press, 1991), especially 234–238.

officials and scientists recognized it, the battle against superstition was inevitably a battle against popular culture.

Conclusion

The top-down project of science dissemination and the bottom-up one of mass science—while both in some sense "popular science"—lay in profound contradiction. Science dissemination was in practice the far stronger of the two, but it was inherently hierarchical and even elitist in character. Though on the surface far more compatible with Maoist ideology, mass science suffered from a relative lack of organization and consistent political backing. It was at times effective in undermining scientists' claims to authority and promoting popular participation in scientific research and technological development. On the whole, however, it was the masses' superstition, rather than their potential contributions, that occupied scientists and others involved in scientific research and dissemination. Mass science was most influential in areas of technology, where workers and peasants had obvious and unproblematic experience to contribute. As a field science, paleoanthropology, too, lent itself to a certain amount of mass participation, but the intellectual and political problems surrounding dragon bones complicated the issue considerably. Throughout the 1950s and 1960s, popular contributions to the interpretive side of the study of human evolution, moreover, were out of the question: what the masses had to contribute was too likely to be tainted with superstition.

It is tempting to dismiss these struggles over the relationship between class and knowledge as artifacts of a flawed political ideology that resulted only in persecuting scientists and distracting them from their research. To do so, however, would be to ignore the problems of class and knowledge already existing before 1949 and lasting to this day. Reformers in the republican era had wrestled with the difficulty of how to write about science in ways accessible and meaningful to the general population. At the same time, their efforts had served to cast popular forms of knowledge as superstition. Science thus helped translate the age-old gap between the literati and the illiterate majority into modern terms. Neither should we overlook the historic and enduring dilemmas surrounding science in the United States and other countries. We continue to grapple, for example, with the responsibility of scientists to educate *and* to listen to the public, the question of whether academics should receive as much credit for writing popular works as for writing professional ones, the resistance of patients to medical knowledge that ignores their own understandings of their physical experiences, and the struggles of people with limited access to education or political connections to find a voice in the scientific decisions that affect their lives. The socialist context in China demanded that issues of class and knowledge be addressed in more explicit and radical ways than they have been in other places, but as we have seen, such efforts were not always very successful. Before reaching judgment, however, let us turn to the latter half of the Cultural Revolution, from 1971 to 1978, the period perhaps most favorable to a resolution of the contradiction between science dissemination and mass science.

FIVE

"Presumptuous Guests Usurp the Hosts": Dissemination and Participation, 1971–1978

Cultural Revolution Science on Its Own Terms

Historical accounts of science in twentieth-century China typically have little good to say of the Cultural Revolution—and often little at all. Perhaps even more than in other fields, the Cultural Revolution in science is seen as a ten-year gap, a time when political struggles interfered with or even put a stop entirely to scientific work. I would suggest, however, that the history books themselves contribute to this gap: they give the Cultural Revolution scant coverage because they do not recognize its priorities. If we broaden our understanding of science to include popular science, a key Cultural Revolution concern, the gap can be closed considerably.

The approach taken here thus differs significantly from that found in existing literature on science in the Cultural Revolution. It is in some ways most similar to accounts published during the late Cultural Revolution by Western visitors to China, in that it takes seriously the stated goals and methods of Cultural Revolution–era "mass science." Foreigners invited to China during this period often had positive views of science as it was then being conducted. The discovery of the Mǎwángduī archaeological site, the delivery of primary healthcare, the development of integrated techniques for controlling insect pests, and advances in earthquake prediction were among the touted examples of recent Chinese scientific achievements. Many of these new accomplishments, moreover, were explicitly linked to contemporary Chinese ideas about mass participation in science.[1] American paleoanthropologists were no exception to this trend. In 1975 they visited the Institute of Vertebrate Paleontology and Paleoanthropology in Běijīng and expressed their positive impressions in a lengthy report, which included a full chapter on the importance of "public archaeology" in China.[2] Political circumstances, however, prevented the authors of such studies from seeing more than what their hosts wished (or dared) to show them. Their accounts thus lack a critical perspective on the very real negative effects of the Cultural Revolution on science and scientists—from the massive amounts of time wasted on political infighting to the violent abuse of scientists as authority figures.

The fall of the Gang of Four in 1976 and the implementation of the Four Modernizations under Dèng Xiǎopíng in 1978 opened the floodgates on criticisms of Cultural Revolution policy toward science and scientists.[3] Chinese as well as foreign accounts began to portray the story as one of scientists persecuted and science stultified, making it almost impossible for scholars and journalists to return to the previous, rosier view. Scientists and other intellectuals were common protagonists in what became known as "scar literature"—personal stories of people psychologically oppressed, physically tortured, and not infrequently killed during the Cultural Revolution.[4] Western scholars have taken up their

---

1. Victor Sidel and Ruth Sidel, *Serve the People: Observations on Medicine in the People's Republic of China* (New York: Josiah Macy, Jr. Foundation, 1973); Gordon Bennett, "Mass Campaigns and Earthquakes: Hai-Ch'eng, 1975," *China Quarterly* 77 (March 1979): 94–112; Science for the People, *China: Science Walks on Two Legs* (New York: Discus Books, 1974); C. K. Jen, "Science and the Open-Doors Educational Movement," *China Quarterly* 64 (December 1975): 741–747; Robert van den Bosch, *The Pesticide Conspiracy* (Garden City, N.Y.: Doubleday, 1978), 147–148.

2. W. W. Howells and Patricia Jones Tsuchitani, eds., *Paleoanthropology in the People's Republic of China: A Trip Report of the American Paleoanthropology Delegation* (Washington, D.C.: National Academy of Sciences, 1977).

3. The Four Modernizations represented a state commitment to prioritizing the development of agriculture, industry, science and technology, and national defense. Although the concept had been a part of state discourse for many years, Dèng's approach differed markedly from the earlier dominant view—found especially during the Cultural Revolution—that development was inextricable from revolutionary politics.

4. For examples of articles published in mainstream Chinese magazines soon after the Cultural Revolution criticizing Cultural Revolution–era treatment of scientists, see Yingqian Guan, "The Rocky Road to Science," *China Reconstructs* 28 (1979): 72–74; and Zhong Lu, "A Woman Chemical Engineer," *Women of China* 7 (1980): 17–21. For an example of "scar literature" with a scientist protagonist, see Feng Jicai, *Voices from the Whirlwind: An Oral History of the Chinese Cultural Revolution* (New York: Pantheon Books, 1991), 224–243.

stories and written about their experiences undergoing forced agricultural labor and becoming targets for political criticism and violence.

These and other narratives, both foreign and Chinese, have also recast the Cultural Revolution as a period in which political campaigns and censorship made scientific achievement in most fields virtually impossible. Such writings all move away from the view, dominant in the Cultural Revolution, that science is or should be a socialist enterprise in which the means (including mass participation and revolutionary spirit) are as important as the ends. Rather, they focus on science as a professional endeavor in which economic development and scientific truth (in that order) are virtually the only clear priorities. While there are many good reasons to hold Cultural Revolution science to the same standards as science in other regional and historical contexts, it is sometimes helpful to put these standards temporarily aside in order to gain a better understanding of the historical period on its own terms—in this case, to reexamine science in the Cultural Revolution by taking seriously one of the stated goals of the time, the promotion of popular science. At the same time, however, we will benefit from access to information and oppositional perspectives unavailable to writers during the Cultural Revolution. In the end, we will be able to assess the degree to which popular science in the Cultural Revolution lived up to its goals.

A Favorable Time for Popular Science

Between the two spheres of popular science in socialist China—science dissemination and mass science—lay both a profound contradiction and a possibility for resonance. The contradiction was general to popular science in the People's Republic. How could scientists be expected to "rely on the masses" (靠群众) when the masses were officially held to be superstitious and ignorant? To look at it in the other direction, how could the project of science dissemination—which followed a top-down model of bringing knowledge from the experts to the masses—be expected to function when the very notion of expert knowledge repeatedly came under attack? For the 1950s and 1960s, the answer to this paradox was that—with the notable exception of agricultural and industrial technology, particularly during the Great Leap Forward—educating the masses took precedence over relying on them. The far more radical politics of the late Cultural Revolution (1971–1978), however, offered new opportunities for "mass science" to flourish.[5]

5. It is conventional to consider 1976 the end of the Cultural Revolution, since in that year

The resonance, on the other hand, was specific to paleoanthropology, although similar points of connection may well be found in other scientific fields. Both the content of popular science materials on human origins and the philosophy undergirding popular participation in paleoanthropology centered on labor. Popular science materials proclaimed that labor created humanity, that labor was the fundamental driving force in human evolution and social development, that labor continued to define what it meant to be human and not animal, and that the laboring masses thus carried the torch of humanity. Rhetoric accompanying calls to promote class struggle in science similarly trumpeted that labor had created science, and that science was best practiced under the guidance of the laboring masses of "workers, peasants, and soldiers" (工農兵). Paleoanthropology was especially easy to portray as a science based on labor, since it included so much dusty work in the field, where shovels, picks, and wheelbarrows figured as prominent tools of the trade.

The late Cultural Revolution was the time most likely to see this resonance produce interesting and productive combinations. The chaos and violence of the early years of the Cultural Revolution, beginning in 1966, had subsided. The spectacular demise of Máo's chosen successor Lín Biāo in 1971 and the rapprochement with the United States in 1971 and 1972 signaled a shift in party policy and "touched off an unheralded movement of intellectual liberation at the height of the Cultural Revolution."[6] Classes had already resumed in most schools in 1968, and after 1971 museums reopened and publishing presses recommenced operation. In 1972, Zhōu Ēnlái began working toward a new scientific initiative, with the urging and support of scientists, that would place more emphasis on basic science and scientific theory, a direct challenge to the radicals' conviction that "scientific knowledge could be acquired only through practice" and only applied science (technology) was of use to society.[7] While it was an uphill battle until Máo's death and the fall of the Gang of Four in 1976, Zhōu's efforts, along with those of Dèng Xiǎopíng, Hú Yàobāng, and others, offered new hope to scientists.

Chairman Máo died and, soon after, the Gang of Four fell. For the history of science at least, however, it makes more sense to cap the period in 1978 with Dèng Xiǎopíng's implementation of the Four Modernizations and the formal reacknowledgment that science was a productive force rather than part of the "superstructure" and that intellectuals could thus be understood as part of the working class.

6. Yan Jiaqi and Gao Gao, *Turbulent Decade: A History of the Cultural Revolution*, trans. and ed. D. W. Y. Kwok (Honolulu: University of Hawai'i Press, 1996), xxiii.

7. Merle Goldman, *China's Intellectuals: Advise and Dissent* (Cambridge, Mass.: Harvard University Press, 1981), 162–163.

These years saw a new burst of activity in disseminating scientific knowledge of human evolution. In 1971, at the urging of Yáo Wényuán (arrested in 1976 as a member of the Gang of Four), Máo authorized the retranslation and republishing of Thomas Huxley's *Man's Place in Nature* (1863) and *Evolution and Ethics* (1893).[8] The Máo quotations chosen for the frontispiece of *Evolution and Ethics* both related to the theme of learning from the past and from the West, a significant break from the early years of the Cultural Revolution. The year 1971 also saw the opening of new and colorful exhibits on the theme "Labor created humanity" in museums across China, and in 1972 IVPP launched the popular magazine *Fossils* (化石).

Nonetheless, radical politics still held sway: scientific and educational institutions remained under the control of "worker propaganda teams," and political campaigns continued to focus the country's attention on class struggle. Even more than in previous decades, labor became the central focus of both science dissemination and scientific research in paleoanthropology. Moreover, the two were more closely integrated. Popular science materials more explicitly emphasized the class politics of scientific research and dissemination; and research practices more often included a dissemination component to facilitate the participation of nonscientists. Thus, the late Cultural Revolution produced the most likely climate in which the two previously separate spheres of science dissemination and mass science could be brought together.

Dissemination: *Fossils* Magazine Strikes a Blow for Popular Science

In 1972, IVPP published a trial issue of a new popular magazine, entitled *Fossils*, that began semiannual publication the following year. The initiative was spearheaded by several of IVPP's younger scientists who were eager to heed Máo's call to bring scientific knowledge to the workers, peasants, and soldiers. Still published today, *Fossils* reached its highest subscription rates in the early years, when it had little competition from other popular science magazines: its peak was 228,000 in 1982.[9] Based on

---

8. According to paleoanthropologist Wú Xīnzhì, the propagandist Yáo Wényuán read *Man's Place in Nature* and wrote a positive review of it for Máo. Máo agreed with Yáo's assessment, and subsequently the paleoanthropologist Wú Rǔkāng and the paleontologist Zhōu Míngzhèn retranslated it along with *Evolution and Ethics*. Wú Xīnzhì, interview with the author, 17 January 2002.

9. Readership was undoubtedly much higher, since friends, families, and even entire work units shared subscriptions. Precise subscription data is not available. Former editor Zhāng Fēng remembers the figure of 228,000. Zhāng Fēng, interview with the author, 6 July 2005.

article submissions and letters to the editor, distribution appears to have been widespread throughout China. Museum workers, employees of cultural centers, and teachers were probably best represented, but readers also included factory workers, students, and others. The trial issue and subsequent issues solicited articles written by "the broad masses of workers, peasants, and soldiers, revolutionary cadres, and revolutionary intellectuals" for the purpose of "spreading Marxism–Leninism–Máo Zédōng Thought, exchanging experience in paleontology and paleoanthropology, disseminating knowledge from these scientific fields, enabling paleontology and paleoanthropology to better serve proletarian politics, and putting into practice the three great revolutionary movements [of class struggle, the struggle for production, and scientific experiment]."[10] The readership was imagined to consist of "the broad masses of workers, peasants, and soldiers, revolutionary cadres, stratigraphy workers, museum workers, and youth," and writers were thus urged to use clear, lively language that was "popular and easy to understand."[11]

Considering the special focus of the magazine, its offerings were diverse. The covers sported colorful photographs and paintings of science workers in the field, exhibits, and reconstructions of prehistoric animals. Inside, readers found poems on evolution, explanations of current research, debates on theoretical issues, and of course criticisms of creationism using Engels's theory that labor created humanity. In keeping with the time, quotations from Marx, Engels, and Máo often peppered the articles in bold print. Many of the authors were scientists and other workers from IVPP, but famous popular science writers like Gāo Shìqí, scientists and museum workers from other institutions, and lay readers also contributed materials.

From the beginning, IVPP recognized that publishing a popular magazine required more than just scientific expertise, and the institute staffed *Fossils* with people possessing substantial writing and editing experience. The first general editor of *Fossils*, Liú Hòuyī, had worked for

10. "Zhēng gǎo jiǎnzé" [Guidelines for Solicited Articles], *Huàshí*, trial issue (1972): 32. The origin of the "three great revolutionary movements" was a selection from an article Máo wrote in 1963 that became part of the famous "little red book" of the Cultural Revolution. "Class struggle, the struggle for production and scientific experiment are the three great revolutionary movements for building a mighty socialist country. These movements are a sure guarantee that Communists will be free from bureaucracy and immune against revisionism and dogmatism, and will for ever remain invincible." See Mao Tse-tung, *Quotations from Chairman Mao Tse-tung* (Peking: Foreign Languages Press, 1966), 40.
11. "Zhēng gǎo jiǎnzé," 32. "Stratigraphy workers" (地層工作者) presumably refers to all "science workers" participating in stratigraphy, geology, paleontology, paleoanthropology, and any other study related to geological strata.

the translation and editing department of the Academy of Sciences and for the Science Press before becoming a graduate student at IVPP in 1957. By the time he began work on *Fossils*, Liú had also written many short essays on science for newspapers and a widely read book entitled *Quick Calculation* (算得快).[12] Right away Liú brought in an outsider to help with the magazine, someone he knew from the Science Press named Zhāng Fēng.[13]

Although Zhāng's colleagues at IVPP respected his writing and editing abilities, he quickly crossed swords with them over the institute's handling of the magazine. Zhāng felt frustrated by what he saw as IVPP's unwillingness to devote significant resources to *Fossils* magazine in comparison with their professional journal, *Vertebrata PalAsiatica* (古脊椎動物與古人類), and other publications. Zhāng had received more than one hundred letters from readers expressing their great interest in the magazine and entreating IVPP to publish it more frequently, but his attempts to see these requests satisfied were rebuffed. To Zhāng, this reflected a larger problem: China's scientific institutes were elitist and so failed to take popular science seriously. On 27 August 1975, however, an unexpected event planted the seed of an idea in Zhāng's mind. Chairman Máo had recently recovered from cataract surgery and wanted something to read. The Central Propaganda Department ordered IVPP to run a special large-print copy of the most recent issue of *Fossils*. Knowing that Máo would now be familiar with the magazine, on the evening of 5 September Zhāng Fēng began writing a letter to the Chairman. He finished as dawn was breaking, and at eight o'clock in the morning with more than a little trepidation he dropped it into a mailbox.[14]

It was a long letter (about 3,200 Chinese characters) in which Zhāng was careful not to point fingers at specific people but nonetheless rained criticism on "leaders" and "authorities" for neglecting science dissemination. He identified a number of concerns at all levels—from the failure of the Chinese Academy of Sciences to make science dissemination a priority to the lack of sufficient workers to set type at the presses. Most relevant for IVPP was his call for *Fossils* to expand to become a monthly

12. Yáng Jiǎo, *Zhōngguó kēpǔ zuòjiā cídiǎn 1* [Dictionary of Chinese Popular Science Writers, Vol. 1] (Hā'ěrbīn [Harbin]: Hēilóngjiāng kēxué jìshù chūbǎnshè, 1989), 26–27.

13. Zhāng Fēng had studied in the biology department at Běijīng University before his assignment to the Science Press (under the Chinese Academy of Sciences) from 1963 to 1972. In 1985, he transferred to the Agricultural Science and Technology Press. Personal communication, 16 May 2005. This chapter relies extensively on interview data. Where I am unable to provide documentary evidence, I will either discuss the events in relatively general terms or indicate that readers should treat given information with appropriate caution.

14. Zhāng Fēng, interview with the author, 6 July 2005.

natural history magazine and his complaint that "certain work units" were rigid in preventing scientists from straying from the research plan to write popular materials.[15]

Máo appeared to take the issue seriously. He personally read Zhāng's letter (or more likely had his secretary read it aloud), appended his own comment, and wrote a note to Dèng Xiǎopíng and Yáo Wényuán, dated 16 September 1975, asking them to review it and consider circulating it to Central Committee members. Unfortunately, his comment, consisting only of the phrase "a speak-bitterness letter" (一封訴苦的信), was ambiguous to say the least.[16] Did he mean that the letter was only a common sort of complaint? Or did he mean that Zhāng's grievance was worthy of the same kind of respect and attention granted to poor peasants when they "spoke bitterness" about their oppression at the hands of landlords?[17]

While people then were hard put to decide how to implement Máo's wishes on the issue, people today have difficulty remembering what was actually done in the end. My interview data on this subject illustrate both the uncertainties of the time and the fragmented character of institutional memories of the Cultural Revolution. One person remembers hearing that Máo called a mass meeting at Workers' Stadium that tens of thousands of people attended.[18] Another rejects this story and says instead that Hú Yàobāng, then director of the Chinese Academy of Sciences and thus ultimately responsible for IVPP itself, was forced to come to IVPP to be criticized.[19] A third person agrees that Hú Yàobāng came to IVPP in response to Zhāng Fēng's letter but does not remember his being criticized. Rather, she recalls that he gave a carefully worded speech intended to encourage the beleaguered scientists, saying that as

15. Zhāng Fēng very kindly shared most of the letter with me. He kept back one part (which he summarized orally) to prevent further harm to what he now sees was a "vulnerable population" of authority figures.

16. A summary of Zhāng Fēng's letter appears along with Máo's memo to Dèng and Yáo in Máo Zédōng, Jiànguó yǐlái Máo Zédōng wéngǎo [Máo Zédōng's Manuscripts since the Founding of the PRC] (Běijīng: Zhōngyāng wénxiàn chūbǎnshè, 1998), 13:468.

17. "Speak bitterness" meetings date from revolutionary times. At such meetings, the Chinese Communist Party encouraged peasants to narrate their grievances against landlords and others who had treated them unfairly.

18. This interview subject began working at IVPP several years after the Cultural Revolution. He remembers hearing about the incident at home from a family member who was then an employee of the Chinese Academy of Sciences. He originally said about one hundred thousand people participated, but then revised it down to ten thousand.

19. The interviewee is a scientist who joined the institute in the 1950s. He was in the field when the events took place, but he remembers hearing about them from his colleagues. Hú Yàobāng may be familiar to Western audiences as the official whose death and memorial sparked the 1989 student protests in Tiananmen Square.

vertebrate paleontologists they should all have "strong backbones" to withstand the trials they were undergoing.[20]

The last of these accounts, from someone who attended Hú Yàobāng's talk at IVPP, tallies best with the notes Zhāng Fēng took at the time.[21] Hú visited IVPP on 19 and 20 September—just days after Máo commented on the letter—and delivered addresses first in a small meeting and then in a larger one. Zhāng was also asked to give a short speech on the use of dialectical materialism in science work. In the smaller meeting on the nineteenth, Hú tackled the obvious questions: Why did Máo comment on this letter? What was his comment meant to teach them? And, how should they firmly implement Máo's comment? In his lecture the following day, Hú admitted he was still not wholly clear about these questions, but he forged ahead nonetheless in offering an interpretation of Máo's interest and the implications for the Academy.

Of all the letters the chairman received from the masses, Hú said, there were three in recent years that had roused his interest. The first was Lǐ Qìnglín's famous letter on behalf of his son and other educated youth suffering in the countryside.[22] The second was Zhāng Tiānmín's letter on the need for greater attention to literature and the arts. The third was Zhāng Fēng's. In commenting on this letter, the chairman was "using the case of Comrade Zhāng Fēng in order to promote science propaganda, science publishing, scientific experiment, and the science and technology work of the entire Academy and indeed the entire country." The party had made mistakes in the past, Hú went on, in not listening to the valuable opinions of people like Zhāng Fēng. In expressing such opinions, Zhāng was a pacesetter that others should follow. Of Zhāng's specific concerns related to popular science, however, Hú was vague, saying instead that "our fundamental goal is to raise up scientific research as the chairman hopes we will."

Hú's visit to IVPP was part of a larger effort on the part of Dèng Xiǎopíng and Hú Yàobāng to promote science—and not popular science, but rather professional scientific work. Máo's cryptic comment on Zhāng Fēng's letter helped them in this endeavor. In late 1975, Hú Yàobāng gave speeches at a number of scientific institutes in which he supported intellectuals and warned against excessive populism in scientific

---

20. All three of these interviews were held while I was in Běijīng in 2002.

21. The following account is based on these notes, which Zhāng Fēng generously shared with me after organizing and rewriting them on 12 February 2005.

22. On this episode, see Jun Zhang, "To Be Somebody: Li Qinglin, Run-of-the-Mill Cultural Revolution Showstopper," in *The Chinese Cultural Revolution as History*, ed. Joseph Esherick, Paul Pickowicz, and Andrew Walder (Stanford, Calif.: Stanford University Press, 2006).

research.[23] Shortly after Máo circulated Zhāng Fēng's letter, Dèng Xiǎopíng mentioned *Fossils* at least twice in ways that suggest he consciously sought to use this episode to promote professonal science. On 19 September, Dèng referred to *Fossils* in a conversation with Hú Qiáomù, then in charge of the State Council's Office of Political Research. A week later, he mentioned the magazine while commenting on a report Hú Yàobāng had made to the State Council about scientific issues. In each case, Dèng interpreted Máo's interest in *Fossils* as evidence of the chairman's support for two of Dèng's own priorities. First, he encouraged both Hú Yàobāng and Hú Qiáomù to step up their involvement in more theoretical periodicals. If Máo was concerned "even about periodicals like *Fossils*," how much more would he be concerned about other science periodicals like *Laws of Natural Dialectics* (自然辯證法) or a general-interest periodical published by the Ministry of Education?[24] Second, he suggested that Máo highlighted the *Fossils* letter to demonstrate his support for the development of scientific research, and even his "high regard for basic theoretical science" (基礎理論科學的重視).[25] This position reflected the interests of research scientists over Cultural Revolution radicals who favored applied science along with popular science. Whatever Máo's intent, Dèng's interpretation thus flew in the face of Zhāng Fēng's populist stance.

While Dèng and Hú were attempting to steer science back onto a professional track, the attention from above nonetheless encouraged IVPP to put more energy into popular science. In preparation for increasing the publication frequency of *Fossils* from two to four issues per year and for sending the magazine to foreign as well as domestic audiences, the editors drafted a detailed plan for the 1976 issues and circulated it along with a cover letter to other work units for feedback.[26] The editors suggested a number of areas ripe for criticism, including "excusing 'excep-

23. "Gēnjù Zhōngguó kēxuéyuàn qúnzhòng jiēfā dàzìbào zhāilù" [Extracts of Big Character Posters in which the Masses Expose the Chinese Academy of Sciences], *Zhōngyāng shǒuzhǎng jiǎnghuà děng* 122-2 (Box 5, JC188Y), Red Guard Collection, Fairbank Center, Harvard University.

24. *Hú Qiáomù zhuàn* biānxiězǔ, ed., *Dèng Xiǎopíng de èrshísì cì tánhuà* [Twenty-Four Speeches of Dèng Xiǎopíng] (Běijīng: Rénmín chūbǎnshè, 2004), 76; "Guówùyuàn lǐngdǎo tóngzhì zài tīngqǔ Hú Yàobāng tóngzhì huìbào shí chāhuà (chuánchāo gǎo)" [State Council Leaders Comment on Hu Yaobang's Report (Private Draft)], *Zhōngyāng shǒuzhǎng jiǎnghuà děng* 122-3 (Box 5, JC188Y), Red Guard Collection, Fairbank Center, Harvard University.

25. *Hú Qiáomù zhuàn* biānxiězǔ, *Dèng Xiǎopíng de èrshísì*, 76. See also "Guówùyuàn lǐngdǎo tóngzhì."

26. I found a copy of the six-page plan, entitled "Huàshí yījiǔqīliù nián xuǎn tí jìhuà" [Plan for the Selection of Topics for *Fossils* in 1976], and the letter (dated February 1976) in an unmarked box among Yáng Zhōngjiàn's books and papers in the IVPP library.

tionalism' in science and technology," "denying that workers, peasants, and soldiers were the main forces in the army of scientific research," and "opposing the policy of making scientific research serve proletarian politics and the workers, peasants, and soldiers and [making scientific research] unite with productive labor." Such criticisms were all aimed at eliminating the privileged status of science as intellectual work and that of intellectuals within science.

The issues for 1976 were to include first and foremost materials in celebration of the hundredth anniversary of Engels's treatise "The Part Plated by Labor in the Transition from Ape to Human." The editors also proposed a range of other topics to satisfy the requirements of being a popular science magazine in a politically charged period: everything from natural history in the service of stamping out superstition, to strategies in science dissemination, to such Cultural Revolution staples as "in agriculture, learn from Dàzhài."[27] Many of the topics—for example, the "struggle" between natural science and religious theology—had been core themes at least since 1949.

There was much in *Fossils* that people who had encountered popular materials on human evolution from the 1950s would find familiar. The difference lay in the higher intensity of the political rhetoric and the greater emphasis on the class politics of knowledge. While many books and other materials had been published in earlier years for popular audiences, *Fossils* was a medium that sought to bring people from wide-ranging backgrounds together between the same covers. Articles were written by and for everyone from the "broad masses" to "revolutionary intellectuals." The magazine further sought to portray science as an endeavor in which all classes participated, and articles sometimes even foregrounded the contributions of workers. Finally, with its vibrant covers and variety of content, *Fossils* was an effective emblem of the importance attributed to popular science during the Cultural Revolution.

What was new was the amount of explicit attention given to the class politics of knowledge. For example, under the category "introductions to scientists" (meaning, biographies of scientists), the editors recommended focusing on "the contributions of laboring people and 'small characters' in paleontology and paleoanthropology," explaining that "science originates in practice and comes from the people," and highlighting the "great truth that 'the lowliest are the smartest.'" Similarly,

---

27. Dàzhài was a model production brigade that had reportedly greatly improved the productivity of its land through the hard work and revolutionary fervor of its members.

when discussing Engels's theory that labor created humanity, the editors suggested authors use this opportunity to criticize the ideas that "intellectual education is most important" and that "in conducting scientific research, one cannot open the doors [to the masses]." While not all the topics had such an explicitly political focus, the editors evinced an awareness that class and education issues were applicable even in the most technical of articles. For example, they emphasized that articles on fossil identification should be clearly written and devoid of overelaboration to facilitate their use by workers, peasants, and soldiers.

Dissemination: Dinosaurs and the Masses at Zhōukǒudiàn

Fossils was not the only venue for such new emphases: exhibits at the Peking Man site at Zhōukǒudiàn and elsewhere were part of the same trend in popular paleoanthropology. The exhibits created in the early 1970s added color and energy to the familiar slogan "Labor created humanity," and emphasized more than ever before the "popular" in popular science. The new public exhibit at Zhōukǒudiàn had its origins in efforts beginning in late 1968 or early 1969 to honor the twentieth anniversary of the People's Republic and the fortieth anniversary of the discovery of the first Peking Man skullcap. The Institute of Vertebrate Paleontology and Paleoanthropology sent a team of scientists, political advisers, and artists to tourist sites around Běijīng and as far away as Hángzhōu and Kūnmíng in order to learn how to create better exhibits for popular audiences.[28] The team then put together a small exhibit (not open to the public) on evolution in the reception hall at Zhōukǒudiàn in late 1969. Guō Mòruò, the famous archaeologist and then president of the Chinese Academy of Sciences, visited the exhibit and recommended that it receive funding for expansion, replacing the existing exhibition hall that had been established in 1953 and had remained relatively unchanged since 1959. The new exhibit opened to the public in 1972 on the anniversary of the founding of the People's Republic (figure 11).

Preparations for the new exhibit at Zhōukǒudiàn coincided with preparations for similar exhibits all around the country. In 1976 IVPP took a traveling exhibit, based on the one at Zhōukǒudiàn, to Tibet. This mission held particular political importance: Engels's lesson that labor created humanity would help assault the stronghold of "superstition"

---

28. I have no documentary confirmation for these dates, and the people who participated are unsure when the events took place.

11 The president of the Chinese Academy of Sciences, Guō Mòruò, visiting the new exhibition at Zhōukǒudiàn, probably in 1972. He is followed by his vice president, the physicist Wú Yǒuxùn. (Photograph courtesy of IVPP archives.)

that the Tibetan religion represented.[29] Two hundred thousand people reportedly viewed this exhibit.[30]

29. Interview data. See also Xīzàng zìzhì qū wénhuàguǎn, "Bù yào shénquán yào kēxué, bù xìn 'tiānmìng' gàn gémìng—xīkàn Lāsà 'láodòng chuàngzào le rén' zhǎnlǎn" [We Need Science, Not Theocracy; We Make Revolution and Don't Believe in Fate—Enjoying the Lhasa 'Labor Created Humanity' Exhibit], Huàshí, no. 1 (1977): 5–6, 2. Also see the inner front cover of Huàshí magazine, 1977, no. 1, for photographs of the exhibit and accompanying activities. The exhibit apparently did not acknowledge the Tibetan belief that Tibetans were descended from a Bodhisattva-ape. In the story, a Bodhisattva-ape married a rock-ogress, and their descendants became the first Tibetan people. For a translation, see Per K. Sorensen, The Mirror Illuminating the Royal Genealogies: Tibetan Buddhist Historiography (Wiesbaden: Harrassowitz, 1994), 127–133. I thank Shěn Wéiróng for suggesting this text.

30. Běijīng bówùguǎn xuéhuì, Běijīng bówùguǎn niánjiàn (1912–1987) [Yearbook of Běijīng Museums (1912–1987)] (Běijīng: Běijīng Yànshān chūbǎnshè, 1989), 509.

As in so many other areas during the Cultural Revolution, young people played a central role in the new exhibits. Both IVPP and the Xī'ān Bànpō Museum (and probably other institutions) brought in young people to serve as guides.[31] In 1971, ten graduating students (six women and four men) from a secondary school near Zhōukǒudiàn began assisting with the creation of the new exhibit and taking classes from IVPP scientists to prepare for their work as guides when the exhibit opened. They were fortunate: serving as guides was pleasant work, particularly compared with the hard agricultural labor many of their peers experienced when they were sent down to serve in the countryside.[32] Moreover, for some this was just the first step in what were to become fruitful careers at these scientific institutions.

The content of the Zhōukǒudiàn exhibit was chosen by a "renovation small group" led by the party branch secretary, several scientists, an artist, and the "military representative" (軍代表) obligatory in all revolutionary committees at that time. With support from Guō Mòruò, they were able to expand the exhibit to one thousand square meters—three times its previous size. They also greatly enlarged its scope. Whereas it had been a "site museum" dedicated specifically to preserving and presenting the relics of Peking Man and related fauna, the politics of the Cultural Revolution demanded its transformation into something akin to a natural history museum with broader popular appeal and greater instrumentality as a vehicle for the dissemination of a materialist view of biological and human evolution. Rather than limiting the content to the site itself or even to the story "from ape to human," the small group sought to convey the whole of biological evolution from a materialist perspective. They even decided to include dinosaurs in the new exhibit despite the lack of dinosaur fossils at Zhōukǒudiàn, a clear sign that popular rather than professional interests had come to the fore. Looking back, the renovation small group member Lù Qìngwǔ calls the dinosaurs "presumptuous guests who usurped the hosts" (喧賓奪主, a Chinese saying), the "hosts" being Peking Man and the other true Zhōukǒudiàn fossils.[33] Scientists at the time may also have seen the masses themselves as "presumptuous guests" whose lowbrow interests had usurped the right-

---

31. Xī'ān Bànpō bówùguǎn "Niánjiàn" biānjí wěiyuánhuì, ed. *Xī'ān Bànpō bówùguǎn niánjiàn, 1958–1998* [Xī'ān Bànpō Museum Yearbook, 1958–1998], 118; Gāo Qiáng, interview with the author, 13 April 2002; Duàn Shūqín, interview with the author, 28 January 2002; Zhāng Lìfēn, interview with the author, 4 February 2002.

32. Zhào Zhōngyì, interview with the author 29 January 2002. Gāo Qiáng, interview with the author, 13 April 2002.

33. Lù Qìngwǔ, interview with the author, 3 June 2002.

ful position of paleoanthropologists and paleontologists at the scientific site of Zhōukǒudiàn.

Even at the time, some opposed this transformation. One was Jiǎ Lánpō, the influential paleoanthropologist and author of many popular materials on human evolution. The previous year (1970) had been different, Jiǎ said, since at that time all other museums were closed. Now that the museums were opening again, an exhibit at Zhōukǒudiàn seeking to encompass all evolutionary history risked redundancy. To avoid this, Jiǎ suggested that the prehuman part be kept to the minimum necessary to illustrate Máo's theories, from his famous 1937 essay "On Contradiction," of internal and external causes and of the "new superseding the old." With this exception, the emphasis should remain on the Zhōukǒudiàn site and its distinctive fossil discoveries. On the other hand, Jiǎ agreed that the "pre–Cultural Revolution exhibit" was sadly deficient in that "workers, peasants, and soldiers" could not understand it. Not only was it "academic for the sake of being academic," but it failed to present even the basic story of evolution. The solution, according to Jiǎ, was for the exhibit to provide a clear explanation of human origins and development based on the political thought of Chairman Máo. In this, at least, he was in accord with the final decision.[34] Zhōukǒudiàn was on the road to becoming a site explicitly oriented to popular interests and political propaganda.

Dissemination: Learning about Humanity at Zhōukǒudiàn and Beyond

The new exhibit at Zhōukǒudiàn opened to the public on 1 October 1972, the twenty-third anniversary of the People's Republic. The exhibit encompassed three parts. The first, titled "The Gestation of Humanity," sought to demonstrate that humans were the product of a long process of unceasing development from the simplest organisms through the stages of fish, amphibians, reptiles, and mammals. The last presented the achievements of Chinese paleoanthropology since 1949. In between was the core of the exhibit: "Labor Created Humanity." Here visitors learned that a century earlier Darwin had "demonstrated that humans had evolved from apes and refuted the fallacy that God had created humanity." But Darwin had suffered from the influence of idealist philosophy

34. Jiǎ Lánpō wrote these opinions in an informal document prepared for a meeting of the small group, and his son has preserved it among his other papers, which are ordered chronologically and provided with page numbers. Jiǎ Lánpō, "Duì Zhōukǒudiàn xīn chénlièguǎn fāng'àn de yìjiàn" [Opinions on the Plan for the Zhōukǒudiàn Exhibition Hall], 16 April 1971, 117–120.

and had not fully comprehended the difference between humans and other animals. This was left to Engels, who explained how humans had evolved their special characteristics through engaging in labor.[35] This account was consistent with the story of human origins told by science disseminators throughout China since 1949, and especially during the Cultural Revolution. The notion that God had created humanity represented a kind of "religious superstition" that the "oppressor classes" used as "opiates" to "paralyze the laboring people."[36] Thanks to the combined forces of Darwin and Engels, however, a materialist and scientific understanding of human evolution was now possible.

Not only the content but also the format of the Zhōukǒudiàn exhibit was greatly elaborated and made to serve mass audiences. The original exhibit had been very simple. Fossils had lain exposed on stands without even any glass to protect them.[37] Signs had provided little interpretation beyond identification of the species displayed.[38] A previous renovation completed in 1959 had added only a few pieces of interpretive art.[39] The Cultural Revolution renovation, on the other hand, used IVPP resident artists and visiting artists from other institutions to create reconstructions of the society and environment of Peking Man and Upper Cave Man in the form of large oil paintings and sculptures that brought the old bones to life.[40] For example, Lǐ Róngshān, who came to IVPP with an art degree in 1965, used the Czech artist Zdeněk Burian's depiction of Java Man as the model for a colorful reconstruction of Peking Man.[41] One of the new guides, Zhāng Lìfēn, had greatly enjoyed the small diorama at the site when she had visited as a child; in 1971 she had

35. Information on the content of the exhibit provided here comes from a pamphlet sold at the exhibit. Zhōngguó kēxuéyuàn gǔjǐzhuī dòngwù yǔ gǔrénlèi yánjiūsuǒ, *Běijīng yuánrén yízhǐ jiǎnjiè* [A Brief Introduction to the Peking Man Site] (Běijīng: Zhōngguó kēxuéyuàn yìnshūchǎng, 1972), 5–6.

36. Shànghǎi zìrán bówùguǎn, *"Cóng yuán dào rén" zhǎnlǎn jièshào* [Introduction to the "From Ape to Human" Exhibit] (n.p. [undoubtedly from the 1970s]), 2.

37. Zhōukǒudiàn Peking Man Site Exhibition Hall Guest Book, 14 June 1956.

38. Zhào Zhōngyì, interview with the author, 29 January 2002.

39. For example, Wú Xīnzhì designed one display which set casts of fossils within a painted silhouette to show the anatomical relationship of the bones and another depicting fossil evidence of fire with a painted backdrop representing smoke. A partial view of the former can be seen in figure 12 on the left. Wú Xīnzhì, interview with the author, 21 February 2002. Reports on when Zhōukǒudiàn first used statues of Peking Man conflict, but there was certainly at least a small diorama.

40. The Upper Cave Man remains belong to the anatomically modern, paleolithic humans found at Zhōukǒudiàn.

41. See Josef Augusta and Zdeněk Burian, *Prehistoric Man* (London: Paul Hamlyn, 1960). At that time, there were no special classes or degrees available in scientific art. Lǐ Róngshān had no training in science and no particular interest in science before being placed ("by chance," as he recalls) at IVPP. Lǐ Róngshān, interview with the author, 24 January 2002.

the opportunity to assist in creating a much larger diorama for the new exhibit.[42] And, in keeping with many other cultural productions of the Cultural Revolution, the exhibit displayed a quotation from Chairman Máo in enormous characters, flanked by a portrait. In this case, as with most books on human evolution, the quotation selected characterized human history as "a continuous development from the realm of necessity to the realm of freedom" and emphasized the unceasing progress and change found in both the social and the natural worlds.[43]

Most visitors came in large groups organized by their schools, army units, and work units (figure 12).[44] One of the young guides, Zhào Zhōngyì, recalls that during the high tide, they had one to two thousand visitors every day, and he once gave a lecture to four hundred visitors in the large lecture hall.[45] Visitors were often passionately interested in the disappearance of the Peking Man fossils in 1941, presumably at the hands of imperialist nations.[46] It is less clear how many felt strongly about the core materialist, antireligion, antisuperstition message that labor, and not God or the Chinese legendary figure Nǚwā, had created humanity. Another guide, Duàn Shūqín, recollects, "Lots of people didn't know before they visited that people came from apes. After we explained it to them, they believed it. Maybe some people didn't completely believe, but they still saw that this explanation had a definite logic to it."[47] Visitors often had more practical questions for the guides, such as why all these early humans had died in one place (had the cave collapsed on them?), how scientists knew the artifacts were tools and

42. Zhāng Lìfēn, interview with the author, 4 February 2002. Zhāng worked as a guide at Zhōukǒudiàn until 1978, after which she transferred to IVPP proper to become a technician who creates casts and prepares fossils. The casts she and others make at IVPP have been used for research by scientists all over the world and have been widely displayed in schools, in museums, and even at the United Nations.

43. The quotation, originating in 1964, was included in *Quotations from Chairman Mao Zedong*. Mao Tse-tung, *Quotations from Chairman Mao*, 203–204.

44. Rules at Zhōukǒudiàn were probably similar to those at Tiānjīn Natural History Museum, which required visitors to submit letters of introduction from their work units. Tiānjīn zìrán bówùguǎn, ed., *Rénlèi de qǐyuán zhǎnlǎn jiǎnjiè* [Introduction to the Exhibit on Human Origins] (n.p. [undoubtedly published between 1971 and 1978]), 16.

45. Zhào Zhōngyì, interview with the author, 29 January 2002. Zhào Zhōngyì now works at IVPP producing reconstructions of prehistoric fauna and busts of modern-day scientists for the exhibition halls. Members of the American Paleoanthropology Delegation who visited IVPP and Zhōukǒudiàn in 1975 heard that Zhōukǒudiàn then had "close to one thousand visitors" every day. Kwang-chih Chang, "Public Archaeology in China," in Howells and Tsuchitani, *Paleoanthropology in the People's Republic of China*, 136.

46. Zhāng Lìfēn, interview with the author, 4 February 2002.

47. Duàn Shūqín, interview with the author, January 2002. Duàn worked for many years in science dissemination at Zhōukǒudiàn; she transferred to IVPP's Paleoanthropology Museum after it opened in 1994.

12 Guide providing interpretation for a group of soldiers visiting the Peking Man site museum at Zhōukǒudiàn. Reproduced from Xīnhuá tōngxùn shè, *Běijīng yuánrén zhī jiā* (Běijīng: Běijīng rénmín chūbǎnshè, 1973), 9.

not just rocks, and how they identified which scraps of bone were from males and which from females.[48] For many visitors, the real excitement lay in the prospect of finding fossils themselves. So many visitors dug holes in the floors in search of fossils that IVPP was eventually compelled to lay cement.[49]

A 1976 movie, *China's Ancient Humans*, captured the spirit of popu-

48. The idea that the cave had collapsed on the early humans may have sounded silly to the guides, but it actually anticipated an important controversy about Peking Man and Zhōukǒudiàn. Some foreign scientists have suggested that Peking Man did not live in the cave, and that the bones accumulated there because hyenas dragged the bodies to *their* home to consume. Noel Boaz and Russell Ciochon, "The Scavenging of 'Peking Man': New Evidence Shows That a Venerable Cave Was neither Hearth nor Home," *Natural History* 110, no. 2 (2001): 46–51.

49. Zhāng Lìfēn and Zhào Zhōngyì, interview with the author, 4 February 2002.

lar science at Zhōukǒudiàn.[50] The director took every opportunity to show crowds of people clustering around a scientist at a makeshift fossil exhibit or pouring through the Zhōukǒudiàn exhibition hall and fossil localities. Such sequences highlighted the mass-oriented character of the new exhibit. The movie was also a means to portray the masses engaged in labor—not, in this case, the ancient labor of stone tool production, but the labor of modern field science. Footage of excavations in progress depicted scores of laborers digging with picks and shovels and moving dirt in baskets and wheelbarrows, while groups of people who appeared to be scientists and workers (although the similarity in dress made it difficult to tell) sat together discussing some of the fossils excavated. This picture of science harmonized with popular written accounts of the discovery of Peking Man published during this period, which emphasized the contributions of laborers.[51]

This, then, was the other side of the popular science coin: at the same time that *Fossils* magazine and the new exhibits were seeking to disseminate scientific knowledge about human evolution to an ever-wider audience, science itself was also ostensibly being transformed into an activity in which workers had as great an acknowledged role as scientists did. The increasing number of popular science materials that emphasized the ideals of "mass science" is one indication that the two spheres of popular science—dissemination and participation—were becoming more tightly integrated during the Cultural Revolution.

Mass Participation: Laborers and Hobbyists

If dinosaurs had been one kind of "presumptuous guest usurping the host" at Zhōukǒudiàn, the "great masses of workers, peasants, and soldiers" were a far more serious kind. The "open-door schooling" movement (開門辦學) welcomed workers, peasants, and soldiers into universities and research institutions while compelling students, teachers, and researchers to set up shop in factories and other places of production.[52] In 1976, *Fossils* printed an article that extended this concept to "open-door science" (開門辦科研). It reported on a new relationship between

50. Shànghǎi kēxué jiàoyù diànyǐng zhìpiànchǎng, *Zhōngguó gǔdài rénlèi* [China's Ancient Humans] (1976).

51. Jiǎ Lánpō, *Zhōukǒudiàn: "Běijīng rén" zhī jiā* [Zhōukǒudiàn: The Home of "Peking Man"] (Běijīng: Běijīng rénmín chūbǎnshè, 1975), 21.

52. For a sympathetic foreigner's view, see C. K. Jen, "Science and the Open-Doors Educational Movement." Jen had lived half his life in China, but having moved to the United States after World War II, he was a foreigner to the People's Republic.

the Chóngqìng Museum in Sìchuān Province and IVPP. In previous years, the Chóngqìng Museum had been required to send newly discovered fossils to IVPP in Běijīng, where for several years only a few people studied them in a "cold and sterile" (冷冷清清) manner. Now that IVPP was taking the "open-door science road," technicians and researchers from IVPP went to Chóngqìng to study the new dinosaur finds and help create a public display for the masses. In the meantime, workers and technicians from the Chóngqìng Museum had several opportunities to visit and study at IVPP.[53]

"Open-door science" was closely related to the concept of "mixing sand" (掺沙子), which tackled hierarchies in the production of scientific knowledge in a very direct way.[54] The idea was that scientists and workers could learn from one another if their work spaces in the scientific institutions were better integrated. This would contribute to the radical goal of "making laborers into intellectuals and intellectuals into laborers," an extention of the Great Leap–era utopian vision of the "all-round Communist" who participated equally in intellectual and physical labor.[55] Technicians grinding rocks to prepare embedded fossils for research, on one hand, and scientists studying fossils, reading, and writing reports, on the other, were moved into the same rooms. While innovative, this strategy greatly impeded the scientists, who found it difficult to think amid the noise and dust of fossil preparation. Although it is possible that each side learned a bit more about the other's contribution to science, the more pronounced effect was likely simply to reinforce the leveling of status distinctions between mental and manual labor, as opposed to actually eliminating the division of labor.

"Mixing sand" also referred to the longer-lasting policy of allowing workers, peasants, and soldiers from outside the institute to enter IVPP for work, study, and criticism.[56] In one case related by an IVPP scientist, a soldier came to the institute to work as a technician making casts. During the Cultural Revolution, the soldier cum technician was assigned to head an excavation team, following the policy of scientists' "standing

---

53. Zhāng Yìhóng, "Kāi mén bàn kēyán jiù shì hǎo" [Open-Door Science Is a Good Thing], *Huàshí*, no. 2 (1976): 5.

54. Máo used this phrase in August 1971 to describe his diversification of the Military Affairs Committee to include people not affiliated with Lín Biāo. Máo Zédōng, *Jiànguó yǐlái*, 13:246–247.

55. 勞動人民知識化, 知識分子勞動化. For a discussion of the "all-round Communist" in the Great Leap era, see Maurice Meisner, *Marxism, Maoism, and Utopianism: Eight Essays* (Madison: University of Wisconsin Press, 1982), 127–128, 192–193.

56. Shuping Yao very briefly defines "mixing sand" as it related to science: "While scientists were leaving the institutes, a host of workers and farmers were invited into them. This was called *chan shazi* (mixing sand)." Shuping Yao, "Chinese Intellectuals and Science: A History of the Chinese Academy of Sciences (CAS)," *Science in Context* 3, no. 2 (1989): 465.

aside" to let workers, peasants, and soldiers lead. In actuality, the technician knew too little about fieldwork to offer any direction, and the scientists in the team had to provide guidance in a discreet manner.

There were positive forms of popular participation to be had in the field. However, the members of "the masses" who took part were local peasants, rather than workers and soldiers, and their roles were ones they knew how to play. While scientists had been relying for decades on local people to lead them to sources of dragon bones, solicitation of local assistance became much more direct and formal during the Cultural Revolution. One of the best examples comes from Yuánmóu, in the southwestern province of Yúnnán, home of China's oldest *Homo erectus*. A geologist first discovered a fossilized human tooth in Yuánmóu on International Labor Day in 1965, leading IVPP to send a small team to investigate in 1967. Full-scale excavations took place between 1971 and 1973. In a 1974 article in *Yúnnán Cultural Relics Bulletin*, the IVPP scientist and *Fossils* magazine editor Liú Hòuyī wrote of the "power of the masses." He said that whenever the team uncovered a number of fossils, the county party committee instructed them to hold an exhibition on biological and human history. Team members provided interpretation of the reason for the investigations, the significance of the discoveries, and the evidence they supplied for historical materialism and the theory that labor created humanity.[57]

A photograph of one of these makeshift exhibitions appeared on the cover of *Fossils* magazine in 1974 (figure 13). The man providing interpretation was Jiāng Chū, a local who graduated from the county's only secondary school in 1966. He was assigned in 1968 to the county's Culture Bureau, and it was through this position that he became involved in the Yuánmóu excavations and later became director of the Yuánmóu Man Exhibition Hall. He recalls that each time they held an exhibition, seven or eight thousand people gathered. Commune leaders typically mandated their participation, but this was the kind of activity people were happy to stop work to attend.[58]

These exhibits were not only opportunities to teach local people about evolution; they also served to encourage locals to assist in the discovery of fossils. Liú Hòuyī noted that many people had provided clues leading to new fossil discoveries or even sent fossils they had discovered to the excavation team's field station.[59] Jiāng Chū remembers this as one of the

57. Liú Hòuyī, "'Yúnnán rén' kǎochá jì" [Notes on the Investigation of 'Yuánmóu Man], *Yúnnán wénwù jiǎnbào* 4 (October 1974): 14–15.

58. Jiāng Chū, interview with the author, 18 May 2002.

59. Liú Hòuyī, "'Yúnnán rén' kǎochá jí," 15.

13 Jiāng Chū of Yuánmóu providing interpretation of fossils and evolution to local people. The women in colorful dress were of the Yí minority nationality, placed in the center to enhance the visual appeal of the photograph. The photo appeared on the cover of *Fossils* magazine in 1974.

important outcomes of the dissemination work. "The people" (peasants and miners) covered a lot of ground in their daily work, whereas the excavation team could only hope to investigate a small area. Once they had learned "a little basic popular science knowledge," they became very dedicated in reporting anything fossil-like they found.[60]

Similar stories are told about Níhéwān, a basin that straddles the border between western Héběi and eastern Shānxī provinces and is home to "the earliest well-documented Paleolithic occurrences containing large artifact assemblages in eastern Asia."[61] The area had been known to paleontologists since 1924. In 1965, local people reported finding fish fossils. Wáng Cúnyì, an IVPP technician whose contributions to paleoanthropology had been noted since the 1930s at Zhōukǒudiàn, went to investigate. Wáng had a passion for archaeology and had found many Paleolithic sites in the past. He solicited the assistance of local people, asking them if they had ever found pieces of flint nearby. A young man named Wáng Wénquán, a peasant then about twenty years old, immediately went and found him several pieces, including a fine microlith. The first years of the Cultural Revolution interrupted this work, but in 1972 a team from IVPP returned. The team promptly found Wáng Wénquán again, and soon several other young peasant men stepped forward to help.[62] Figure 14 shows local people gathering to see fossils collected in 1972. The particularly impressive mass participation achieved at sites in the Níhéwān Basin made print in a 1975 Fossils article entitled "The Road of Mass Science Gets Broader the More It Is Traveled" and in a 1978 book, Conversations about Fossils.[63]

The 1970s also saw a dramatic increase in the number of letters IVPP received from "workers, peasants, and soldiers" reporting fossil finds. An article in the 1972 trial issue of Fossils applauded that in recent years many "workers, peasants, and soldiers" from all over China had sent letters with important clues for researchers, some had sent fossils via post, and a few had been so devoted that they had brought fossils to the capital in person. "Their concern for the scientific work of the ancestral country can teach and inspire us specialists; we should learn from

60. Jiāng Chū, interview with the author, 18 May 2002.

61. Xinzhi Wu and Frank E. Poirier, *Human Evolution in China: A Metric Description of the Fossils and a Review of the Sites* (New York: Oxford University Press, 1995), 17.

62. Except where noted, the information presented here about Níhéwān comes from Wèi Qí, interview with the author, 7 October 2002.

63. Héběishěng Yángyuán xiàn wénhuàguǎn, "Qúnzhòng bàn kēxué de dàolù yuè zǒu yuè kuānguǎng" [The Road of Mass Science Gets Broader the More It Is Traveled], *Huàshí*, no. 2 (1975): 4–5; Xià Shùfāng, *Huàshí màntán* [Conversations about Fossils] (Shànghǎi: Shànghǎi kēxué jìshù chūbǎnshè, 1978), 207–208.

14 Local people of the Níhéwān Basin brought together in 1972 to view recently unearthed fossils and to be photographed. Wèi Qí has developed long-lasting personal relationships with many local people from areas where he has conducted research. (Photograph courtesy of Wèi Qí.)

their love and care for the cultural treasures of the ancestral country," the author commented.[64] With the publication of *Fossils* magazine, the number of such letters greatly increased. Of course, the majority of these reports were dead ends—either fossils of little scientific interest or everyday rocks that appeared to be only the remains of some ancient life form. Yàn Défā, who for many years was responsible for reviewing and replying to the letters, recalls many reports of such geological improbabilities as "fossilized eyes." Nonetheless, in at least a few cases, letters from "the masses" led to significant discoveries.[65]

64. "Qúnzhòng bào huàshí" [The Masses Report Fossils], *Huàshí*, trial issue (1972): 26–27.
65. Yàn Défā, interview with the author, 2 December 2001.

Writing these letters and bringing fossils to IVPP in person represented a radically different kind of participation from that of local people guiding scientists to troves of "dragon bones," being employed as workers on a dig, or even learning about fossils and then reporting subsequent finds to locally stationed researchers. These letters mark the emergence of what may loosely be termed "fossil hobbyists"—people who developed a personal interest in fossils as scientific objects nurtured through such activities as reading *Fossils* magazine, hunting for fossils, and reporting fossil discoveries. One hobbyist began fossil hunting in 1973 when in his late thirties, around the same time he started reading *Fossils* magazine. As a pharmacologist, his interest in fossils overlapped with an interest in herbs used in Chinese medicine, and he hunted for both simultaneously.[66] Another such hobbyist was Lǐ Xùwén, a tax collector in Lìjiāng, Yúnnán, who put his job's peripatetics to a new use when he began hunting for fossils in 1971. He took an active and critical interest in excavations there in the 1970s, differing with IVPP scientists on issues of methodology and interpretation. At seventy-five in 2002, he remained an avid fossil collector respected for his amateur contributions by local employees of the cultural center and scientists in the provincial capital.[67]

It is tempting to see these forms of popular participation as evidence that paleoanthropology in the Cultural Revolution was open to people of all classes, and to attribute this openness to the socialist political context. Such a conclusion, however, requires several qualifications. First, the most active "hobbyists" were almost undoubtedly relatively well educated, as were the pharmacologist and tax collector just discussed. Second, popular participation and especially cooperation between scientists and local people are characteristics commonly found in paleoanthropology, and field sciences as a whole, around the world. Paleoanthropologists and many other kinds of scientists all rely heavily on local people to provide "expert" knowledge about local environmental and social conditions. Even mass science has counterparts outside the communist world. The Audubon Society's Christmas Bird Count, held annually since 1900 in the United States and Canada, is a good example. Amateurs report the numbers and species of birds observed, and their data become the basis for population estimates. Other examples abound. Even if Chinese paleoanthropological discoveries have involved nonprofessionals either

---

66. Survey no. 39, 2002. I mailed a survey to people who responded to an advertisement I placed in *Fossils* magazine. Fifty-four people returned completed surveys.

67. Lǐ Xùwén, survey no. 33, 2002. Lǐ Xùwén, interview with the author, 21 May 2002. Jí Xuépíng, interview with the author, 15 May 2002.

as first finders or as valued assistants more often than those in the West, this is probably due to the relatively small number of scientists in China and the large number of farmers and miners digging over a significant portion of the land.

Whatever the scale and type of actual popular involvement in socialist China compared with that in other social and historical contexts, the meanings ascribed to such participation were nonetheless specific to socialism—and meanings are important. The Christmas Bird Count is not celebrated as a victory of the uneducated, working class over the ivory tower. Far from it: bird-watching, at least in the cultural imagination, is the hobby of intellectuals or even aristocrats.[68] In contrast, the rhetoric surrounding popular participation in science in China during the Cultural Revolution hailed nonscientists' contributions to science as examples of the great benefits of traveling "the mass-science road."

Mass Participation: Criticism of Scientists

The formal criticism of scientists by the masses, on the other hand, was an area in which not just meanings, but social relations were different for Cultural Revolution–era China. Inviting "workers, peasants, and soldiers" into scientific institutions was a case in point. Several workers and soldiers—apparently no peasants—took advantage of this opportunity to study and criticize at IVPP, although they did not participate in research.[69] Others did not physically enter the institution, but studied by themselves and published articles in *Fossils* and sometimes in *Vertebrata PalAsiatica* and *Science Bulletin* (科學通報). In general, scientists then at IVPP remember these people and their actions as disruptive, domineering, ignorant, and generally useless. As one scientist recalls, "There were no good ideas [in what they wrote]. . . . When we work, we look at fossils, compare them with other fossils, and then generate some ideas. They weren't like this. They just invented ideas out of thin air." A few people, however, are more sympathetic. One of the former guides at Zhōukǒudiàn says, "They cared about [evolutionary theory]. Also, I have to say they studied a little and knew a little bit about it." Where these two members of IVPP agree is that the workers and soldiers immersed

68. See Gary Trudeau's depiction of bird-watchers in his comic strip *Doonesbury*. Gary Trudeau, *Doonesbury Dossier: The Reagan Years* (New York: Holt, Rinehart and Winston, 1984).

69. Such "workers" and "soldiers" may have been people who came from intellectual backgrounds but who had the necessary connections to obtain these desirable labels. In any case, anyone able to read and write about Marxist theory had almost certainly at least attended secondary school.

themselves only in theory—especially the writings of Engels and Máo—and did not engage in empirical research. Nonetheless, their theoretical disputations were sometimes quite lively, and twice provoked extended debates in the popular forum presented by *Fossils* magazine.

An article by the IVPP scientist Zhōu Guóxīng, printed in a 1973 issue of *Fossils*, ignited the first of these debates. His topic was the question "Can modern apes become human?" This was a common theme in Chinese and Soviet materials on human evolution that seems to have originated in a section of Mikhail Ilin's *How Man Became a Giant*.[70] The question was based on the work of Ivan Pavlov and other Soviet animal trainers who taught human behaviors—like eating with silverware and sleeping in a bed—to chimpanzees.[71] The answer for Ilin and Zhōu Guóxīng was no: while apes can learn certain human behaviors, others—like speech—are physically beyond them. Human evolution had occurred under specific environmental conditions that no longer existed, and modern anthropoid apes had evolved considerably since the times of our common ancestor. Zhōu concluded, "History cannot run backwards. For modern apes to return to the ancestral state of not yet being specialized, and for them then to become human under a 'specific environment' of your creation, is impossible."[72]

The first to take issue with the article was a worker from a low-pressure boiler factory in the Yangtze River city of Wǔhàn named Yuán Hànxīng, who together with a Wǔhàn secondary school student wrote a letter published in a 1975 issue of *Fossils*.[73] The editors introduced the

70. The chapter section is titled "Can a Chimpanzee Be Turned into a Man?" in the English translation. M. Ilin and E. Segal, *How Man Became a Giant*, trans. Beatrice Kinkead (Philadelphia: J. B. Lippincott Company, 1942), 41–45. See Chéng Míngshì and Fāng Shīmíng, *Cóng yuán dào rén tōngsú huà shǐ* [A Simple History from Ape to Human in Pictures] (Shànghǎi: Rén shì jiān chūbǎnshè, 1951), 29; Wáng Shān, *Láodòng chuàngzào rénlèi* [Labor Created Humanity] (Shànghǎi: Shànghǎi qúnzhòng chūbǎnshè, 1951), 14; and Wú Rǔkāng, *Rénlèi de qǐyuán hé fāzhǎn* [Human Origins and Development] (Běijīng: Kēxué pǔjí chūbǎnshè, 1965), 33. Many more such examples exist, and they have persisted even into recent years.

71. Ilin and other Soviet authors dealt at length with Vladimir Leonidovich Durov's work training chimpanzees for a theater production (opened in 1912) that featured many animal acts in addition to a natural science museum and a working animal psychology laboratory. The theater is still running (http://www.ugolokdurova.ru/eng_museum.html, 6 March 2007).

72. Zhōu Guóxīng, "Xiàndài de yuán néng biànchéng rén ma?" [Can Modern Apes Become Human?], *Huàshí*, no. 2 (1973): 10. The notion of history as a fixed entity whose tape could not be rolled back did represent the historical materialism of socialist China well, but it was by no means unique to that time and place. The tape of history that cannot be rewound is found also, for example, in Stephen Jay Gould's *Wonderful Life: The Burgess Shale and the Nature of History* (New York: Norton, 1989).

73. Yuán Hànxīng and Zhāng Jiànpéng, "Wǒmen duì 'Xiàndài yuán néng biànchéng rén ma?' Yī wén de yīdiǎn kànfǎ [A Few of Our Views on the Article "Can Modern Apes Become Human?"], *Huàshí*, no. 1 (1975): 25.

letter by celebrating this departure from the past practice of debating issues of human evolution only in the "cold and sterile" manner of a small number of researchers addressing a specialist audience. The critics' principal concern was that Zhōu had focused too heavily on the question of the "specific environment" in which human evolution had occurred. This interpretation did not "fit scientific facts," they argued. As evidence, they produced a line from Máo's "On Contradiction": Dialectical materialism **"holds that external causes are the condition of change and internal causes are the basis of change, and that external causes become operative through internal causes"** (bold in original, as with all quotations from Máo, Marx, Engels, Lenin, and Stalin). The "external causes" of human evolution included environmental conditions, while the more important "internal cause" of human evolution was 20 or 30 million years of undergoing labor in a struggle for survival. Moreover, they objected to the term "specific environment" (特定環境), which in Chinese can also be understood as "specified environment" or "specially designated envrionment." They contended that the changes in the earth's crust and climate were due **"chiefly to the development of the internal contradictions in nature**, and not to someone specially arranging it that way."[74]

The next issue of *Fossils* printed two letters from workers defending Zhōu Guóxīng. One, from a "young worker" at a brick and tile factory in Tàiān, suggested that Zhōu's original article did acknowledge the primary importance of labor in human origins, and that Zhōu's critics had failed to acknowledge the importance of the role played by the "specific environment" as the external cause of human evolution.[75] The other was a full-page letter from another "young worker," Chén Chún, at a Shànghǎi iron foundry.[76] Chén agreed that an answer to the question of why modern apes cannot become human had to address both internal and external causes. Nonetheless, he defended Zhōu Guóxīng's use of the term "specific environmental conditions" and said that it referred not to conditions "specially arranged by someone," but rather to "a period of time with special features manifested from out of the long, slow river of history." He further restated Zhōu's main points and argued that, far from "obscuring the dialectical relationship between internal

74. Rather than retranslating these quotations, I have relied here on Mao Tse-tung, *Selected Works of Mao Tse-tung* (Peking: Foreign Languages Press, 1967–1971), 1:314.

75. Zhèng Hóng, "Guānyú tèdìng de huánjìng jí qí zài rénlèi qǐyuán zhōng de zuòyòng" [On Specific Environment and its Role in Human Origins], *Huàshí*, no. 2 (1975): 23.

76. Chén Chún, "Zěnyàng lǐjiě rénlèi qǐyuán zhōng de 'tèdìng huánjìng'" [How to Understand the "Specific Environmental Conditions" of Human Origins], *Huàshí*, no. 2 (1975): 22.

and external causes," Zhōu had "properly analyzed dialectically the relationship."

As "workers," these two letter writers were able to help Zhōu in ways his colleagues could not: they gave the appearance that Zhōu's theories had support from members of the "broad masses." In fact, however, Chén Chún at least had only recently gained worker status. The son of a politically unfortunate teacher of English and Russian, Chén was one year short of finishing secondary school when the Cultural Revolution broke out in 1966. Thanks in part to his homeroom teacher's recommendation, he was able to secure a position in the iron foundry along with nine other students. Although the factory conditions were primitive and the work pouring molten iron was hot and grueling, they counted themselves lucky to have escaped being sent to the countryside. Chén began reading *Fossils* because his friend had subscribed and there was precious little else to read to while away the time. Soon he became engrossed in the articles, especially those on human evolution. Although he knew he was no expert, when he saw Yuán's criticism of Zhōu, Chén felt he could use his own judgment to formulate a reply. He was pleased when a response came from Liú Hòuyī asking him to revise it for publication.[77]

While reassuring, the letters of support did not prevent Zhōu Guóxīng from being labeled a counterrevolutionary and having to write a self-criticism. Nor did Zhōu's rocky relationship with Yuán Hànxīng end with the publishing of the letter. Yuán was one of the "workers, peasants, and soldiers" invited to come to IVPP to help "lead" the institute. He spent his time at IVPP reading and writing, and he published other articles in *Fossils* and in *Vertebrata PalAsiatica*.[78] Zhōu Guóxīng remembers Yuán coming to his house to eat and discuss his ideas about paleoanthropology, at which time Zhōu told Yuán that, as Máo had said, knowledge required practice, and Yuán was short on practice. Yuán then allegedly reported that Zhōu had used food and capitalist ideology against him. When the tables turned after the fall of the Gang of Four in 1976, however, Yuán still hoped to make a career in paleoanthropology. He wrote letters to Zhōu asking to become his student, an honor Zhōu refused.[79] Yuán returned to factory work, where he apparently remained until he

77. Chén Chún, interview with the author, 21 July 2005.
78. Yuán Hànxīng, "Cóng rénlèi qǐyuán kàn fùzhì tiāncáilùn de fǎndòngxìng" [A Look from Human Origins at the Reactionary Character of Reproducing the Theory of Innate Genius], *Huàshí*, no. 3 (1976): 16, 5; Yuán Hànxīng, "Duìyú rén yuán huàfēn wèntí yīdiǎn yìjiàn" [An Opinion on the Question of the Division between Humans and Apes], *Gǔjǐzhuī dòngwù yǔ gǔrénlèi* 13, no. 2 (1975): 77–80; Yuán Hànxīng, "Guānyú dìsìjì mìngmíng yǔ huàfēn tàntǎo" [Inquiries into the Naming and Divisions of the Quaternary], *Gǔjǐzhuī dòngwù yǔ gǔrénlèi* 14, no. 4 (1976): 222–227.
79. Zhōu Guóxīng, interview with the author, 7 January 2002.

retired.[80] In the end, then, it was the worker and not the academic whose intellectual ambitions suffered the more crushing defeat, a reversal that was undoubtedly far more common than has been conveyed by the literature on suffering in the Cultural Revolution.

On the other hand, for Chén Chún, the *Fossils* magazine debate served as a launching pad to fulfill scientific dreams. Zhōu was impressed by Chén's self-taught knowledge of paleoanthropology. He was also undoubtedly moved to help a stranger who had supported him in a difficult situation. They remained in touch, and Zhōu sought Chén out when he visited Shànghǎi for business. In 1978, when IVPP began soliciting applications for graduate study, Zhōu helped Chén prepare for the examination. Of seventy-five candidates, Chén alone was selected to study with Jiǎ Lánpō.[81] Chén proved a deserving student and went on to study in Canada before taking a position at Fùdàn University in Shànghǎi.

Zhōu Guóxīng was not the only IVPP scientist criticized in the pages of *Fossils*. From 1975 to 1977, *Fossils* published articles and letters debating the IVPP scientist Wú Rǔkāng's attempts to deal with the paradox of *Australopithecus*: was it ape or human? This issue had originally arisen among scientists in the early 1960s and had been resurrected in *Vertebrata PalAsiatica* in 1974 before spilling over into the more popular arena of *Fossils* the following year. In all, *Fossils* published eleven letters and articles on the subject (seven criticizing Wú, four defending), and these were only a few of the many submissions the editors received.[82] The debate centered on Wú Rǔkāng's use, in a 1974 article published in *Vertebrata PalAsiatica*, of Engels's concept of "both this and that" from *Dialectics of Nature*.[83] Engels had deemed "hard and fast lines" of "either/ or" characteristic of a metaphysical outlook insufficient for understanding the process of biological evolution. He suggested instead a dialectical "outlook on nature where all differences become merged in intermediate steps, and all opposites pass into one another through intermediate links," thus replacing "either/or" with "both this and that." The quint-

---

80. Efforts to contact Yuán Hànxīng have failed.

81. Chén Chún, interview with the author, 21 July 2005. Chén notes that being the son of an English teacher gave him an advantage in the English portion of the exam. His close relationship with Zhōu undoubtedly also helped his candidacy.

82. See the editors' note introducing Dǒng Tiěhàn and Liú Guójùn, "Guānyú cóng yuán dào rén guòdù jiēduàn de tǎolùn" [Debate on the Stages in the Process from Ape to Human], *Huàshí*, no. 1 (1976): 25, which included two letters. For the other nine letters and articles, see *Huàshí*, no. 2 (1975): 23; no. 2 (1976): 31; no. 2 (1976): 31, 26; no. 3 (1976): 31 (three letters); no. 4 (1976): 14–16, 21; no. 1 (1977): 28–29; and no. 3 (1977): 29–30.

83. Wú Rǔkāng, "Rén yǔ yuán de jièxiàn wèntí" [The Problem of the Boundary between Human and Ape], *Gǔjǐzhuī dòngwù yǔ gǔrénlèi* 12, no. 3 (1974): 181–182, 186.

essential example for Engels was the transition between the birdlike reptile *Compsognathus* and the reptilelike bird *Archaeopteryx*.[84] To Wú, it seemed logical to extend the theoretical concept of "both this and that" to address the problem of *Australopithecus*, who appeared to represent an intermediate stage which was "both ape and human."

Unfortunately for Wú, not everyone agreed. The lengthiest criticism came from a soldier, Lài Jīnliáng, in the rear service department of the People's Liberation Army in Nánjīng. Even before his polemic was published in the last 1976 issue of *Fossils*, Lài had already sent it to *People's Daily*, which printed an article summarizing his points and celebrating Lài's work as evidence against "Dèng Xiǎopíng's reactionary fallacy that workers, peasants, and soldiers do not understand science."[85] Lài focused on the way Wú allegedly downplayed the chief difference between ape and human, namely, labor. Making *Australopithecus* a midway point between ape and human, contended Lài, split the qualitative change of ape to human into two parts, leading to the preposterous conclusion that "the long period of struggle with the natural world and the long period of labor had only created half a human!"[86] At the end of the article, Lài somewhat feebly made a link to the pressing political issues of the day—criticizing capitalism, revisionism, and eclecticism (折東主義)—and, of course, the *People's Daily* article took this link as its focus. In September, an expanded version of Lài's article appeared in *Science Bulletin*, which had already published several articles on the debate over the previous year and a half.[87]

Wú Rǔkāng rode the waves fairly successfully, despite all this negative attention. He continued to publish: for example, he had the honor of authoring the chief article commemorating the hundredth anniversary of Engels's "Labor created humanity" in the spring 1976 issue of *Fossils*.[88] And in 1977, *Fossils* published both a new article in which Wú defended

84. Frederick Engels, *Dialectics of Nature*, in *Karl Marx, Frederick Engels: Collected Works*, ed. E. J. Hobsbawm et al. (Moscow: Progress Publishers, 1975–), 25:493.

85. Dèng was then undergoing his second round of formal criticisms. Lài Jīnliáng, "Píng 'yì yuán yì rén' jiēduàn lùn" [Criticizing the 'Both Ape and Human' Stage Theory], *Huàshí*, no. 4 (1976): 14–16, 21. Xīnhuá shè Nánjīng, "Xiàng gǔrénlèixué zhōng zīchǎn jiējí guāndiǎn yǒnggǎn tiānzhàn" [Bravely Challenging the Capitalist Viewpoint in Paleoanthropology to Battle], *Rénmín rìbào*, 18 August 1976, 3.

86. Lài Jīnliáng, "Píng 'yì yuán yì rén,'" *Huàshí*, 15.

87. Lài Jīnliáng, "Píng 'yì yuán yì rén' jiēduàn lùn" [Criticizing the 'Both Ape and Human' Stage Theory], *Kēxué tōngbào* 21, no. 9 (1976): 404–409.

88. Wú Rǔkāng, "Láodòng chuàngzào rénlèi: Jìniàn 'Láodòng zài cóng yuán dào rén zhuǎnbiàn guòchéng de zuòyòng' xiě zuò yībǎi zhōu nián" [Labor Created Humanity: Commemorating the Hundredth Anniversary of "The Part Played by Labor in the Transition from Ape to Human"], *Huàshí*, no. 2 (1976): 6–8.

himself on the subject of "both ape and human" and an article by a worker and a shop employee asserting that although Wú's notion of "both ape and human" was "not complete enough," it was "basically tenable."[89]

What, then, do we make of these passionate debates on human origins? First, these were "real" debates in the sense that both positions were given space, and the editors did not explicitly endorse either side. Given the fervor of political campaigns then underway, the letters were also remarkably polite: even when using political rhetoric to criticize scientists' theories, they fell significantly short of labeling the scientists themselves revisionists or capitalists. Second, *Fossils* (and to some extent professional journals as well) certainly succeeded in creating a forum for nonspecialists to sink their teeth into important scientific questions. It is true, as a scientist quoted above noted, that the workers, soldiers, students, teachers, and others who contributed to this forum had no empirical evidence with which to anchor their theories. However, the original articles whose positions they debated were largely theoretical themselves, and what empirical data they presented were well within the grasp of many nonspecialist readers. Third, while Chén Chún had been born into an intellectual's household, he was a foundry worker with a high school education when he discovered in the pages of *Fossils* a passion for human prehistory. The debates opened a door for him to "enter science" in a meaningful and lasting way. If this had been a more common outcome, the judgment on "mass science" might be more positive than it is today.

On the other hand, this kind of mass participation constrained scientists by compelling them to debate in terms readily accessible to nonspecialists. Moreover, it was an ugly experience for the scientists whose work became the target of criticism. It is very possible that some of those who made the criticisms—particularly residents at IVPP like Yuán Hànxīng—were coached by others at the institute who bore grudges against the targeted scientists. To be criticized by lay people was a particularly difficult burden to bear: the intellectual authority scientists wielded had to bow before the political authority of the "workers, peasants, and soldiers." Their only hope was that some of these privileged members of "the masses" would come to their defense, and fortunately for them, some did.

89. Wú Rǔkāng, "Guānyú cóng yuán dào rén de guòdù jiēduàn zhōng 'yì yuán yì rén' de xìngzhì wèntí [On the Problem of the Nature of "Both Ape and Human" in the Stages in the Process from Ape to Human], *Huàshí*, no. 1 (1977): 28–29. Wáng Jiànjūn and Wáng Ānzhèng, "Zěnyàng rènshí 'fēi cǐ jí bǐ' yǔ 'yì cǐ yì bǐ' de biànzhèng guānxì" [How to Identify the Dialectical Relationships between "Either This or That" and "Both This and That"], *Huàshí*, no. 3 (1977): 29.

The Missing Link

In 1978, Dèng Xiǎopíng moved into position to become China's next leader. One of the key stepping-stones in this process was the National Conference on Science (全國科學大會) held in March 1978. There Dèng gave a speech in which he asserted his commitment to the Four Modernizations (agriculture, industry, science and technology, and national defense). He further promoted science and technology as a "productive force," rather than part of the "superstructure," as it had been considered during the Cultural Revolution. This in turn signified that scientists' intellectual work could again be acknowledged as labor, and thus scientists could be considered members of the working class rather than people fundamentally divorced from the proletariat.[90] The same year, IVPP assigned Huáng Wèiwén to overhaul the Peking Man site exhibit again; it reopened in 1979 on the fiftieth anniversary of the discovery of the first Peking Man skullcap. No longer did the exhibit display dinosaurs or other fossils foreign to Zhōukǒudiàn. Neither did it take on the whole of biological evolution in order to educate the masses about materialism. Zhōukǒudiàn once again became a place that celebrated scientific research, past and present, rather than popular science.[91] Professional interests had risen again over popular ones: the hosts had returned to expel the "presumptuous guests."

Fossils itself underwent a transformation in the late 1980s into a more specialized magazine geared toward more educated audiences. In 1982, the editors still identified their audience as "mainly young workers and students . . . lacking in basic knowledge" and proclaimed their intention to increase the readability of the magazine.[92] The editors of *Fossils* have continued until the present day to receive letters from readers calling for simpler language and more attention to basic scientific knowledge. However, competition from the many other popular science magazines (which by 1983 numbered more than sixty)[93] has caused *Fossils* subscription numbers to plummet to about 4,000 from its peak of 228,000, and

90. This same point had been made in the post–Great Leap Forward reevaluation of science and scientists, and it was a key part of Dèng's remarks after Hú Yàobāng's September 1975 report to the State Council (see note 24 above).

91. Huáng Wèiwén, interview with the author, 27 December 2001. See also Jia Lanpo and Huang Weiwen, *The Story of Peking Man: From Archaeology to Mystery*, trans. Yin Zhiqi (Beijing: Foreign Languages Press, 1990), 217–218.

92. *Huàshí* biānjí zú, "*Huàshí* shí nián," *Huàshí*, no. 4 (1982): 1.

93. See page 31 of the no. 4 issue of *Fossils* magazine for 1982 for a complete list of popular science magazines available for subscription in 1983.

the editors have responded by focusing on serving a more specialized and better-educated audience.[94]

The late Cultural Revolution was the closest China came to a productive union between the top-down model of science dissemination and the bottom-up model of mass science. New, colorful popular media like the "Labor created humanity" exhibits and *Fossils* magazine emphasized the role of the masses in paleoanthropological research. Dissemination in the field encouraged local people to become active participants in the location and protection of fossil sites. *Fossils* magazine also encouraged the emergence of a new kind of participation: that of "hobbyists" who hunted for fossils out of an interest in their scientific value, as opposed to their medicinal and commercial value as dragon bones. Moreover, *Fossils* provided a forum for laypeople to participate in debates on issues of human evolution alongside professionals.

Nonetheless, the union of science dissemination and mass science fell far short of its radical potential. The notion that labor created humanity was central to all dissemination materials. At the same time, laborers were entering scientific institutions, leading paleontological digs, and challenging scientists on their theories of human evolution. These two kinds of labor—that which defined humanity and that which provided the backbone of contemporary science and society—were closely related. They were both understood specifically as physical labor: labor preceded the development of the brain in Engels's account of human evolution, and intellectual work was excluded from the definition of "labor" during the Cultural Revolution. With the two spheres of dissemination and participation becoming more integrated, and the concept of labor being foundational to both, it would seem an easy jump to say, "Because labor created humanity, laborers are uniquely qualified to interpret evidence of human origins." Indeed, paleolithic archaeologists in China and elsewhere have often engaged in "experimental archaeology," producing primitive stone tools in their attempts to understand paleolithic technologies (figure 15). In the West, archaeologists have also turned to modern flint knappers to learn these skills and gain insight into early tool manufacture.

And yet, it appears no one ever made this link. Ironically, it was in the mental realm of theory, rather than the manual realm of practice, that "mass scientists" like Yuán Hànxīng and Chén Chún had the most play. One might well have expected to find materials—even if they were only the most flagrant propaganda—suggesting that those who worked with

94. Liú Shífān, interview with the author, 6 February 2002.

63. 为了进一步探索中国猿人制造工具的过程，我们用很多方法，来做各种不同的打制石器的试验。

15 Image from a serial picture book based on the 1959 film *Zhōngguó yuánrén* (Peking Man). The caption reads: "To explore further Peking Man's tool-manufacturing processes, we conduct experiments using many different methods of flaking stone tools." Note the white coat of the person conducting the experiment. The authors do not seize the obvious opportunity to highlight the potential contributions of laborers to science. From *Zhōngguó kēxuéyuàn gǔjǐzhuī dòngwù yánjiūsuǒ*, *Zhōngguó yuánrén* [Peking Man] (Běijīng: Kēxué chūbǎnshè, 1972), 63.

their hands had an important insight into the first labor, the production of tools. One might have expected there to be an explicit connection made between knowledge about human origins on the one hand and the people who produced that knowledge and the practices through which they produced it on the other. The lack of such an explicit connection indicates the absence of a serious, class-based reappraisal of the structure of scientific knowledge.

An example of what such a reappraisal could have looked like in popular paleoanthropology of the Cultural Revolution can be found among a few Marxist geologists in postwar Japan. Nakayama Shigeru has written on the philosophy of science and "grass-roots geology" that geologist Ijiri Shōji proposed and implemented in the early decades after World War II.[95] Ijiri was the leader of the Society for Corporate Research in Earth

95. Nakayama Shigeru, "Grass-Roots Geology: Ijiri Shōji and the Chidanken," in *Science and*

Science, a part of the Association of Democratic Scientists. Through the society, he helped organize amateur geologists around Japan whose fieldwork genuinely contributed to ongoing geological research. The society's "bible" was Ijiri's *On Science—Centering on Paleontology* (*Kagaku ron—koseibutusgaku o chūshin to shite*), in which Ijiri described (geological) science as based on personal experience, hypothesis venturing, and fieldwork—all equally accessible to amateurs and to professional scientists. Indeed, Ijiri saw hypothesis venturing in particular as dependent on a "denying or dissenting spirit" identified with "class-consciousness itself."[96] With respect to fieldwork, as Nakayama interprets Ijiri, "those whose labors involve endless walking and exhaustive observations, whether amateurs or semi-professional geologists, know more and know it better than an established top-notch geologist settled in Tokyo."[97]

A similar perspective can be found in a far more unlikely place: the writings of Robert Ardrey, an American playwright with strong anti-communist commitments whose influential 1961 book, *African Genesis*, highlighted murderous aggression as the key to understanding human origins and evolution. Ardrey criticized professional scientists for ignoring ideas outside the mainstream and questioned whether they had as much insight into human nature as the lay public. His approach implied that "the more untrained the observer, and the closer to the humanities, the better their intuitive grasp of the findings and implications of a new science of human nature. So 'poets and peasants' know these truths better than professional scientists."[98]

In contrast, the *Fossils* article that in 1975 celebrated the "road of mass science" as "traveled" in Níhéwān said nothing about what "the masses" as "the masses" brought to science. Rather, it was only "after the masses had grasped science and understood science" that they became useful to science by "hunting for, reporting, and protecting fossils."[99] In short, despite ostensibly being about mass science, the article paid much greater homage to science dissemination. It even suggested that as a result of participating in mass science the local people had begun to have "preliminary knowledge" about the "natural geographic features of their own village." Specifically, local people had learned from scientists that the area, dry today, was home to a large lake 2 million years ago.

Society in Modern Japan: Selected Historical Sources, ed. Nakayama Shigeru, David L. Swain, and Yaga Eri (Cambridge, Mass.: MIT Press, 1974), 270–289.

96. Nakayama, "Grass-Roots Geology," 274.

97. Nakayama, "Grass-Roots Geology," 277.

98. Nadine Weidman, "Popularizing the Ancestry of Man: Ardrey, Dart, and the Killer" (paper presented at the annual meeting of the History of Science Society, 4 November 2006).

99. Héběishěng Yángyuán xiàn wénhuàguǎn, "Qúnzhòng bàn kēxué," 5.

This account conflicts strangely with the IVPP scientist Wèi Qí's memory of working with local people for whom the lake's past existence was old news: local legend said the same.[100] Had the authors of the article been serious about "mass science" in the sense of a class-based claim to knowledge and knowledge production, the local legend come true would have been an easy way to discuss what local peasants knew by virtue of their being peasants—that is, by virtue of their connection to the land and their inheritance of knowledge in the form of legends—as opposed to what they knew only after being taught by scientists.

Rhetoric and policy on popular science in the People's Republic of China contained a deep-seated and unresolved contradiction between reverence of the masses on the one hand and concern about their lack of scientific knowledge on the other. Much of science dissemination throughout the 1950s, 1960s, and 1970s was almost synonymous with "stamping out superstition." What has become evident through an examination of the Cultural Revolution is that even in the most radical period, the notion that the masses were superstitious outweighed any idea that they might have had special access to knowledge. While propaganda asserted time and again the emptiness of "the fallacy that workers, peasants, and soldiers do not understand science," their understanding was seen only as the outcome of successful science dissemination rather than the result of knowledge forms or mental orientations they possessed by virtue of their class position. It is possible that such a rigorous, class-based philosophy of science—a Marxist standpoint epistemology—was more explicitly elucidated in other scientific fields, for example, industry, agriculture, or even seismology.[101] Paleoanthropology represents a "harder case," since its connection to popular forms of knowledge is less conspicuous. However, since apparently no one made the obvious connection that, because labor created humanity, laborers were uniquely suited to study human evolution, such a philosophy of science could not have been well entrenched in the natural sciences as a whole.

100. Wèi Qí, interview with the author, 7 October 2002. Wèi Qí published an article on Níhéwān in which he included a discussion of the lake legend. Wèi Qí, "Zǒuxiàng shèhuì, zǒuxiàng qúnzhòng: Níhéwān yánjiū yīng xiàng zhīshí jīngjì zhuǎnhuà" [Go to Society, Go to the Masses: Níhéwān Research Should Turn toward the Knowledge Economy], Huàshí, no. 1 (2001): 33–35.

101. Peasant knowledge of farming was an obvious well to tap in agriculture. Earthquake prediction efforts in the 1970s relied extensively on peasants' observations of livestock behavior. For something close to such a Marxist standpoint epistemology for industry, see Máo's memo of 1958 cited in chapter 4, note 36.

SIX

"Springtime for Science," but What a Garden: Mystery, Superstition, and Fanatics in the Post-Máo Era

Some Other Spring

Zhōu Guóxīng has become one of China's foremost producers of popular materials on human origins in the post-Máo period. In addition to publishing a great number of books and articles, he became responsible for the anthropology wing of the Běijīng Natural History Museum at the end of the Cultural Revolution.[1] Despite his rationalistic approach firmly rooted in the established, superstition-squashing form of science dissemination, Zhōu sees himself as a maverick and cultivates that image by consistently pushing scientists, the state, and society to stretch their tolerance for new ideas about humanity. In the last chapter, we saw him criticized in the pages of *Fossils* magazine for his alleged neglect of "internal causes" and his emphasis on the "specific environment" in which apes evolved into humans. Today, his public explorations of marginal scientific theories and his willingness to engage with nonexperts

---

1. Originally a scientist at IVPP, he obtained a transfer to the Natural History Museum because of political tensions arising from the positions he and others took during the Cultural Revolution.

continue to generate letters from China's growing number of fossil hobbyists and amateur evolutionary theorists.

Some of these people are what Tián Sòng, a scholar at Běijīng University, calls "one-person institutes"—that is, people without formal scientific education who have a passion for a particular scientific field and develop and promote their own scientific theories.[2] One of Zhōu's most enthusiastic fans has a pet theory that the key characteristic distinguishing humans from other animals is not labor, but front-to-front sex.[3] The behavior is almost unknown in the animal world, but Zhōu has cautioned this lay expert that it is not quite unique to humans: bonobos (also known as pygmy chimpanzees) do it too.

The theory itself is not quite unique, either, but rather reflects movements in international science to consider the kinder, gentler side of human nature and efforts within China to treat sex in a more open manner. Zhōu himself is no stranger to these trends. The opening of the new anthropology exhibit he created for the Natural History Museum in 1988 drew crowds of excited visitors after being delayed for weeks by state officials. In addition to softening the force of Engels's theory that labor created humanity, Zhōu created a section on the human reproductive and life cycles that included detailed drawings and a photograph of a naked woman and man.[4] As shocking as such images were, rumors exceeded them. People said, for example, that visitors could drink the milk of a lactating woman to reexperience this primal aspect of human, mammalian life.[5]

The experience of meeting Zhōu and listening to him discuss such issues can almost be captured in a single Chinese syllable: *mí*. Three kinds of *mí* have defined popular science in post-Máo China. The first is mystery (謎, *mí*): always a common component of popular science writings, mysteries have come to dominate in the post-Máo period. Whether cracking the mystery of human origins, or indulging in the mystery of human sexuality, audiences have flocked to topics that challenge the

2. Tián Sòng, paper presented at the Tenth International Conference on the History of Science in East Asia, Shànghǎi, 20–24 August 2002.

3. Zhōu Guóxīng, interview with the author, 7 January 2002. Tián Sòng, personal communication.

4. Jǐng Xīng, "'Rén zhī yóulái' zhǎnlǎn héyǐ bùnéng rúqī kāifàng" [Why the "Human Origins" Exhibit Cannot Open on Schedule], *Běijīng qīngnián bào*, 27 August 1988, 1. An opinion piece published in the same newspaper three days later called for the long-awaited exhibit to open without further delay. Yì Yùnwén, "Xīwàng 'Rén zhī yóulái' jǐnkuài zhǎnchū" [I Hope "Human Origins" Will Open as Soon as Possible], *Běijīng qīngnián bào*, 30 August 1988. The exhibit finally opened on 8 October after the photograph of the naked woman and man was replaced with an anatomical drawing. I thank Zhōu Guóxīng for providing these and other articles on the issue.

5. This rumor is remembered by a woman born in the early 1980s.

stability of knowledge and politics. The second is superstition (迷信, *míxìn*): state officials and scientists have remained deeply concerned about the general population's adherence to all sorts of "superstitious beliefs"—a situation exacerbated by the growing sensationalism of popular science materials themselves. Finally there are the hobbyists, or fans (迷, *mí*): having been encouraged in the late Cultural Revolution by the publication of *Fossils* magazine, fossil collectors and amateur evolutionary theorists have since grown wildly in number and commitment.

The three terms are etymologically very close. *Mí*, 迷, has a cluster of meanings: lost, confused, bewitched, fascinated. *Míxìn*, 迷信, is a belief (*xìn*, 信) that is founded in confusion, thus a superstition. A fan is someone who is fanatically interested, who is fascinated in a given subject, hence *mí*, 迷. In the *mí* of "mystery" (謎) a language classifier is added to the left side of the character, but its meaning is nonetheless related, and in fact it is not infrequently written as 迷. It originally referred to verbal riddles; the sixth-century literary theorist Liú Xié defined 謎 as "having the words scrambled so as to make one confused [昏迷]."[6] Its meaning has since expanded to denote anything that has no clear explanation.

The post-Máo era is often described in Chinese writings as "springtime for science"—with the chaos and persecutions of the Cultural Revolution largely put to rest, scientists have been able to focus on their work with less fear of political reprisal and more freedom from forced participation in political activities. Many credit Dèng Xiǎopíng's adoption of the Four Modernizations (agriculture, industry, science and technology, and national defense) as the centerpiece of his political agenda.[7] But for popular science, spring has worn a somewhat different face. The "garden of science" in which the "hundred flowers" of scientific ideas bloom has indeed become more diverse, but not always in ways considered positive by scientific and political elites. The pressure to sell stories on the market has cultivated many garish flowers, and the hope of liberalization has even encouraged the kind of politically threatening ideas and practices decried in the Máo era as "poisonous weeds."[8]

6. Wú Zéyán, Huáng Qiūyún, and Liú Yèqiū, eds., *Cí yuán, dàlù bǎn* [Origin of Words, Mainland Edition] (Táiběi: Táiwān shāngwù yìnshūguǎn, 1993 [1989]), 2:2907.

7. Promoting advances in the four key sectors of "agriculture, industry, national defense, and science and technology" was part of party rhetoric in the early decades of the People's Republic. See, for example, Zhōngguó kēxué jìshù xiéhuì, "Guānyú kēxié zǔzhī gōngzuò ruògān wèntí de jiěshì" [Regarding the Resolution of Certain Questions about the Organization of Work in CAST], 30 August 1960, Běijīng Municipal Archives, 10.1.157: 8–10.

8. The unintended consequences of reform are an important theme in Deborah Davis and Ezra Vogel, eds., *Chinese Society on the Eve of Tiananmen: The Impact of Reform* (Cambridge, Mass.: Council on East Asian Studies, Harvard University, 1990).

Tensions of Reform

The enormous changes witnessed in the post-Máo era should not blind us to the ways that trends begun under Máo have continued to exert influence. Some of the key institutions—for example, the Chinese Association for Science and Technology and the local cultural centers—were established in the 1950s. The second half of the Cultural Revolution also generated new or newly emphasized approaches to popular science that served as a foundation for later years. The seeds sown by *Fossils* magazine and other pioneers of the early and mid-1970s gave rise to the profusion of science dissemination materials in the post-Máo era. The late Cultural Revolution's brightly colored magazine covers that displayed energetic people, dramatic landscapes, and appealing animals, and museum exhibits that featured dinosaurs and large oil paintings, testified to the importance placed on catering to mass tastes, a priority not unlike that of post-Máo market-oriented popular science. The cultivation of fossil hobbyists in the early 1970s also marks a Máo-era beginning for what became a serious phenomenon in the post-Máo period. Such forms of precedence require special attention because they are often elided in official Chinese historiography, which shies away from recognizing apparently progressive trends during the supremacy of the Gang of Four.[9]

Chinese people and foreigners alike have watched the changes of the post-Máo era with amazement. After 1978, many new magazines competed with one another for readers, and in recent years corporations have begun marketing their own popular science magazines.[10] Colorful magazine covers with pictures of glamorously made-up women decorate street corners, and the abundant consumer goods advertised in their pages are within the economic reach of an increasing number of China's newly legitimate middle class. The experience of this transformation in material culture and social relations has its roots in the policies of liberalization, market economics, and "opening" to other countries. But the admittedly sweeping character of the changes does not reflect a consensus on the direction Chinese society and politics should take. Rather, opening has bred fears of "spiritual pollution" (精神污染), liberalization has paved the way for dissent, the market has encouraged hedonism, and all of these factors together have allowed for the reemergence

9. Rudolf G. Wagner has made the same point with respect to the historiography of science fiction. See his "Lobby Literature: The Archaeology and Present Functions of Science Fiction in China," in *After Mao: Chinese Literature and Society, 1978–1981*, ed. Jeffrey C. Kinkley (Cambridge, Mass.: Harvard University Press, 1985), 40.

10. See, for example, *China Science and Technology Global Knowledge* (中國科技縱橫知識).

of ideas, beliefs, and practices the state broadly labels "superstition."

Science dissemination was an obvious and important element in the Four Modernizations program, as it had been throughout the republican and Máo eras. As with other sectors, formal institutions responsible for science dissemination resumed normal operations or formed for the first time in the late 1970s. For example, the China Science Dissemination Writers Association (中國科普作家協會) was founded in August 1979 and soon began publication of a bimonthly periodical, *Science Dissemination Writings* (科普創作). Local branches of this association also formed and launched new science dissemination magazines, as did the revitalized natural history museums in Běijīng and Shànghǎi.[11] Writers and editors pronounced their commitment to disseminating scientific knowledge in order to further the goals of the Four Modernizations. And in 1980, the renowned science writer Gāo Shìqí convinced the State Council to allow the founding of the China Research Institute for Science Popularization (中國科普研究所, CRISP) under the umbrella of the Chinese Association for Science and Technology.[12]

As part of the strategy for achieving the Four Modernizations, Dèng and other Chinese leaders in 1978 embraced market economics. The market has had two very different consequences for popular science. On one hand, with market pressures sometimes outstripping political pressures, producers of science dissemination materials have engaged in sensationalism and other strategies to appeal to the greatest possible number of readers. On the other hand, the legitimation of economic stratification, understood to serve the greater purpose of cultivating a needed scientific elite, has allowed some science disseminators to cater to those with the most spending power and the greatest access to educational privileges. For the editors of *Fossils*, this has meant, for example, incorporating articles on the "strange rocks" (奇石, rocks of unusual and beautiful shapes and patterns, rarely including fossils) collected by scholars in imperial times and enjoying a renaissance today thanks to a renewed spirit of acquisition and the related cultivation of Chinese traditional elite practices. From 1998 through 2001, *Fossils* devoted a

---

11. See the Shànghǎi Natural History Museum's *Natural History* (博物), begun in 1979, which later changed its name to *Nature and Man* (自然與人); the Běijīng Natural History Museum's *Nature* (自然), begun in 1980; the Sìchuān Science Dissemination Writers Association's *Science Hobbyist* (科學愛好者), also begun in 1980; the Yúnnán Science Dissemination Writers Association's *Science Window* (科學之窗), begun in 1980, and its comic-book-style supplement *Mysteries Illustrated* (奧秘畫報).

12. CRISP has been responsible for studies of science literacy in addition to surveys of influential science writers and writings in China and abroad.

section of each issue's contents to "strange rocks." Recent years have also seen greater attention among science disseminators to enhancing extracurricular science education for children at elite schools. Scientists of the IVPP have been active in organizing field trips and even summer camps for the fortunate few. Such programs have moved away from earlier, socialist models of science dissemination, which explicitly prioritized educating the "broad masses" in basic science and eradicating their superstitions. Instead, they now focus on inspiring and educating a new generation of scientists whose intellectual work will make the economic dreams of the Four Modernizations a reality.

Also critical to Dèng's plan to achieve the Four Modernizations was an "open-door policy" in international economic relations, coupled with a general opening to cultural interaction with foreign countries. With the stage set for increased international cooperation with the Máo-Nixon talks in 1972, these programs and the specific policies they generated were among the main forces behind the dramatic changes in Chinese society in the post-Máo period. They led to the emergence of an expanding class of consumers and a growing variety of manufactured goods for them to consume. Moreover, "opening" provided increased access to knowledge about foreign countries while also whetting Chinese people's appetites for more such knowledge.

Dèng's support for market economics and international cultural exchange did not extend to support for some of the consequences of these policies. Beginning in 1982, Dèng and other leaders attacked a growing body of art and literature that cast a bleak light on recent Chinese history. Blaming the phenomenon on damaging influences from the West, they called for a new political campaign against "spiritual pollution." As a blanket political term, "spiritual pollution" is almost as flexible and powerful as "superstition": it easily stretches to cover sex and consumerism in addition to political dissidence and unorthodox theories. Tellingly, in 1979 Wáng Ruòshuǐ labeled Máo's cult of personality a form of superstition and four years later lost his position at the *People's Daily* during an attack on humanism in the campaign against "spiritual pollution."[13] This same bogeyman appeared in official explanations of the democracy movement that culminated in the June Fourth Incident at Tiānānmén Square in 1989. Dèng characterized the incident

---

13. David A. Kelly, "The Emergence of Humanism: Wang Ruoshui and the Critique of Socialist Alienation," in *China's Intellectuals and the State: In Search of a New Relationship*, ed. Merle Goldman with Timothy Cheek and Carol Lee Hamrin (Cambridge, Mass.: Council on East Asian Studies of Harvard University, 1987), 159–161.

as having arisen from "bourgeois liberalization and spiritual pollution," and he reaffirmed the need to foster openness and economic growth without succumbing to such noxious Western influences.

It was in this spirit that he followed the political crackdown after June Fourth with clear statements about the separation of political and economic liberalization. For the former, he promised strict controls to secure the nation against destabilizing forces, while the latter he pledged to encourage with a new set of economic policies. In 1992, Dèng embarked on his famous "southern tours," where he laid out new policies designed to spearhead rapid growth through market economics. However, ten years after June Fourth came another challenge to political orthodoxy, the Fǎlún Gōng movement, which provoked the state, now headed by Jiāng Zémín, into unleashing all its weapons against "spiritual pollution," superstition, and political dissidence.

The tensions of reform in the post-Máo period have been strongly felt in the realm of science, particularly where science meets the public. Best known to Western audiences is the astrophysicist Fāng Lìzhī, who has been vocal in linking scientific progress, freedom of scientific inquiry, and more general democratic values. As early as 1980, Fāng began publicly questioning Dèng's reform program, which provided only technocratic and no democratic solutions. In 1987, he was expelled from the party, and after being accused of helping to instigate the democracy movement that ended in the June Fourth Incident, he is now living in exile in the United States.[14] While foreign observers applaud his political daring in the cause of democracy and human rights, he is not without his detractors even in the West. Some see much of his activism as motivated by a desire to secure special privileges for scientists, an already privileged social group.[15] It is telling that in Dèng's China economically undemocratic attempts to cultivate a scientific elite were politically acceptable or even desirable, while speaking for democracy in the political sphere was the mark of a dangerous dissident.

These tensions go far beyond highly visible figures like Fāng. Science in the post-Máo period has been characterized by a set of related "breakdowns," and by the anxieties that they have in turn produced. A breakdown of control over the economy has been necessary for the

14. James Williams, "Fang Lizhi's Big Bang: Science and Politics in Mao's China" (Ph.D. diss.: University of California, Berkeley, 1994). A collection of Fāng's essays is available in English as Fang Lizhi, *Bringing Down the Great Wall: Writings on Science, Culture, and Democracy in China* (New York: Norton, 1990).

15. Richard C. Kraus, "The Lament of Astrophysicist Fang Lizhi: China's Intellectuals in a Global Context," in *Marxism and the Chinese Experience: Issues in Contemporary Chinese Socialism*, ed. Arif Dirlik and Maurice Meisner (Armonk, N.Y.: M. E. Sharpe, 1989).

introduction of market forces, but along with the new abundance of state-approved consumer goods has come the emergence of a black market in fossils and sensationalism in popular science media. The breaking down of barriers to foreign knowledge and culture has breathed new life into scientific disciplines and allowed the reading public greater exposure to inspirational accounts of scientists and scientific discovery around the world, but it has also been blamed for the sensationalist and even pornographic trend in science dissemination materials. The breakdown of central control over scientific research and dissemination agendas has encouraged greater innovation among regional scientific and cultural institutions, but members of these institutions have also begun to push for more self determination. And amid all of this has come a more general breakdown in the perceived stability of knowledge itself, which has permitted analytical frameworks about human origins other than Engels's "Labor created humanity" to claim space on the stage of popular science but has also encouraged scientifically and politically dubious claims about the role of extraterrestrials and supernatural forces in human evolution. None of these breakdowns has come close to being total, but in each case a highly controlled system has become less highly controlled, and Chinese leaders remain uncertain how to cultivate their economic benefits while limiting the challenges to political order that they engender.

"Opening"

Máo-era scientists and laypeople were by no means entirely cut off from scientific activities outside China. But this was nothing compared with the scale of such exposure once China "opened" to the world. Beginning in the immediate post-Máo period, scientists and others voraciously read foreign materials and translated or adapted them for mass consumption. As in other countries, accounts of the work of women primatologists— Jane Goodall's with chimps, Dian Fossey's with mountain gorillas, and Penny Patterson's with her sign-language student, Koko the gorilla— were especially popular.[16] Also common were accounts of "new theories" on established scientific questions. Most such articles were adapted and translated from Western sources, and many directly contradicted the story of human evolution as constructed in Máo-era China. Human intelligence arose in connection with solar flares; spoken language emerged

---

16. In addition to a great many articles on the individual research projects and the women who run them, see Shān Fán, "Wèihé nǚ língzhǎnglèi xuézhě rúcǐ duō" [Why Are There So Many Women Primatologists?], *Huàshí*, no. 1 (1996): 17–19.

when human ancestors moved into cold climates and lived in dark caves where they could not see hand signals; mysterious artifacts may be evidence that a highly developed civilization existed on earth prior to the beginnings of human history; walking erect was an adaptation to keep early hominids cool in a hot climate; and still more radical, walking erect did not emerge in conjunction with tool use but rather as a social adaptation to assist in the gathering, transport, and distribution of food.[17] Such "new theories" implicitly challenged the dominance of labor as the fundamental cause of human origins and in the process also broadened the range of discourse on humanity. Humans could be said to be special because of our intelligence, human society could be said to be prior to human labor, and the emergence of such key human characteristics as language or walking erect could be traced to the most mundane causes and said to be mere accidents of nature.

Fossils in particular also began to highlight international cooperation in science, in 1994 going so far as to place on the magazine's cover a visually uninteresting but politically potent picture of a Chinese scientist talking with three foreign scientists.[18] Over time, even discussions of international participation in the discovery and disappearance of the Peking Man fossils became more conciliatory toward the foreign scientists. In 1979, an article published in Scientific Experiment on Péi Wénzhōng had little to say about foreign participation at Zhōukǒudiàn except that "in the 1920s [Dragon Bone Hill] inspired some geologists and paleontologists to come to China along with the foreign aggressive forces in order to 'visit' and 'investigate.'"[19] In 1982, an article in the popular Chinese magazine Nature, also on Péi Wénzhōng, wrote favorably on Davidson Black's positive influences and on the contributions of the American Rockefeller Foundation.[20] And the same year, Huáng Wèiwén published a celebratory article in Fossils on Davidson Black himself.[21] Subsequent

17. Zhōu Lìmíng, "Rénlèi zhìnéng qǐyuán xīn shuō" [New Theory on the Origin of Intelligence], Zìrán yǔ rén, no. 4 (1987): 46–47; Wáng Hóngyuǎn, "Rénlèi yǔyán qǐyuán de xīn xuéshuō" [New Theory on the Origins of Human Language], Bówù, no. 1 (1983): 32; Yú Jié and Wáng Lìlí, "'Qiánrénlèi' xīn shuō" [New Theory on "Prehumans"], Àomì huàbào, no. 7 (1990): 30–32; Zhāng Mínghuá and Chén Wéimín, "Rénlèi wèishénme zhílì xíngzǒu" [Why Do Humans Walk Erect?], Zìrán yǔ rén, no. 2 (1989): 14–15; Chén Chún, "Xīn de rénlèi jìnhuà lǐlùn" [New Theory on Human Evolution], Bówù, no. 2 (1980): 41. Note that this last author is the worker from chapter 5 who defended Zhōu Guóxīng and went on to study at IVPP.

18. See the cover of issue no. 3 for 1994.

19. Liú Hòuyī and Liú Qiūshēng, "Huànxǐng 'Zhōngguó yuánrén': Jì Péi Wénzhōng jiàoshòu" [Awakening "Peking Man": On Professor Péi Wénzhōng], Kēxué shíyàn 1979.12, no. 12 (1979): 1.

20. Zhèng Shuònán, "Běijīngrén de fāxiànzhě: Yì Péi Wénzhōng jiàoshòu" [The Discoverer of Peking Man: Remembering Professor Péi Wénzhōng], Dà zìrán, no. 4 (1982): 8.

21. Huáng Wèiwén, "Gěi Běijīngrén mìngmíng de rén: Jì Jiānádà rénlèi xuéjiā Bù Dáshēng" [The

published materials tended to highlight the cosmopolitan and cooperative character of the Zhōukǒudiàn excavations and have often taken an
agnostic or even kind position on American culpability in the loss of the
Peking Man fossils.[22]

Opening has also been credited with, or blamed for, the increased
sexiness of Chinese cultural productions and social practices in the post-
Máo era. Following trends set by Chinese fashion magazines, popular
science magazines like the Shànghǎi Natural History Museum's publication *Nature and Man* began in the 1980s to sport photographs of beautiful women, often Western and increasingly bikini clad, on their covers.
And in 1988 Zhōu Guóxīng's new anthropology exhibit at the Běijīng
Natural History Museum provoked official censure and temporary closing, followed by enormous popular interest, in large part because of
the new section on human reproduction. One of the concerns was that
the photographs depicted naked *foreign* people. Why, asked one official critic, did Zhōu not use Chinese models? Evidence from magazines
suggests that Chinese audiences, or at least Chinese magazine editors,
are more comfortable seeing nude or explicitly sexual photographs of
visibly foreign people than of visibly Chinese people.[23] Zhōu, however,
notes that at the time there were no suitable photographs of Chinese
people, or any viable ways of obtaining such photographs.[24] While the
exhibit gained political legitimacy through its usefulness in supporting
education about birth and population control, Zhōu himself has a much
broader and deeper interest in sex as fundamental to an understanding
of the nature of humans.

Scientists and other science disseminators were thus active agents in
the incorporation of sex into popular science materials on human origins. Nonetheless, some also expressed concern over these trends. Zhōu
Guóxīng won his struggle to exhibit graphic sexual education materials, but anxiety about the increasingly sexual character of media forms,
including science dissemination, continued. A comic published in 1996

Person Who Gave Peking Man Its Name: On the Canadian Anthropologist Davidson Black], *Huàshí*,
no. 4 (1982): 6–8.

22. Jiǎ Lánpō and Huáng Wèiwén, *Zhōukǒudiàn fājué jì* [The Excavation of Zhōukǒujiàn] (Tiānjīn:
Tiānjīn kēxué jìshù chūbǎnshè, 1984); Lǐ Míngshēng and Yuè Nán, *Xúnzhǎo "Běijīngrén"* [The Search
for "Peking Man"] (Běijīng: Huáxià chūbǎnshè, 2000), especially 442. The Japanese do not get off as
easy in Lǐ and Yuè's account.

23. Julia Andrews and Kuiyi Shen, "The New Chinese Woman and Lifestyle Magazines in the late
1990s," in *Popular China: Unofficial Culture in a Globalizing Society*, ed. Perry Link, Richard P. Madsen,
and Paul G. Pickowicz (Lanham, Md.: Rowman and Littlefield, 2002), 146.

24. Zhōu Guóxīng, interview with the author, 7 January 2002. He obtained the photographs
from a museum in France.

with the title "A Wily Rabbit Has Three Burrows" (狡兔三窟, a Chinese adage) showed a rabbit labeled "obscene content" hopping from general books and magazines to video cassettes, and finally into science dissemination materials.[25] An official in the science dissemination department of the Chinese Association for Science and Technology today blames the increasingly sexual content of science dissemination materials on foreign influences, a clear indication of the continued potency of the "spiritual pollution" perspective on undesired trends in the media.[26]

If opening to foreign cultures has had an enormous impact on popular science, the inward-looking interest in regional diversity has been another post-Máo trend of great influence. The Máo-era state generally discouraged regional identities for fear they would detract from the sense of national unity so dearly bought during the wars of the early twentieth century. Significantly, when in 1951 the Culture Ministry of the central government released guidelines for local museums, it specified the function of such institutions as "furthering revolutionary patriotic education" and "making the masses accurately understand history and nature and love the ancestral country." The content was to be "intimately connected with the locale," but museum workers should also "pay attention to coordinating the national character with the local character so as to avoid emphasizing the locale and overlooking the nation."[27]

In the case of paleoanthropology, and probably other subjects, consistency in establishing the national characters of local exhibits was accomplished through practical assistance and the provision of materials from central institutions like IVPP and the national-level Science Dissemination Association to provincial and local organizations. The Institute of Vertebrate Paleontology and Paleoanthropology has remained in the post-Máo period a phenomenal influence over the dissemination of knowledge on human origins throughout the country. In the 1980s, scientists developed at Zhōukǒudiàn what they jokingly called their own "Whampoa Academy." Like the academy of the early twentieth century that trained the great military leaders of republican-era China, IVPP's program educated many museum workers who then brought what they had learned back to China's increasing number of regional and local museums.[28]

Nonetheless, researchers and science dissemination workers outside the capital have also enjoyed a growing authority to define their own

25. "Jiǎo tù sān kū" [A Wily Rabbit Has Three Burrows], *Dāngdài diànshì*, no. 5 (1996), 58.
26. Interview, August 2002.
27. Guójiā wénwù shìyè guǎnlǐ jú, ed., *Xīn Zhōngguó wénwù fǎguī xuǎnbiān* [Selected Laws on Cultural Property in New China] (Běijīng: Wénwù chūbǎnshè, 1987), 18.
28. Yóu Yùzhù, interview with the author, 22 July 2002.

agendas.[29] In the 1980s and 1990s, museums sprang up all over the country to display local human fossils, for example, Lántián Man in Shǎnxī, Liǔjiāng Man in Guǎngxī, Yuánmóu Man in Yúnnán, and Mǎbà Man in Guǎngdōng. Such museums typically emphasize a regional identity. Regional popular science magazines, such as Yúnnán's *Science Window* (科學之窗), which began publication in 1980, have similarly focused on nearby natural history and scientific activities and have addressed their readers in regional terms. Such regionalism is sometimes compatible with nationalism. For example, an article celebrating Guǎngxī's abundant human fossils and the museum erected at Liǔzhōu to display them quoted Péi Wénzhōng as saying, "China can be called the world's center for paleoanthropology; Guǎngxī is the center of the center."[30] But other times regional identity bumps up against central authority, as in Lántián, where local people resent that their human fossils are less famous than the more recent Peking Man fossils and that their local museum can display only casts of the Lántián fossils because the originals are housed in Běijīng.[31]

As with opening to foreign cultural influences, the rise of regionalism has expanded the ways Chinese people in the post-Máo period talk about humanity. A case in point is the Yuánmóu Man Exhibition Hall and the nearby excavation site in Yuánmóu, Yúnnán. Both the exhibit, created in 1989 and located in downtown Yuánmóu, and a memorial erected at the excavation site in 1995 treat visitors to a story of human evolution with a distinctively Yúnnán stamp. The exhibit and memorial connect the Yuánmóu Man fossils, reportedly aged 1.7 million years, with the identity constructed for Yúnnán as a part of the post-Máo fascination with primitiveness. While in the Máo era the primitive qualities of the minority nationalities were stages of development that, with the party's help, the nationalities were to leave behind as quickly as possible, in the post-Máo period this conceptualization has been joined by more explicit yearnings for the apparently simple and free existence enjoyed by members of "primitive cultures."[32]

The memorial is a towering pillar of rough-hewn bricks carved with figures meant to evoke primitive art (figure 16). At the exhibit, an introductory panel calls Yuánmóu itself a "museum of nature and human primitive society." Illustrations meticulously point out the continuities

29. Jí Xuépíng and Gāo Fēng, interview with the author, 15 May 2002.

30. Yuán Lìng, "Péi Wénzhōng jiàoshòu yǔ tā de 'zhōngxīn' zhī shuō de yóulái" [Professor Péi Wénzhōng and the Origin of His "Center" Statement], *Huàshí*, no. 1 (1994): 20.

31. Interviews conducted in Lántián, Shǎnxī, April 2002.

32. Ralph Litzinger, *Other Chinas: The Yao and the Politics of National Belonging* (Durham, N.C.: Duke University Press, 2000), 231.

16 Memorial, erected at the Yuánmóu Man excavation site in 1995, evoking the celebrated "primitivism" of Yúnnán's minority nationalities. Photograph by the author.

in technology from Neolithic to modern minority times. A painting depicting a Paleolithic-era woman with large hoop earrings and posed in an unconsciously seductive manner further brands Yuánmóu's remote archaeological past with the vibrant beauty and sexuality post-Máo Chinese people see in Yúnnán's ethnic minority women (figure 17).

It is easy, proper, and necessary to be alarmed at the ways such imagery stereotypes Yúnnán's people and promotes harmful practices like sex tourism.[33] However, it is also possible to see more positive ramifications.

33. Dru C. Gladney, "Representing Nationality in China: Refiguring Majority/Minority Identities," *Journal of Asian Studies* 53, no. 1 (1994): 92–123.

17 Painting at the Yuánmóu Man Exhibition Hall depicting two women and a man from the
Paleolithic era. Note especially the woman in the upper half of the picture. The large hoop
earrings and suggestive pose reinforces a stereotype that southwestern minority women are
exotic and sexually uninhibited. Photograph by the author.

18 Painting at the Yuánmóu Man Exhibition Hall depicting three pairs of Neolithic-era people
engaged in fishing and hunting, with each pair consisting of one woman and one man.
Compared with figure 17, this painting suggests a more positive spin on the same common
stereotypes about southwestern cultures and their strong, active women. Photograph by the
author.

The Yuánmóu Man Exhibition Hall portrays Neolithic and even Paleo-lithic women in highly unusual and powerful ways, a characteristic almost certainly arising from the commonly celebrated significance of women in many of Yúnnán's cultures. In a painting of the Neolithic period, three pairs of people engage in fishing and hunting activities, and each pair includes a woman and a man (figure 18). Still more strik-ing is the diorama of Paleolithic society, in which a full-breasted woman wields a spear to help bring down a hunting party's quarry. A fascination with the primitive, like opening to foreign influence, bears with it dan-gers but also new possibilities for liberating discourse.

The Strange and the Mysterious

Primitivism was itself a part of a more general surge of interest in the "strange" (怪) and the "mysterious" (謎) in the post-Máo period. Fasci-nation with primitivism was not limited to China's ethnic minorities, but also, particularly in the beginning of the period, included foreign peoples and practices. *Fossils* magazine in particular printed a great many photographs and articles on foreign "primitive" peoples and their "strange" customs, such as facial piercing, tattoos, and head-hunting. Attention to tribal peoples occasionally had a political punch familiar from socialist times—for example, in a discussion of the role of impe-rialism in the extermination of the Tasmanians and in an article on a "primitive" white-skinned tribe in Kashmir as evidence against ranking peoples according to racial characteristics.[34] More often, however, such articles blended the Marxist/Stalinist approach to historical stages with an unabashed curiosity about exotic peoples and places that reportedly appealed greatly to many readers.[35]

Readers of popular science magazines in the post-Máo era also gained much exposure to marginal subjects in foreign science, for example, UFOs, lake monsters, Bigfoot, and the *Guinness Book of World Records*. Popular science magazines, books, and films in the late 1970s and 1980s teemed with accounts of extrasensory perception and images of people with interesting deformities, such as extra limbs or thick hair all over their bodies.[36] In this context, the "hairy people" (毛人) born with unusual

34. Chén Wéi, "Bǎinián jiān xiāoshì de mínzú" [A People Exterminated in One Century], *Huàshí*, no. 1 (1985): 19–21; Xú Xiǎomíng, "Shàng chǔyú yuánshǐ shèhuì fāzhǎn jièduàn de báizhǒng rén" [White People Still in the Developmental Stage of Primitive Society], *Huàshí*, no. 1 (1988): 12–13.

35. *Huàshí* biānjí zǔ, "*Huàshí* shí nián," *Huàshí*, no. 4 (1982): 1.

36. A great many exist, some published by state-run presses, and others by semiprivately run presses. See, for example, Liú Míngyù et al., *Zhōngguó máorén* [The Chinese Hairy Person] (Liáoníng

amounts of body hair (who had appeared regularly in materials on human evolution since the republican era) became more explicit as sensational symbols of the freakish mysteries of nature, despite scientists' efforts to treat them as rational evidence of human evolution from apes.

As with sex, excessive fascination with the strange has often been blamed on foreign influence. For example, the author of a 1986 *Fossils* article blamed the "blind" following of foreign ideas for the excitement over certain "strange theories" on scientific subjects circulated in the Chinese press, including the widely trumpeted "new theory," generated by a British person, that humans had evolved not from apes but from marine mammals. He complained, "Some people think that any foreign new theory should be unthinkingly disseminated."[37] A review of early post-Máo popular science literature bears him out. Eager to overcome decades of relative isolation from foreign scientific developments, some scientists and political leaders appeared ready to embrace the most far-out foreign ideas as the most cutting edge and thus the most likely to propel China to become a leader of scientific innovation.

Other critics traced the fascination in the strange to indigenous sources. In late 1983, as the antispiritual pollution campaign got under-way, an article in the magazine *Science Dissemination Writings* criticized the resurgence of "strange writings" (志怪) in contemporary media. Trac-ing the genre back to the Six Dynasties (A.D. 220–A.D. 589), the author concluded that accounts of such phenomena as ghostly voices and con-joined twins were acceptable as science dissemination only if presented from the proper perspective. A spirit of "curiosity and delight in the strange" had long prompted people to investigate nature's mysteries and thus forward scientific discoveries. But too often, "strange writings" only "wore the clothing of science dissemination" to conceal their "antisci-entific" (違反科學) nature. Rather than providing a solidly materialist explanation of the phenomena in question, they fostered an idealist out-look by encouraging readers to believe in supernatural forces as explana-tions of anomalies.[38] This type of analysis had a distinguished history in

---

kēxué jìshù chūbǎnshè, 1982); and a widely viewed film on the same child: Běijīng kēxué jiàoyù diànyǐng zhìpiànchǎng, *Máohái* [Hairy Child] (1978). A more recent example is Xíng Wànlǐ, ed., *Rénlèi shénmì xiànxiàng quán jìlù* [Complete Records of Mysterious Human Phenomena] (Běijīng: Dàzhòng wényì chūbǎnshè, 1999).

37. Zhāng Fǎkuí, "Rénlèi qǐyuán yú hǎiyáng ma?" [Did Humans Originate in the Ocean?], *Huàshí*, no. 2 (1986): 19. A paleontologist at IVPP, Zhāng wrote another essay criticizing the maga-zine for encouraging superstition. Zhāng Fǎkuí, "Chì 'Zhōnghuá zǔxiān shì wàixīng rén' xiéshuō" [Denouncing the Heretical Theory That "Chinese Ancestors Are Extraterrestrials"], *Huàshí*, no. 2 (2001), 35–36.

38. Chén Bǐngxī, "Zhìguài yǔ kēpǔ" [Strange Writings and Science Dissemination], *Kēpǔ chuàng-zuò*, no. 4 (1983): 23–24. See also Zhēn Shuònán, "Lièqí zhìguài zhī fēng bù kě zhǎng" [Do Not

Chinese communist writings on science: in 1934 Ài Sīqí published an article in the popular science magazine *Tàibái* warning against science dissemination efforts that focused too much on shocking and strange facts and so "mythologized science" and promoted superstition.[39]

Concerned officials and intellectuals sometimes also point to the transition away from socialism as the cause of the allegedly unhealthy fascination with strange phenomena. For example, in 1998 a series of articles titled "Exploring the Mystery of Human Origins" began with an experience the author (a scientist at IVPP) had recently had in a sleeper car on the No. 13 Express from Shànghǎi to Běijīng. As the train passed rural people performing rites at the graves of their ancestors, the author overheard two fellow passengers discussing the practice. "The economy is growing, but the spiritual civilization is not growing with it. Instead, we have a resurgence of feudal superstitions," one said. His companion agreed, saying, "The spiritual realm is never a vacuum. If materialism does not occupy it, theories of spirits and ghosts will step in to fill the gap."[40]

Some critics pin the blame even more squarely on the market economy. In an interview with employees of the science dissemination department of the Chinese Association for Science and Technology, I asked what they thought of the ubiquitous term "mystery" in popular science materials. The more senior employee quickly cut off her junior colleague as he waxed poetic about the potential for mysteries to inspire scientific curiosity: she thought the term encouraged superstition and was a sensationalist tactic employed by private companies to make money.[41]

This characterization is not entirely fair. Official institutions also compete on the market and have participated heavily in sensationalist storytelling. Moreover, one of the chief proponents of scientific investigations into strange phenomena has been none other than Qián Xuésēn,

---

Encourage Hunting for the Peculiar and Writing on the Strange], *Kēpǔ chuàngzuò*, no. 2 (1983): 58–59. For a thorough account of "strange writings" in the Six Dynasties period, see Robert Ford Campany's *Strange Writing: Anomaly Accounts in Early Medieval China* (Albany, N.Y.: State University of New York Press, 1996). Tián Sòng has written on the distinct but often conflated concepts of "scientism," "antiscience," and "pseudoscience" and has argued with respect to "antiscience" that criticisms of science (for example from the environmentalist movement) can positively influence science. "Wéikēxué, fǎnkēxué, wěikēxué" [Scientism, Antiscience, and Pseudoscience], *Zìrán biànzhèng fǎ yánjiū* 16, no. 9 (2000): 14–20.

39. Liú Wéimín, *Kēxué yǔ xiàndài Zhōngguó wénxué* [Science and Modern Chinese Literature] (Héféi: Ānhuī jiàoyù chūbǎnshè, 2000), 173.

40. Xú Qìnghuá, "Tànsuǒ rénlèi qǐyuán zhī mí (yī): Gǔyuán de chūxiàn yǔ fúshè fēnhuà" [Exploring the Mystery of Human Origins (1): The Emergence and Radiation of Ancient Apes], *Huàshí*, no. 1 (1998): 2.

41. Interview, Chinese Association for Science and Technology, August 2002.

the renowned physicist who directed the research responsible for the creation of the Silkworm missile. He has advocated teaching about UFOs in classes on geoscience, and most famously he has supported systematic research into "human body science" (人體科學) focusing on the use of *qìgōng* for extrasensory perception and telekinesis.[42] When he began in the late 1970s to speak on such issues, Qián's influence was at something of a low ebb. He had participated actively in the 1976 criticisms of Dèng Xiǎopíng, and having consequently fallen somewhat in political standing, he found himself increasingly pushed to the periphery of the physics community as well.[43] Although still officially revered as one of China's most important scientists, Qián's increasingly marginal role in his profession may help explain his enthusiasm for such seemingly unorthodox scientific research. Yet he was hardly the only one: others influential in science and politics did much to help promote both the research itself and its image in the popular media.[44] In the case of official support for research on *qìgōng* in particular, the potential for national glory was certainly an important factor: if *qìgōng* could be proven to offer powers beyond those already recognized by science, this would be a tremendous feather in China's national cap. But scientific and political elites are people, too, and not immune from the excitement of investigating the strange and the mysterious. Many of those who promoted strange science probably did so for the same reasons the general public eagerly consumed it.

Still, not everyone saw the promise of the supernormal in *qìgōng*, and some considered scientific attention to the question legitimate only insofar as it helped counter dangerous trends emerging from the "*qìgōng* fever" sweeping the nation. To such people, bringing science to the study of *qìgōng* was necessary to combat this new wave of superstition. In the mid-1990s official opinion began to turn increasingly against people who claimed to have supernormal *qìgōng* powers. In 1995 two *qìgōng* masters were arrested on charges of swindling.[45] By 1999, the Chinese state had

---

42. Qián Xuésēn, "Réntiānguān, réntǐ kēxué yǔ réntǐxué" [Man in Cosmic Environment: Anthropic Principle, Somatic Science, and Somatology (English title provided with abstract)], *Dà zìrán tànsuǒ*, no. 4 (1983): 15–22.

43. Iris Chang, *Thread of the Silkworm* (New York: Basic Books, 1995), 254–257.

44. In addition to numerous articles in popular science magazines, *People's Daily* printed very positive articles on the research in the mid-1980s. See, for example, Liú Bǐngyàn and Zhèng Xītóng, "Dǎkāi shēngmìng àomì zhī mén—'shēngmìng zhìliáo yí' chuàngzàozhě de jīnglì" [Opening the Door to the Mysteries of Life—the Experience of the Creator of the "Life Information Treatment Instrument"], *Rénmín rìbào*, 17 December 1986, 3.

45. Jian Xu, "Body, Discourse, and the Cultural Politics of Contemporary Chinese Qigong," *Journal of Asian Studies* 58, no. 4 (1999): 965–966. See also Nancy N. Chen, *Breathing Spaces: Qigong, Psychiatry, and Healing in China* (New York: Columbia University Press, 2003), 145–153.

identified a very specific target for its anxieties over such superstition and its social and political consequences. The government instituted a formal ban on *qìgōng* sects, including Fǎlún Gōng, whose following had increased alarmingly.

A blend of Buddhism, Daoism, millenarian prophecies, and marginal scientific ideas culled from around the world, Fǎlún Gōng challenges the state's monopoly on truth and encourages followers to put their faith instead in the sect's leader, Lǐ Hóngzhì. Lǐ and his followers, like many other members of religious sects, consider themselves to be contributing to science and science education. They generate and promote marginal scientific theories and attack establishment science. Lǐ, for example, has criticized Darwinism for persuading people to identify with and thus cultivate their "bestial nature," and he has advanced theories of extra-terrestrial influences on human history and civilizations far older than the earliest human fossils.[46] The senior IVPP scientist Wú Xīnzhì has doubted Lǐ's claims to have found evidence for such theories in an American scientist's book.[47] While flattering to Americans, this particular criticism is unfounded: the theories did indeed originate in Michael Credon and Richard Thompson's influential *The Hidden History of the Human Race: Major Scientific Coverup Exposed*. Credon (a writer) and Thompson (a mathematician) are members of the Bhaktivedanta Institute, a branch of the International Society for Krishna Consciousness.[48]

The Chinese government's response to Fǎlún Gōng has generated criticism not only abroad but also within China itself. A communications (信息) scholar at the Shànghǎi Academy of Social Sciences who has also been an active producer of science dissemination writings on human origins, Zhū Chángchāo has conducted substantial research on the movement and has concluded that the government's "nationwide, Cultural Revolution–style mass criticism" has been both unreasonable and counterproductive. His research demonstrates that the majority of sectarians are from the "lowest levels of society" and that their participation is not politically motivated. He particularly opposes labeling as "heretical" (邪

---

46. Mínghuìwǎng, "Jìnhuà lùn duì rénxìng de pòhuài" [The Theory of Evolution's Destructiveness to Humanity], http://minghui.ca/mh/articles/2002/5/29/31006.html (viewed 21 March 2007) (an English version is available at http://www.clearwisdom.net/emh/articles/2002/6/7/22831.html); Lǐ Hóngzhì, "Zài Niǔyuē jiǎng fǎ" [Teaching the Fa in New York], http://www.falundafa.org/book/chigb/mgjf_1.htm, viewed 21 March 2007; Lǐ Hóngzhì, *Zhuàn fǎlún* [Spinning the Law Wheel] (Hong Kong: Fǎlún fó fǎ chūbǎnshè, 1997), 18–25.

47. Wú Xīnzhì, "Rénlèi lìshǐ yǒu duō cháng?" [How Long Is Human History?], *Huàshí*, no. 3 (1999): 2.

48. Michael A. Cremo and Richard L. Thompson, *The Hidden History of the Human Race: Major Scientific Coverup Exposed* (Badger, Calif.: Govardhan Hill Publishing, 1994). The book criticizes at length the excavation and interpretation of the Peking Man fossils.

教) a movement with such large numbers and such a strong foundation among "the basic masses of workers and peasants." He considers the root problem to be one of belief, and therefore the appropriate measures to be educational, and foremost "dissemination of scientific knowledge and materialist dialectics."[49]

Science dissemination actually has been a significant component of the Chinese government's attack on Fǎlún Gōng, although it has been combined with many more drastic and invasive measures that Zhū understandably opposes. When I visited the Chinese Association for Science and Technology in 2002, employees of the science dissemination department were busily producing posters and other materials designed to challenge medical and scientific theories the movement espouses. They and others involved in science dissemination note that funding has become more plentiful since the beginning of the Fǎlún Gōng crisis.[50] The Science and Technology Dissemination Law, passed 29 June 2002, was also probably motivated at least in part as a response to Fǎlún Gōng. Article 1, section 8, requires science dissemination workers to "uphold a scientific spirit and resist antiscience"; it forbids "any group or individual to use the name of science dissemination to engage in activities harmful to society and the public interest."

It is clear from both the content of the Fǎlún Gōng teachings and the government and party's response that the obvious crisis of political authority is embedded in a somewhat less obvious crisis of scientific authority related to the perceived stability of knowledge itself. In the Máo era, nature followed established laws interpreted by clear authorities, mainly Marx, Engels, Lenin, Stalin, and Máo. The challenge for scientists was to fit their new and sometimes problematic data into the frameworks they inherited. The post-Máo era has seen effective challenges to authorities both political and scientific. The state acknowledges Máo to have made mistakes, and magazines have filled their pages with accounts that contradict established Marxist theories and challenge mainstream science. The post-Máo fascination with the strange and the mysterious is thus part of a general decline in the effectiveness of authoritarianism (here understood in a very specific sense), and the

---

49. "Guānyú tiáozhěng duì Fǎlún gōng zhèngcè de jiànyì . . . (Shànghǎi) Zhū Chángchāo" [An Opinion by Zhū Chángchāo (of Shànghǎi) on Revising the Government's Policy toward Fǎlún gōng], *Běijīng zhī chūn*, no. 7 (2003), http://beijingspring.com/bj2/2003/180/2003712171012.htm (viewed 21 March 2007).

50. Léi Qǐhóng and Shí Shùnkē (of the China Research Institute for Science Popularization), interview with the author, 26 July 2002; Duàn Shūqín, interview with the author, 28 January 2002; Wáng Huìméi (of the Chinese Association for Science and Technology, Department of Science and Technology Dissemination), interview with the author, August 2002.

government's response to Fǎlún Gōng represents an attempt to explain away the mysteries that threaten to undermine not only its control over the people but also its authority to define the workings of the world.

"Labor Created Humanity" and Its Post-Máo Fate

Along with the fading influence of Marxist authority figures in science, the theme of labor quickly lost the dominant position it had held since the revolution. The 1986 movie *The Peking Man Site: Discovery and Research*, which replaced the Cultural Revolution–era *China's Ancient Humans*, reduced "Labor created humanity" to a single sentence tucked away in the middle of the film.[51] One article published in 1989 began, "In our discussions of questions about our own human origins and development, we have focused mostly on causes of human evolution such as the significance of the appearance of tools, the role of labor, etc.," before dismissing these overworn themes and turning instead to the relatively fresh question of fire in human evolution.[52] Many other articles published in popular science magazines did not mention Engels or his theory at all, even when there were obvious opportunities for them to do so.

Yet this Máo-era mantra has by no means entirely disappeared. A few books and articles, for example, Huáng Wèiwén's 1985 *Labor Created Humanity: The World-Famous Peking Man*, could almost be mistaken for earlier publications.[53] Many others still include explanations of the theory apparently as a matter of course. And some authors have continued to find it a productive paradigm for generating new ideas about human evolution. For example, a 1992 article in *Nature and Man* used Engels's theory as the background for a discussion of the new conclusion reached by some Chinese scientists that women were responsible for the majority of the small stone tools (microliths) found at Zhōukǒudiàn.[54] In 2002, the director of the China Engineering Institute, Sòng Jiàn, made use of a slightly modified version of "Labor created humanity" in an article published in *Science Times*. Writing against recent talk of an economy based on knowledge rather than material production, he produced a narrative

51. Zhōngguó rénmín dàxué, *"Běijīngrén" yízhǐ de fāxiàn yǔ yánjiū* [The Peking Man Site: Discovery and Research], 1986.

52. Cài Shùxióng, "Huǒ zài rénlèi jìnhuà zhōng de zuòyòng" [The Role of Fire in Human Evolution], *Huàshí*, no. 1 (1989): 8–9.

53. Huáng Wèiwén, *Láodòng chuàngzào le rén: Wénmíng shìjiè de "Běijīngrén"* [Labor Created Humanity: The World-Famous "Peking Man"] (Běijīng: Shūmù wénxiàn chūbǎnshè, 1985).

54. Huáng Xiànghóng, "Běijīng yuánrén yánjiū de xīn jìnzhǎn" [New Advances in Research on Peking Man], *Zìrán yǔ rén*, no. 2 (1992): 24–25.

of human history beginning with "Manufacturing created humanity" and ending with the continued critical importance of industry to modern society.[55]

A few venues visited in 2002 and 2005 still retained introductions to human evolution indistinguishable from those produced in the first three decades of the People's Republic. A prime example is the Běijīng zoo, where a large display near the great apes exhibit, created in the mid-1980s, began with a quotation from Engels: "Labor . . . is the primary basic condition for all human existence, and this to such an extent that, in a sense, we must say that labor created humanity itself." A subsequent panel provided the familiar trajectory from Darwin's refutation of creationism to Marx and Engels's critique of Darwin's idealism and Engels's theory that labor created humanity. It further added Máo's 1938 statement that humans' "conscious dynamic role" (自覺的能動性) is what separates them from other living things. It concluded, echoing Wú Rǔkāng's influential 1965 book, "If we say that Darwin liberated us from the hands of God, then our great leaders separated us from the animals."[56]

School textbooks have also continued to discuss Engels's theory, although its proper interpretation and application is now often debated. Until about 1985, the phrase "Labor created humanity" remained a very common opening for textbooks on Chinese history beginning with Peking Man. After that point, many textbooks altered the phrasing to say instead that labor had a "definite role" in human evolution. Political science textbooks on the history of social development have been more consistent in their use of the phrase "Labor created humanity." However, for a brief period after 1988, some such textbooks also adopted the more limited "definite role" wording. Zhōu Guóxīng attributes this change to the influence of his new anthropology exhibit at the Běijīng Natural History Museum, which had come under political attack for a similar toning down of the "Labor created humanity" theme. After the exhibit received official approval to open, the editors of certain political science textbooks contacted Zhōu and solicited advice on how they should change their texts.[57]

55. Sòng Jiàn, "Zhìzàoyè yǔ xiàndàihuà" [Manufacturing and Modernization], Kēxué shíbào, 23 September 2002, 1 and 3. I thank Lín Shènglóng for providing me with this article.

56. Wú Rǔkāng, Rénlèi de qǐyuán hé fāzhǎn [Human Origins and Development] (Běijīng: Kēxué pǔjí chūbǎnshè, 1965), 11.

57. Rénmín jiàoyù chūbǎnshè, ed., Shèhuì fāzhǎn shǐ (shàng běn) [History of Social Development, Vol. 1] (Běijīng: Rénmín jiàoyù chūbǎnshè, 1989 [1988]), 3; and Rénmín jiàoyù chūbǎnshè, ed., Chūjí zhōngxué kèběn (shìyòng běn): Shèhuì fāzhǎn shǐ (shàng běn) [Lower Secondary School Textbook (Trial Edition): History of Social Development (Vol. 1)] (Běijīng: Rénmín jiàoyù chūbǎnshè, 1990), 3. I thank Zhōu Guóxīng for providing these sources.

Zhōu has not been alone. The relative freedom from an authoritarian attitude toward scientific and political theories has inspired other scientists as well to reconsider Engels's theory from a more nuanced perspective. In 1982, *Fossils* published two articles that questioned the standard line. The first, by the IVPP scientist Lín Shènglóng, boldly stated, "For many years and until very recently, people often summarized Engels's basic point about human evolution as 'Labor created humanity.' Looking at it today, stating Engels's point in that way is very incomplete, inaccurate, and inconsistent with Engels's original intention." Instead, Lín suggested that the "crux [of Engels's argument] is 'in a sense.'" Engels had not baldly stated, "Labor created humanity," but had said, "In a sense, labor created humanity"—not "in every sense," but "in *a* sense." Lín then quoted other passages to demonstrate that Engels also appreciated that humans were "products of the natural world" and not just of labor. Lín disagreed, however, with a recent article published in the *Dialectics of Nature Bulletin*, which saw a logical contradiction in Engels's understanding of labor as both the creator of humanity and as humanity's distinguishing characteristic.[58]

The author of the second article was Zhū Chángchāo, the communications scholar who has since criticized the government's response to the Fǎlún Gōng crisis. Zhū agreed with Lín that previously people had wrongly "taken a quotation to represent a scientific theory." He also similarly sought to find a way to incorporate both natural selection and labor into an understanding of human evolution. He differed with Lín, however, on the question of Engels's alleged logical contradiction. The contradiction was there, he said, but this did not detract from Engels's contribution any more than Lamarck's theory of the inheritance of acquired characteristics justified a denial of the greatness of his evolutionary thought.[59]

Zhōu Guóxīng's new exhibit, and the materials he wrote to accompany it, were thus a part of a greater trend toward critical evaluation of the use of Engels's theory in the study of human origins. In a 1991 book, Zhōu tackled "Labor created humanity" head-on: "There are now two tendencies. One is to make human labor excessively sacred and mysterious, to reduce the very complex process of human origins into the single

---

58. Lín Shènglóng, "Rén běnshēn shì zìránjiè de chǎnwù: 'Láodòng chuàngzào le rén běnshēn' jǐnjǐn shì 'zài mǒuzhǒng yìyì shàng' shuō de" [Humans Are Inherently Products of Nature: "Labor Created Humanity" Only "in a Certain Sense"], *Huàshí*, no. 2 (1982): 28–30.

59. Zhū Chángchāo, "Rénlèi qǐyuán de dònglì shì yīzhǒng zōnghé dònglì" [The Driving Force of Human Origins Is a Synthetic Force], *Huàshí*, no. 4 (1982): 27–28, 32. Note that this statement cuts at Michurinism/Lysenkoism in addition to making Engels fallible. Zhū incorporated his critique into a popular book he published in 1995. See Zhū Chángchāo, *Rénlèi zhī mí* [The Human Mystery] (Shànghǎi: Shànghǎi yuǎndōng chūbǎnshè, 1995), 13–16.

factor of labor, and to overlook all other factors. The other plays down, to the point of overlooking entirely, the role of labor in the transition from ape to human." Zhōu considered both these positions to be unsatisfactory. He proposed instead that with respect to human evolution, "labor" should be understood as a key "adaptation" in the transition from ape to human.[60] For Zhōu, as for the two *Fossils* authors just discussed, one of the consequences of a simplistic understanding of Engels was that the social context of human evolution overshadowed the biological. Their revisionism was an attempt to correct that bias.[61]

Thus, while labor remained an important explanatory mechanism, its role narrowed and its interpretation broadened significantly. For Zhōu, labor was not just a social activity but also a biological adaptation.[62] For the author of a 1983 article printed in *Fossils*, labor was not just about creating functional tools; it was also responsible for the emergence of "beauty," or for the rise of an aesthetic sense among humans. It was when humans looked upon the products of their labor and felt satisfaction and joy that aesthetics first came about.[63]

Perhaps most important, labor in post-Máo China officially came to represent not just manual work, but also intellectual work. This change was supported, at least for scientists, by the crucial National Conference on Science of March 1978, where Dèng Xiǎopíng had said that "science and technology are a productive force" (科學技術是生產力), thus enabling scientists to become legitimate members of the working class.[64] Ten years later, Dèng promoted science and technology to "the primary productive force" (第一生產力).[65] This transformation has influenced the

60. Zhōu Guóxīng and Liú Lìlì, *Rén zhī yóulái* [Human Origins] (Běijīng: Zhōngguó guójì guǎngbō chūbǎnshè, 1991), 31–32. A previous edition of this book was released in 1986. Zhōu Guóxīng, *Rén zhī yóulái* [Human Origins] (Běijīng: Mínzú chūbǎnshè, 1986), 40.

61. Zhōu Guóxīng, interview with the author, 7 January 2002.

62. This bears some interesting comparison to the move in 1970s paleoanthropology internationally to look at the "human revolution" as a biological rather than social transformation. See Roger Lewin, *Bones of Contention: Controversies in the Search for Human Origins* (Chicago: University of Chicago Press, 1987), 98–100.

63. Zhū Chāng, "Měi de bùfá" [The Pace of Beauty], *Huàshí*, no. 2 (1983): 15–16, 12.

64. Dèng Xiǎopíng, *Dèng Xiǎopíng wénxuǎn, 1975–1982* [Selected Works of Dèng Xiǎopíng, 1975–1982] (Hong Kong: Rénmín chūbǎnshè, 1983), 83–85. The phrase "science and technology are a productive force" had an interesting recent history. In 1975, the Chinese Academy of Sciences, under Hú Yàobāng's direction, drafted a report emphasizing the notion that philosophy should guide science but not substitute for it. The report included a statement Máo ostensibly made in 1963 that "science and technology are a productive force." Upon reading the draft, Máo claimed that he had never said such a thing, so Dèng ordered it removed. *Hú Qiáomù zhuàn* biānxiězǔ, ed., *Dèng Xiǎopíng de èrshísì cì tánhuà* [Twenty-Four Statements by Dèng Xiǎopíng] (Běijīng: Rénmín chūbǎnshè, 2004), 96–99.

65. Dèng Xiǎopíng, *Dèng Xiǎopíng wénxuǎn, dìsānjuàn* [Selected Works of Dèng Xiǎopíng, Vol. 3] (Běijīng: Rénmín chūbǎnshè, 1993), 274–276.

portrayal of labor in materials on human evolution. For example, in the Cultural Revolution the need to emphasize the relative unimportance of intelligence compared with physical labor led the IVPP scientist Wú Rǔkāng to include in the 1976 edition of his book *Human Origins and Development* a section on how the limb bones of our ancestors evolved more quickly than their cranial capacity. The scientific conclusion that labor preceded intelligence thus supported the political attack on the theories of *a priorism* (先驗論) and innate genius (天才論). The 1980 edition did away with this passage entirely.[66] While the 1958 and 1965 editions of Fāng Zōngxī's *How Apes Became Human* defined labor simply as "conscious, goal-oriented productive activities," the 1990 version added the sentence, reminiscent of New Democracy–era politics, "In short, it is the synthesis of physical and mental labor," a definition reiterated six pages later.[67] While Fāng did not go so far as to add scientific, artistic, or otherwise highly intellectual examples of work to his standard examples of agriculture and industry, such a shift can be seen in another post-Máo popular book on human origins with Máo-era precedents. The 1977 edition of the picture book produced by the Shànghǎi Natural History Museum to accompany its exhibit on human evolution used photographs of heavy industrial and military work to illustrate the caption "Laboring Hands, Glorious Hands" (figure 19).[68] In the 1980 edition, the entire page was replaced: the slogan became "Ingenious Hands," and the photographs depicted a piano player and calligrapher in addition to people engaged in light industry (figure 20).[69] In a drastic reversal of Engels's "Labor created humanity," the editors of *Knowledge Is Power* went so far in 1985 as to publish an article titled "Brains: The Creators of Human Civilization."[70]

Engels and his theory that labor created humanity did not disappear in the post-Máo era. The continued relevance of the paradigm and willingness of people to keep discussing it despite its overuse for thirty

66. Wú Rǔkāng, *Rénlèi de qǐyuán hé fāzhǎn* [Human Origins and Development] (Běijīng: Kēxué chūbǎnshè, 1976), 71–72. Compare with pp. 71–86 of the 1980 edition.

67. Note that for the 1958 and 1965 editions, Fāng used a pen name. Fāng Shàoqīng, *Gǔ yuán zěnyàng biànchéng rén* [How Ancient Apes Became Human] (Běijīng: Zhōngguó qīngnián chūbǎnshè, 1958), 15; Fāng Shàoqīng, *Gǔyuán zěnyàng biànchéng rén* [How Ancient Apes Became Human] (Běijīng: Zhōngguó qīngnián chūbǎnshè, 1965), 23; Fāng Zōngxī, *Gǔyuán zěnyàng biànchéng rén* [How Ancient Apes Became Human] (Běijīng: Zhōngguó qīngnián chūbǎnshè, 1990), 29, 35.

68. Shànghǎi zìrán bówùguǎn, *Láodòng chuàngzào rén* [Labor Created Humanity] (Shànghǎi: Shànghǎi rénmín chūbǎnshè, 1977), 43.

69. Shànghǎi zìrán bówùguǎn, *Rénlèi de qǐyuán* [Human Origins] (Shànghǎi: Shànghai kēxué jìshù chūbǎnshè, 1980), 53.

70. Niú Jiànzhāo, "Nǎo: Rénlèi wénmíng de chuàngzào zhě" [Brains: The Creators of Human Civilization], *Zhīshí jiùshì lìliàng*, no. 7 (1985): 26–27.

19 Photographs of military and industrial activities illustrating the caption "Laboring Hands, Glorious Hands." From Shànghǎi zìrán bówùguàn, ed., *Láodòng chuàngzào rén* [Labor Created Humanity] (Shànghǎi: Shànghǎi rénmín chūbǎnshè, 1977), 43.

years testifies to its elegance and significance as a way of understanding humanity. Indeed, officials and intellectuals in the post-Máo era found the theory a useful reference point when they revived a discussion of "human nature" and debated the significance of Marxist humanism. For example, an article entitled "Human Nature Is Human's Social Nature" began: "For a long time, the question of human nature has been forbidden territory. By smashing the Gang of Four, we have broken into this territory and regained the opportunity to explore the question of human nature." In pursuing this exploration, the author wove together Engels's maxim and a story recently circulated by Zhōu Guóxīng and other authors interested in scientific "mysteries." She wrote, "Labor created humanity, and at the same time labor created human society. . . . If a human is removed from social intercourse, he will not develop the full characteristics of humans. The discovery of the wolf children is very good evidence of this." She then proceeded to recount the case of two Indian girls who had been discovered in the wild. Raised by wolves, they did not behave as human children normally do, so demonstrating that society is

199

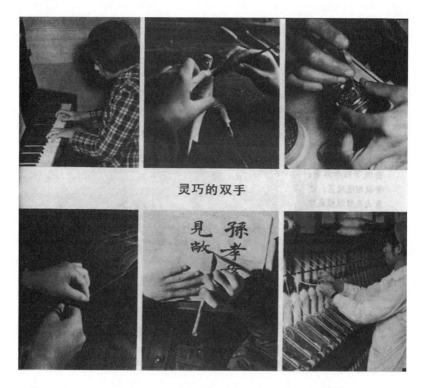

20 "Ingenious Hands" responsible for music, art, and light industry in place of the "Laboring Hands" of just a few years earlier (see figure 19). From Shànghǎi zìrán bówùguǎn, ed., *Rénlèi de qǐyuán* [Human Origins] (Shànghǎi: Shànghǎi kēxué jìshù chūbǎnshè, 1980), 53.

the critical factor in producing human nature.[71] "Labor created humanity" thus meshed easily with post-Máo popular currents. Moreover, the human identity it had promoted throughout the Máo era formed a basis for thousands of post-Máo articles on human nature, which in turn paved the way for continuing, critical Chinese debates over "human rights" and, more recently, intensive discussions under Hú Jǐntāo of the significance of "human-centered" (以人爲本) development."[72]

71. Xú Méifēn, "Rénxìng jiù shì rén de shèhuìxìng" [Human Nature Is Humans' Social Nature], *Shànghǎi shīfàn xuéyuàn xuébào*, no. 2 (1980), 64. For other examples of post-Máo discussions of humanism with respect to "Labor created humanity," see note 15 of the introduction. On wolf children, see Zhōu Guóxīng, *Lánghái, xuěrén, huó de huàshí* [Wolf Children, Yeti, and Living Fossils] (Tiānjīn: Tiānjīn rénmín chūbǎnshè, 1979).

72. A search through the China Academic Journals database (www.cnki.net) for 1994 to 2007 alone reveals 2,582 articles with the keywords "human nature" (人的本質); 129 of those contain the phrase "Labor created humanity" (勞動創造了人). The phrase was probably much more common in articles published in the 1980s, but the database is very limited for those years. On "labor

If Engels and his "Labor created humanity" remain significant, what has largely disappeared is—to borrow a phrase from Máo—a "slavish mentality" with respect to this Marxist authority and his celebrated text. Writers now have far more freedom to reinterpret, challenge, or even ignore Engels. With their heightened status in the post-Máo era, it is no wonder that scientists have shifted the discourse to emphasize the role of mental labor in the process of human evolution and have used Engels's theory to promote the exploration of humanism that had been off limits for so many years.

Mass Science and Its Post-Máo Fate

What, then, of the Máo-era insistence on the authority of the masses in science? How has that fared in the post-Máo period? Despite much abstract rhetoric during the Máo era on "following the masses" in science there was little serious elaboration of these principles in the practice of paleoanthropology. The post-Máo era has seen a decline in, although not a disappearance of, the rhetoric of mass science, and it has witnessed the elimination of some of the more radical and often highly unproductive attempts to achieve it. But participation itself has not declined. Rather, some of the previous types of participation are still active, and other kinds are now possible. Although lacking the strong class politics of previous decades, some of these forms of participation still have the potential to challenge or at least complicate the authority scientists wield.

A careful eye can discern some legacies of the Máo-era emphasis on the roles of labor and laborers in science. In the main hall of the paleontology museum at IVPP, constructed in 1996, a large glass window opens onto an enormous room where technicians assemble dinosaur fossils or other prehistoric animals. This focus on the manual and practical aspect of scientific work would have been fitting in Máo-era museum exhibits. Government officials sometimes still talk about the need to cultivate "mass science," and some senior scientists like IVPP's Wèi Qí continue to find value in "relying on the masses."[73] When Wèi, one of the scientists

---

created humanity" and human rights, see Sū Wěi, "Cóng yuán de běnzhì kàn rénquán de duōchóng hányì" [A Look at the Multiple Meanings of Human Rights from the Perspective of Human Nature], *Tànsuǒ*, no. 2 (1992): 57–60 and 64. On the connection to "human-centered" development, see Zhōu Hépíng, Gù Yíhuá, and Zhōu Huì, "Láodòng qúnzhòng shì chuàngzào lìshǐ de yuánshǐ dònglì: Jiān lùn kēxué fāzhǎn guān 'yǐ rén wéi běn' de zhéxué jīchǔ" [The Working Masses Are the Primitive Forces to Create History: On the Philosophical Basis of the "Human-Centered" Scientific Development Concept], *Xiāngnán xuéyuàn xuébào* 26, no. 4 (2005): 1–4.

73. Wèi Qí also encourages scientists to pay attention to the local economies of the areas where they excavate. More than any other scientist I have met, he has cultivated profound, long-lasting

who worked in Níhéwān in the 1970s, was asked to lead an investigation in the Three Gorges in 1994 as part of a last-chance archaeological effort before the dam flooded the area, he insisted on bringing along Wáng Wénquán and several of the other peasants who had actively participated in the 1970s excavations because they far surpassed college students in experience and diligence. Yet things are not just what they were. The museum spotlights the role of technicians, but scientists are not made to conduct their work amid the technicians' dust as in the early Cultural Revolution. And there is no danger that Wèi Qí will be required to "stand aside" and allow his assistants to take charge of the excavations.

Other kinds of participation in the post-Máo era also strongly resemble Máo-era forms. When Chinese scientists explore a new paleontological or archaeological site today, they continue to inquire of local people where dragon bones can be found, and they sometimes also visit apothecaries that sell this item of the Chinese pharmacopoeia. Yet the dragon bone trade has also been responsible for the destruction of many fossils, and the concept of dragon bones is often taken as an indication of peasants' lack of scientific knowledge. The case of a young peasant's 1986 discovery of the Yuánmóu ancient ape fossils is both a testament to the success of science dissemination and also an indication of what peasants who play by the rules have to gain from their participation in science.

Scientists participating in the Cultural Revolution–era excavations of Yuánmóu Man in Yúnnán made special efforts to disseminate knowledge about fossils and human evolution to local people. These efforts probably had a lasting influence on local education and science dissemination practices. In 1986, a thirteen-year-old Yí-nationality girl named Lǐ Zìxiù was cutting grass in the hills while on summer break.[74] When Lǐ came across old bonelike objects, she did not identify them as dragon bones, but rather guessed they were the fossils she had learned about at school. She told her mother, and together they filled a gunny sack with the fossils and sent them to the Yúnnán Provincial Geological Research Institute. Researchers there surmised that one of the fossils was a human tooth; it was later reclassified as an anthropoid ape like

---

personal relationships with rural people. See his "Zǒuxiàng shèhuì, zǒuxiàng qúnzhòng: Níhéwān yánjiū yīng xiàng zhīshí jīngjì zhuǎnhuà" [Go to Society, Go to the Masses: Níhéwān Research Should Turn toward the Knowledge Economy], *Huàshí*, no. 1 (2001): 33–35.

74. Written accounts of the discovery sometimes say that she had attended primary school but that her family did not have the money to provide for further education. Wú Xīnzhì, "Yúnnán Yuánmóu—yánjiū rényuán fēnyě de wòtǔ" [Yuánmóu, Yúnnán: Fertile Soil for the Study of the Dividing Line between Human and Ape], *Huàshí*, no. 1 (1988): 1. My data come partly from such sources and partly from interviews I conducted with Lǐ and others in Yuánmóu.

that of *Lufengpithecus* discovered in nearby Lùfēng in the 1970s. Scientists quickly traveled to Lǐ's village, and she showed them where she had found the fossils. Local people with time to spare participated in the excavations, although the eight yuán per day (plus an extra yuán for each fossil found) was not enough to persuade many local people to forego their agricultural duties. The event had an enormous initial impact on Lǐ and the village, as scientists, reporters, and busloads of tourists came to study, interview, and admire the site and its young discoverer. In the end, however, with the exception of the occasional visitor (including yours truly) life returned to much the way it was. Lǐ had been led to believe that since she had discovered the first fossil, she would be assigned relevant work after graduating from high school. But when she graduated in 1992, there was nothing for her to do, and she returned to her village to become a farmer again.[75]

Lǐ Zìxiù's accomplishment was undoubtedly what the author of an article published in 1981 in the Yúnnán science dissemination magazine *Science Window* had in mind when he called upon "the thousands upon thousands of 'fossil fans' [化石迷] to keep their eyes peeled in the hometown of the ancient ape and the cradle of humanity, Yúnnán, and to stretch their talents to the utmost for the glory of their country."[76] A young, female peasant who knew enough about fossils to recognize them when she saw them, Lǐ is a wonderful example of lay participation, but she has never considered herself a "fossil fan." For the most part, fans or hobbyists have been members of the better-educated classes with white-collar jobs. While hobbyists were certainly encouraged during the late Cultural Revolution by the publication of *Fossils* magazine, the widespread and "fanatical" pursuit of this rarified, intellectual, and (as a form of collecting) even "bourgeois capitalist" avocation has been very much a post-Máo phenomenon. Collecting all kinds of things—from coins to phone cards to Cultural Revolution memorabilia to fossils—has become a recognized and highly popular avocation in the comsumer-oriented post-Máo era.

A survey of fifty-four readers of *Fossils* magazine provides a general sense of the demographics and interests of people who consider themselves "fossil fans." Respondents were overwhelmingly male, with 85 percent identifying themselves as male and 7 percent identifying

75. Yáng Quányì, one of the farmers who in 1974 discovered the terra-cotta warriors of the first emperor in Xī'ān, has received more continued attention. He was signing books at the site in 2002. I thank Suzanne Cahill for loaning me her signed book.
76. Zhāng Xīngyǒng, "Lùfēng gǔyuán fājué sǎnjì" [Notes on the Excavation of *Lufengpithecus*], *Kēxué zhī chuāng*, no. 3 (1981): 2–4.

themselves as female. The average age was fifty-one, and three quarters were evenly spread between forty and seventy years old. They wrote from twenty-three provinces and metropolitan districts, representing a very widespread audience for the highly specialized magazine. Twelve respondents indicated that they were employed in education, five as workers, four as engineers, three as officials, three in medicine, and three as museum workers. Eighty percent of the respondents reported having engaged in fossil-hunting activities. Many had communicated in the past with scientists or scientific institutions, suggesting a confidence in their own social and intellectual status in addition to a degree of openness in Chinese scientific institutions.

Most of the respondents pursued the hobby on their own.[77] Those who did participate in social activities related to fossils tended to have informal networks of friends, colleagues, and family who shared their interests. Only a few had developed relationships with fellow hobbyists through ties to more formal science organizations. One wrote that *Fossils* magazine itself had helped introduce him to someone who became a friend—and this was something that a good number of the respondents clearly hoped would happen with me.[78] A college student reported participating in a fossil hunting trip organized by her campus Astronomy Association. And two respondents referred to a community of collectors—one to friends in "the collecting world" and another to friends who "researched the collection of fossils."

Scientists' encouragement of amateur fossil collectors (*Fossils* explicitly promotes such activities) and their generally permissive attitude toward dragon bones is somewhat problematic. Such activities became illegal in 1950, and under the 1982 Cultural Relics Protection Law (中華人民共和國文物保護法), fossil collection is at best a gray area. Both versions of the law explicitly protect vertebrate fossils "of scientific value." Under the 1950 law, local governments had the responsibility to protect "precious fossils" and to prohibit people from collecting them for their own purposes (嚴禁任意採捕).[79] The 1982 law recognizes the existence of private collectors and seeks to regulate their activities. Most important,

77. Fourteen specified that they did not have friends or relatives interested in fossils; twenty-two did not answer the question, suggesting they most likely did not.

78. This may help explain why so few of the respondents reported having friends who shared their interests. I may have received a disproportionate number of responses from people hoping to connect with someone they thought was a fellow fossil hobbyist. The editor of *Fossils* changed my advertisement to create a more friendly tone. He identified me as an American looking to make friends with Chinese people interested in fossils.

79. Article 3 of the 1950 law. Guójiā wénwù, ed., *Xīn Zhōngguó wénwù*, 4. Note that a 1930 law (discussed in chapter 1) served as an important precedent for the 1950 law.

private collectors are strictly forbidden to sell fossils for profit or to sell to foreigners under any circumstances. They are permitted to sell objects they collect only to appropriate governmental organizations.[80]

The vagueness of the law in protecting only fossils "of scientific value," combined with the allowance in the current version for private collectors who choose not to sell their fossils, may contribute to scientists' continued direct encouragement of amateur fossil collecting and indirect encouragement of the dragon bone trade. However, laypeople cannot be expected to determine the value of fossils for themselves. Moreover, scientists are well aware that fossil collectors in the entrepreneurial economy are very often seeking to profit from their activities. Thus, scientists probably ignore the law because they depend on local people's knowledge of fossils or dragon bones to guide their research, because they recognize that they are powerless to prevent trade in fossils and dragon bones, and also perhaps because they value science dissemination and see fossil hobbyism as a good means toward that end.

Nonetheless, the consequences of such activities are growing increasingly severe, as paleontologists who never had very firm control over the raw materials of their science have found themselves with still less control in the post-Máo era. The increase in entrepreneurial activity, combined with a growing demand for fossils by fossil hobbyists as well as by scientists, has even given rise to a black market for fossils. Scientists have been compelled to buy fossils on the black market, and on occasion have even been fooled by fakes produced by increasingly sophisticated black-market suppliers.[81] At the same time, foreign scientific institutions have purchased fossils in China only to discover the transactions were illegal and the Chinese state continues to claim them as national property.[82] A recent scandal linked these two phenomena: a fossil smuggled out of China and sold to a museum in Utah as *Archaeopteryx*—a missing link between dinosaurs and birds—was discovered to be a forged composite.[83] Scientists at IVPP further report that the letters they receive from people who have found fossils display a decreasing interest in science and nation and an increasing desire to know how much their finds are worth on the market. The rise in fossil hobbyism is thus strongly

80. Article 5 (sections 24–26) addresses private collectors. Guójiā wénwù, ed., *Xīn Zhōngguó wénwù*, 217.

81. Interview with a Chinese scientist, 2002.

82. Anne Carlisle Schmidt, "The Confuciusornis Sanctus: An Examination of Chinese Cultural Property Law and Policy in Action," *Boston College International and Comparative Law Review* 23, no. 2 (2002): 185–228.

83. Rex Dalton, "Fake Bird Fossil Highlights the Problem of Illegal Trading," *Nature* 404, no. 696 (2000).

connected with market economics, which is now in competition with science for social value and authority.

Less troubling than the fossil black market but also challenging to scientists' authority are the fans that Tián Sòng calls "one-person institutes." For a prime example of such a person, we can turn once again to Yúnnán, this time to the small city of Lìjiāng in the northwest corner of the province. Lǐ Xùwén was seventy-five years old in 2002, a proud member of the Nàxī minority nationality, and a retired tax official. He became interested in human evolution and human history as a child, and he began collecting fossils in the early 1970s, a hobby made possible by his itinerant duties as a tax official. He has strong convictions about the significance of the fossils in his area and about the ways they have been excavated, and he is not afraid to express his ideas in letters to scientists, in essays, and in a book he was writing when I interviewed him. Specifically, he disagrees with the contention of certain "experts" and "authorities," as he smirkingly refers to them, that many of the fossil sites were not human dwellings. Moreover, Lǐ believes the human fossils to be the ancestors of the Nàxī people, providing evidence that the Nàxī are indigenous to Yúnnán and not migrants from the north as claimed by some noted anthropologists.

Lǐ's strong opinions and his refusal to kowtow to the "experts" have earned him the respect of his local community and have not prevented him from forming close relationships with some scientific authorities. Employees of the local cultural center visit him regularly, and at least one scientist in the provincial capital of Kūnmíng visits when he is in town and keeps up a friendly correspondence. His social status, strong Nàxī identity, and passion for local fossils have made him a celebrity in Lìjiāng. When I visited his home, we were soon joined by a television crew interested in filming a young foreigner admiring his fossil collection and taking notes at "Professor Lǐ's" knee.

Another kind of authority in post-Máo science has emerged in public opinion as created in the press, which some might say represents a "public sphere." A case in point is the protection of the Peking Man site at Zhōukǒudiàn. Nearby industrial activities, beginning with limestone and coal mining and later including cement manufacturing, have threatened the site ever since its discovery. One scientist at IVPP notes that each time the government takes steps to protect the site, the impetus has come from critical articles in the newspapers.[84] In 1983, *People's Daily*

---

84. The interviewee specifically identified the *Běijīng Evening News* (北京晚報) as the most important influence.

printed a letter from "some employees" at the Peking Man site complaining about pollution from a nearby cement factory and calling upon the leadership to take appropriate action.[85] In 1984, the government ordered the plant and other factories and mines to be shut down by the end of 1985, and in 1986 *People's Daily* triumphantly reported that the task had been accomplished.[86] Yet these measures were not enough to stop the damaging effects of acid rain and other environmental hazards. Pointing out its designation as a UNESCO world heritage site in 1987, *People's Daily* reporters visited Zhōukǒudiàn in 2000 and called for greater efforts to return it to its green past.[87] The following year, the newspaper printed another story on the subject, this time noting that IVPP was diligently attempting to raise funding for protection and had joined with a social organization (社會團體, the most common term for the current Chinese version of nongovernmental organizations, or NGOs) to form a "Committee for the Protection of the Zhōukǒudiàn Site."[88] The article concluded that the Běijīng Department for the Management of Cultural Relics should pay more attention to the issue.[89]

The movement to protect Zhōukǒudiàn can be seen as part of a growing environmentalism that questions humans' dominance over nature and science's omnipotence as a force for good. This trend is evidenced as well in post-Máo writings on human evolution. Fāng Zōngxī wrote in 1981 that it was "incomplete" to say "humans are the transformers of nature," since "humans are also the destroyers of nature."[90] The same year, an article in *Knowledge Is Power* titled "Forests: The Cradle of Humanity" called upon modern people to protect the vanishing places that once sheltered their tree-dwelling ancestors.[91] Naming forests the "cradle of humanity" became a pattern, as evidenced by an article published in

85. Zhōukǒudiàn Běijīng yuánrén yízhǐ guǎnlǐ chù bùfēn gōngzuò rényuán, "Běijīng yuánrén yízhǐ páng bù yīng jiàn shuǐníchǎng" [Cement Factories Should Not Be Built next to the Peking Man Site], *Rénmín rìbào*, 8 January 1983, 5.

86. Wǔ Péizhēn, "Wēihài Zhōukǒudiàn yuánrén yízhǐ huánjìng de yīxiē chǎngkuàng ànqī tíngchǎn" [Factories and Mines Harmful to the Zhōukǒudiàn *Pithecanthropus* Site Environment Shut Down according to Schedule], *Rénmín rìbào*, 2 January 1986, 3.

87. Qián Jiāng and Wú Kūnshèng, "Zěnyàng bǎohù 'Běijīng rén' yízhǐ" [How to Protect the "Peking Man" Site], *Rénmín rìbào*, 17 October 2000, 10.

88. Especially since 1989, Chinese "social organizations" operate under considerably greater restrictions and government oversight than do Western NGOs. See Tony Saich, "Negotiating the State: The Development of Social Organizations in China," *China Quarterly* 161 (2000): 124–141.

89. Wú Kūnshèng and Shī Fāng, "Zhěngjiù 'Běijīng rén' de jiāyuán" [Saving 'Peking Man's' Homeland], *Rénmín rìbào*, 8 May 2001, 6.

90. Fāng Zōngxī, "Rén shì dòngwù, yòu shì rén" [Humans Are Animals, and Humans Are Also Humans], *Dà zìrán*, no. 6 (1981): 17.

91. Yán Xīmíng, "Sēnlín: Rénlèi de yáolán" [Forests: The Cradle of Humanity], *Zhīshí jiùshì lìliàng*, no. 7 (1981): 1.

1996 by the same magazine.[92] An article by Zhū Chángchāo published in 1984 carried a similar message but with a more urgent tone. In his characteristic, politically provocative manner, Zhū called forests "the wet-nurse of civilization" and linked the fall of many great civilizations to a failure to protect forests from invasive human practices.[93]

Many IVPP employees have been anxious to find ways to preserve the environment at Zhōukǒudiàn and are appreciative of the role of public opinion in pushing the issue. Yet media attention to environmental threats provided fodder for Běijīng city officials excited to take the site away from the institute and place it under city management. On 16 August 2002, a plan to transfer to the municipal government all responsibility for the museum, for tourism, and for protection of the site was approved. In an interview, one key figure at IVPP noted that the plan has a logic to it, since the city government has the resources to carry out these tasks. Moreover, he says, the current president of the Chinese Academy of Sciences, of which IVPP is a member institution, considers science dissemination work outside the province of the Academy. The same member of IVPP also suggested that the Běijīng government has a legitimate interest in the site's status as the origins of Běijīng history. On the other hand, IVPP employees cannot help but be saddened by the handover, he added. Zhōukǒudiàn is where IVPP has its "roots"; it is IVPP's "old home."[94]

No one in China today suggests that scientific authority should be vested in the "broad masses of workers, peasants, and soldiers." And the extent to which the physicist and dissident Fāng Lìzhī was able to speak on the independence of science and the need for further reforms before being effectively silenced further testifies to the way that scientists have reclaimed much authority in the post-Máo period. Nonetheless, new and old forms of popular participation in science continue to place scientists in complex relationships with laypeople. The dragon bone trade has been joined by a black market in fossils to reduce even further scientists' control over research materials. Because of market sensationalism or native place identity, theories circulated by "one-person institutes" and other lay experts sometimes attract more attention than those put forward by scientists. The rise of the media as a voice for public opinion at times benefits scientists but also adds another external form of pres-

92. Zhū Chūnquán, "Sēnlín, rénlèi, wèilái" [Forests, Humanity, Future], *Zhīshí jiùshì lìliàng*, no. 12 (1996): 36–37.

93. Zhū Chángchāo, "Sēnlín: wénmíng de nǎiniáng" [Forests: Wet Nurse of Civilization], *Huàshí*, no. 3 (1984): 10–12.

94. Interview with the author, 2002.

sure. "Mass science" may have faded, but scientists in post-Máo China cannot escape the masses themselves.

Conclusion

The decades since the death of Máo and the end of the Cultural Revolution are well spoken of as "springtime for science," and the analogy is easily extendable to popular science. Thawing political restrictions and energizing market reforms have encouraged a "blooming and contending" far surpassing the original Hundred Flowers Movement and its revivals. Yet, as in Máo's China, officials in the post-Máo era have found the proliferation of new voices and new ideas difficult to manage. And once again, many scientists and other intellectuals are firmly on their side. The introduction of market forces, collapse of barriers to foreign influences, encouragement of regional diversity, acknowledgment of "undefined phenomena," and challenges to scientific authoritarianism have together created the right conditions for the spread of ideas that many political and intellectual elites consider superstitious and downright dangerous. But neither do these elites advocate a return to the restrictions of earlier decades. Rather, they recognize the freedom to rethink or even ignore Engels's theory that labor created humanity and to engage in ideas generated in Western countries as a great boon to paleoanthropology. And such freedoms have greatly expanded the range of expressed meanings about humanity—including greater emphases on biological aspects of the human and on brains and intelligence. At the same time, the anti-Darwinian teachings of Fǎlún Gōng, the black market in fossils, and other troubling phenomena have become the new "poisonous weeds" that scientists and government officials believe need to be eradicated for the "garden of science" to become beautiful, productive, and ordered.

SEVEN

"From Legend to Science," and Back Again? Bigfoot, Science, and the People in Post-Máo China

"*Yěrén* Fever"

One of the central concerns of any good popular book, article, movie, or museum exhibit on human origins is to paint a vivid picture of the lives of early humans—in short, to bring the bones to life. For a few people, however, such attempts have hardly been necessary. In 1956, Professors Máo Guāngnián and Wáng Zélín attended the opening ceremonies of the Běijīng Natural History Museum. They were standing in front of a reconstruction of Peking Man when Máo heard Wáng say, "This *yuánrén* [猿人, *Pithecanthropus*, or ape-human], I've seen it before!" Surprised, Máo asked, "Are your nerves all right, or do you have a little bit of a fever? Peking Man is a human ancestor from 500,000 years ago. How could you have seen it with your own eyes?!" Wáng proceeded to explain that in 1940, while on an expedition for the Yellow River Water Control Committee, he and others in his group had come across a half-ape, half-human creature who had been shot dead by local people. Wáng's story was the catalyst for Máo Guāngnián's long and determined investigation of such *yěrén* (野人, "wild people"). Like others after him, Máo was convinced that *yěrén* were living primates, distant relatives of modern humans, and thus should be investigated scientifically.[1] It would be twenty years before knowledge of *yěrén* spread beyond a handful of rural communities and intellectual elites, but when it did, it lit a fire that another twenty years could not quench.

Chinese people have hardly been alone in their wild enthusiasm for creatures that dance on the boundary between the human and animal worlds. The yeti of the Himalayas, yowie of Australia, alma of Mongolia, Biaban-guli of the Caucasus, Mono Grande of the Andes, and sasquatch, Bigfeet, and skunk apes of North America have similarly emerged from legend to populate new forms of folklore and generate heated controversy.[2] Nor are stories about half-ape, half-human monsters a recent phenomenon in China. Accounts of such creatures are easily found in Chinese natural history works like Lǐ Shízhēn's *Systematic Materia Medica* (本草綱目, 1578), the encyclopedia *Synthesis of Books and Illustrations Past and Present* (古今圖書集成, published in 1726–1728), and local gazetteers. Similar monsters had been resurrected periodically in the pages of science dissemination texts of the Máo era. "Hairy people"—humans born with the atavistic trait of thick body hair—had regularly appeared in books and articles on human evolution since the republican era, and so served as a kind of precedent for using such anomalies to investigate scientifically the boundary between human and animal. The 1950s and 1960s saw some small-scale research efforts on *yěrén* by Chinese scientists from the Institute of Vertebrate Paleontology and Paleoanthropology. At the same time, Máo Guāngnián was conducting investigations of his own and theorizing about the possible relationships of *yěrén* to yeti, fossil apes and hominids, and modern humans themselves. Unlike the IVPP scientists, Máo was optimistic that *yěrén* represented an advanced primate unknown to science.

But it was only in the closing years of the Cultural Revolution that widespread interest in *yěrén* began to take off, generating a *"yěrén* fever" (野人熱) in the early post-Máo period that still burns today.[3] Several factors help explain its emergence in this historical period. First, as the

---

1. The source clearly specifies 1956, although the museum did not open until 1959. Liú Mínzhuàng, *Jiēkāi 'yěrén' zhī mí* [Solving the *"Yěrén"* Mystery] (Nánchāng: Jiāngxī rénmín chūbǎnshè, 1988), 51–52. Wáng Zélín was a biologist who graduated from the biology department at Dōngběi University. Máo Guāngnián was a professor at Xúzhōu Professorial Advanced Training Institute (徐州教師進修學院). It is unclear what type of education he had.

2. One of the broadest English-language treatments of these creatures can be found in Myra L. Shackley, *Wildmen: Yeti, Sasquatch and the Neanderthal Enigma* (London: Thames and Hudson, 1983). For "el Mono Grande," see Pino Turolla, *Beyond the Andes: My Search for the Origins of Pre-Inca Civilization* (New York: Harper and Row, 1980).

3. Note that in the previous chapter we encountered a *"qìgōng* fever": the early post-Máo era was a time feverish with fevers of all kinds.

intensity of political campaigns died down, people became less afraid of the consequences of associating themselves with what many identified as "superstition." Second, among the newly available Western writings scientists leapt to read and translate were materials on yeti and Bigfoot, which enjoyed a significant following in those days. Increased publishing freedoms coupled with growing commercialization further led to a proliferation of popular science and science fiction, to which the subject of *yěrén* was well suited. At the same time, a broad primitivist movement in literature and art gave a new significance to these creatures on the border between human and animal.[4]

The person credited with sparking *yěrén* fever was a historian by training, a man named Lǐ Jiàn who soon became known affectionately as "Minister of *Yěrén*" (野人部長).[5] Lǐ was the vice-secretary of a prefectural propaganda department in the mountainous area of Shénnóngjià in Húběi Province. In 1974, he began collecting reports of *yěrén* sightings from local villagers, some recent and some dating back as far as 1945. In 1976, Lǐ's findings attracted the interest of an East China Normal University professor of evolutionary theory, Liú Mínzhuàng—who was to become China's most prolific writer on *yěrén*, dubbed "Professor of *Yěrén*"—and several scientists from IVPP.[6] Teams of scientists and technicians, local officials, and rural people began tracking *yěrén* in earnest and collected dozens more reports, footprints, and even a few hair samples. The first team included the IVPP sculptor (and former Peking Man site guide) Zhào Zhōngyì so that, if they found a *yěrén*, he could sculpt a reproduction.[7] In addition to consulting Soviet and American materials on yeti and Bigfoot, respectively, researchers turned to Chinese historical sources and found consistencies between the accounts found therein and the stories currently in circulation. The *yěrén* of yore and of late had hair

4. On post-Máo "fifth-generation" filmmakers' portrayal of humans as animals and their use of the animal's point of view, see Claire Huot, *China's New Cultural Scene: A Handbook of Changes* (Durham, N.C.: Duke University Press, 2000), 91–125.

5. Lǐ's former positions included vice-director of the Institute of History at the Húběi Academy of Social Sciences. See front flap of Lǐ Jiàn, *Yěrén zhī mí* [The *Yěrén* Mystery] (Wǔhàn: Zhōngguó dìzhì dàxué chūbǎnshè, 1990). Also see Liú Mínzhuàng, *Jiěkāi 'yěrén' zhī mí*, 78; Zhāng Xiǎosōng and Bǎi Hóng, "'Yěrén' shì bàn shuì bàn xǐng de mèng ma?" [Are "Yěrén" a Half-Sleeping, Half-Waking Dream?], *Zhōngguó qīngnián bào*, 23 October 1980, 4.

6. According to the front flap of his *Jiěkāi 'yěrén' zhī mí*, Liú Mínzhuàng was born in 1933 and graduated in 1956 from the biology department of East China Normal University, where he subsequently taught courses in evolutionary theory, natural dialectics, and heredity, among other subjects. He published an article on the need for a middle ground between Morganism and Lysenkoism in 1975: "Tán shēngwù jìnhuà de nèiyīn hé wàiyīn" [On Internal and External Causes in Biological Evolution], *Zìrán biànzhèng fǎ zázhì*, no. 3 (1975): 63–85. Also see Zhāng Xiǎosōng and Bǎi Hóng, "'Yěrén' shì bàn shuì," 4.

7. Zhào Zhōngyì, interview with the author, 5 November 2002.

all over their bodies, often laughed wildly when encountering humans, and occasionally abducted villagers to mate with them.

By 1977 scientists involved in the project were publishing articles in such popular science magazines as *Scientific Experiment* and the IVPP publication *Fossils*. General-interest magazines and newspapers also carried stories about the investigations, and in the 1980s book-length treatments of the *yěrén* "mystery" began appearing. Aside from local eyewitnesses and guides, scientists, technicians, and intellectuals from many fields have taken part in *yěrén* research, and scientists as well as government officials and professional writers have produced materials for popular consumption. As might be expected, *yěrén* fever spread quickly beyond the hard world of scientific facts and investigative reporting into the realm of popular and elite literature, where it became a potent way of exploring the meaning of humanity. Most notable are Nobel laureate Gāo Xíngjiàn's 1985 play *Yěrén* and his 1990 novel *Soul Mountain* (靈 山), in which *yěrén* figure prominently as symbols of the primitive and pure past that modern humans seek to recover. While true of popular science in general, in this case especially, the lines between professional and popular, scientist and lay writer, official and nonofficial, sciences and humanities have been blurry.[8]

The wealth of materials on *yěrén* published in China since 1977 has been staggering, but participation in *yěrén* fever has gone beyond reading and writing. "Minister of Yěrén" Lǐ Jiàn and other enthusiasts attended the inaugural meeting of the Chinese Anthropological Society in 1981 and received endorsement from the board of directors to establish the Chinese Yěrén Investigative Research Association (中國野人考察研究會, hereafter Yěrén Association).[9] From the beginning, it was a "people-run" (民辦) association. The term "people-run" has a revolutionary pedigree dating from the populist movements of the Yán'ān period, but "people-run organizations" (like the "social organizations" discussed in the previous chapter) multiplied more quickly and took on new significance in the relatively liberal political culture of the 1980s.[10] Membership in

8. Examples of "official" pieces of *yěrén* science dissemination include the exhibit at the Shénnóngjià nature reserve and articles published in such newspapers as *People's Daily*. On difficulties delineating "popular" from "professional" science literature, see Stephen Hilgartner, "The Dominant View of Popularization: Conceptual Problems, Political Uses," *Social Studies of Science* 20 (1990): 519–539.

9. Liú Mínzhuàng, *Jiēkāi 'yěrén' zhī mí*, 211–212.

10. Pauline Keating, "The Ecological Origins of the Yan'an Way," *Australian Journal of Chinese Affairs* 32 (1994): 123–153; Edward Gu, "Cultural Intellectuals and the Politics of the Cultural Public Space in Communist China (1979–1989): A Case Study of Three Intellectual Groups," *Journal of Asian Studies* 58, no. 2 (1999): 389–431; Gordon White, "Prospects for Civil Society in China: A Case Study of Xiaoshan City," *Australian Journal of Chinese Affairs* 29 (1993): 63–87.

the Yěrén Association grew to between two and three hundred people—mostly scientists, science teachers, administrators, and people with some personal connection to the subject.[11] Its reach into the general population, however, was far greater: attendance at four exhibitions the association held in the 1980s topped four hundred thousand.[12] Many more have since traveled to Shénnóngjià and other wilderness areas with the hope of spotting, or even capturing, a yěrén.

Legacies of the Máo era are apparent in the recent fascination with yěrén. Eliminating superstition and (at least in the early years) promoting mass science have both been important rationales for pursuing yěrén research. Nonetheless, yěrén fever is decidedly a post-Máo phenomenon. As wild creatures, yěrén have become powerful agents in challenging authorities once considered sacrosanct and in articulating the desires and values of people living in a rapidly industrializing and commercializing society. Elusive, savage, and yet strangely human, they speak the right language for the time.

Replacing Superstition with Science

Much of the literature on yěrén takes the form of popular paleoanthropology. The initial investigations in the late 1970s were led by IVPP, and many of the most prominent people to research and comment on yěrén have been IVPP scientists and their colleagues at other institutions. Moreover, popular science materials on yěrén contain many of the same established themes found in popular paleoanthropology as a whole. Materials on human origins have battled superstition by seeking to replace ideas of divine creation with Engels's theory that labor created humanity. Writings on yěrén have similarly set out to convince audiences that yěrén are creatures of science, not fantasy. In doing so, they have relied on the same interpretations of human evolution promoted in their more established counterparts.

Especially before the 1980s, yěrén were unavoidably associated with superstition. Local legends and imperial-era texts attributed supernatural powers to these creatures, and the stories were often told side by side with stories about ghosts, demons, and monsters. Concerns about fueling superstition undoubtedly lay behind the failure of yěrén research to gain respectability in the early decades of the People's Republic. The Cultural Revolution brought an intensification of such views. Rural people were

11. Yuán Zhènxīn, interview with the author, 19 March 2002.
12. Liú Mínzhuàng, Jiěkāi 'yěrén' zhī mí, 249.

punished for circulating stories about *yěrén* that "spread feudalist super-stition" and threatened production by making workers afraid to tend their fields and harvest forest products.[13] It was only in the closing years of the Cultural Revolution that it became possible for Lǐ Jiàn and others to revisit the subject and begin collecting eyewitness accounts. Whereas previously the tendency had been to suppress reports, Lǐ Jiàn spearheaded the movement to investigate them instead. Yet discomfort remained. The IVPP scientist Wú Rǔkāng reportedly faced censure in 1978 upon express-ing the belief that *yěrén* might exist. According to Liú Mínzhuàng's later account, Wú had been sternly told that "someone who practices natural science shouldn't believe this kind of thing. If someone told you ghosts existed, would you believe them?" Such criticisms did not prevent IVPP scientists from continuing to investigate *yěrén*, but it did cause them to scale back their efforts and take a lower profile for a time.[14]

Yěrén research and writing in the early years was not a bow to supersti-tion, nor was it a concession to the idea that there were limits to what science could explain about the world. On the contrary, "solving the *yěrén* mystery" could and did become a way to further the goal, so impor-tant to Máo Zédōng and his followers, of replacing superstition with science. Such attempts were similar to the efforts of communist activists thirty years earlier to replace superstition with socialism through the famous opera *The White-Haired Girl*, which told audiences that a super-natural creature with human and animal characteristics had a rational explanation rooted in social realities.

Like the authors of *The White-Haired Girl*, many of those who studied and wrote about *yěrén* hoped to correct superstitious views. In fiction especially, many authors continued to blame social evils and inhuman-ity for causing people to flee to the mountains and become *"yěrén."*[15]

13. Zhāng Xiǎosōng and Bǎi Hóng, "'Yěrén' shì bàn shuì," 4. On state fears of superstition, see Steve A. Smith, "Talking Toads and Chinless Ghosts: The Politics of 'Superstitious' Rumors in the People's Republic of China, 1961–1965," *American Historical Review* 111, no. 2 (2006): 405–427.

14. Liú Mínzhuàng, *Jiēkāi 'yěrén' zhī mí*, 139. Whether Wú Rǔkāng actually supported *yěrén* research is questionable. In addition to Liú's testimony (which has been recirculated in books by other authors), Lǐ Jiàn's family has preserved a letter Wú wrote Lǐ in June 1976 in which Wú prom-ised to send the IVPP scientist Huáng Wànbō to Shénnóngjià to investigate. However, interviews with other paleoanthropologists (including those who support *yěrén* research) indicate that Wú was never interested in the subject. Wú ended his 1992 book *Human Origins* with the following assertion: "Many people are interested in '*yěrén*' and 'extraterrestrials.' . . . The press also often reports on them. However, as far as I know, we have only legends and conjectures without a reliable scientific basis." Wú Rǔkāng, *Rénlèi de yóulái* [Human Origins] (Běijīng: Kēxué jìshù wénxiàn chūbǎnshè, 1992), 92. If Liú's account of Wú's 1975 statements is accurate, Wú may for political reasons have expressed a view that he did not actually hold.

15. Hú Xuéchún, "Yěrén zhī mí" [The "Wildman" Mystery], *Jù yǐng yuèbào*, no. 5 (1993): 30–41; Liú Yǒuhuá, "Yěrén zhài" [*Yěrén* Camp], *Dāngdài zuòjiā* 59 (1995): 56–70; Chén Yùlóng, "'Yěrén'

《古今图书集成》第五二二册，第八
十八卷之二十一集上的玃图

21 Image of a *jué* from *Synthesis of Illustrations and Books Past and Present* (Gǔjīn túshū jíchéng, 1726–1728), reprinted in Jiāng Tíng'ān and Yún Zhōnglóng, "*Yěrén*" *xún zōng jì* [Tracking the "*Yěrén*"] (Xī'ān: Shǎnxīshěng rénmín chūbǎnshè, 1983), 102. Interestingly, the creature depicted appears female, while the encyclopedia's text entries identify "*jué*" as consisting solely of males. The creature is standing in a martial arts pose commonly used in illustrations of the famous character the Monkey King.

But most writings on *yěrén* sought answers in natural science rather than social conditions. In the 1960s, Máo Guāngnián explicitly connected his work with an attack on superstition. Máo postulated that *yěrén* were descendants of the Chinese fossil ape *Gigantopithecus*, and could be found listed as *jué* (玃) in imperial-era writings (figure 21). According to him, a key reason for thinking that *Gigantopithecus* still roamed Chinese forests was that "we do not believe in ghosts and spirits."[16] In other words, imperial-era records and recent reports of *yěrén*-like creatures had

---

yìshì" [A "*Yěrén*" Anecdote], *Mángzhòng*, no. 5 (1996): 39–44; Tán Tán, *Yěrén* (Běijīng: Zhōngguó wén-lián chūbǎnshè, 1993).

16. Máo Guāngnián in Wáng Bō, *Yěrén zhī mí xīn tàn* [New Investigations of the *Yěrén* Mystery] (Chóngqìng: Kēxué jìshù wénxiàn chūbǎnshè, 1989), 137. Máo's treatise, "The Question of *Jué* and Human Origins" (*Jué yǔ rénlèi qǐyuán*) is included in an appendix (pp. 117–165).

to have a scientific explanation. Proof of *yěrén*'s existence as an evolutionarily identifiable primate would be useful precisely for its ability to undermine appeals to ghosts as explanations for encounters with curious animals. In Máo's words, "The existence of *jué* can provide a scientific explanation for the origins of Chinese ghost stories and legends."[17]

People in the post-Máo period have similarly seen themselves to be engaged in transforming legends into science. In his 1979 account of the investigation into yeti in the Mount Everest region conducted twenty years earlier, Shàng Yùchāng suggested, "Historical experience tells us that common people [民間] often mistake animals in the natural world for imaginary animals." He went on to conclude that science would someday probably show yeti and *yěrén* to be unexceptional animals, much in the same way scientists had already shown "mermaids" simply to be marine mammals.[18]

The IVPP scientists Yuán Zhènxīn and Huáng Wànbō were more enthusiastic that *yěrén* would turn out to be exceptional animals, at least in the sense that they were as yet unknown to science. However, they acknowledged the legitimacy of imaginary critics, whom they paraphrased to say, "People treat *yěrén* legends like Westerners treat the Bible. Though people's accounts of *yěrén* may seem very clear and logical, so might the tens of millions of superstitious people's [accounts of] ghosts."[19] Yuán and Huáng nonetheless maintained the study of *yěrén* could and should proceed as a serious scientific investigation. They began a 1979 *Fossils* article with a quotation from Thomas Huxley's 1863 *Man's Place in Nature*, which has since graced the introductions to several other Chinese writings on *yěrén* and is also found in Russian literature on cryptozoology: "Ancient traditions, when tested by the severe processes of modern investigation, commonly enough fade away into mere dreams: but it is singular how often the dream turns out to have been a half-waking one, presaging a reality."[20] Yuán and Huáng likened their work to that of early scientists who, as Huxley recounted, transformed strange

---

17. Máo Guāngnián in Wáng Bō, *Yěrén zhī mí*, 164.

18. Shàng Yùchāng, "Cóng Zhū Fēng dìqū 'xuěrén' kǎochá tánqǐ" [Talking about Yeti Investigations from the Mount Everest Area], *Huàshí*, no. 3 (1979): 8.

19. Yuán Zhènxīn and Huáng Wànbō, "'Yěrén' zhī mí xiàng kēxué tiāozhàn" [The "*Yěrén*" Mystery Poses a Challenge to Science], *Huàshí*, no. 1 (1979): 8.

20. Thomas H. Huxley, *Man's Place in Nature* (New York: Random House, 2001 [1863]), 3. As one of the first scientific books republished in the latter half of the Cultural Revolution, *Man's Place in Nature* had enormous influence on scientists, particularly paleoanthropologists. The quotation also appears in Liú Mínzhuàng, *Jiěkāi 'yěrén' zhī mí*, preface, 1; and Zhāng Xiǎosōng and Bǎi Hóng, "'Yěrén' shì bàn shuǐ," 4. Dmitri Bayanov, of the International Center of Hominology in Moscow, used the same quotation as the epigraph for his book on evidence of *yěrén*-like creatures from folklore. Personal communication, 28 February 2005.

stories from Africa and Asia into scientific accounts of living apes.[21]

In language reminiscent of Máo's "On Practice," Liú Mínzhuàng similarly celebrated the "progression" of *yěrén* research "from myth, legend . . . this kind of primitive, beginning level, to the level of scientific investigation."[22] Thus, rather than celebrating *yěrén* as something outside the realm of science, *yěrén* scientists and popularizers justified investigations of *yěrén* as a way of replacing such legends or superstitions with a more scientific explanation. This replacement narrative is well expressed in a 1996 book on *yěrén*: "Today, superstition is the enemy of science. In history, she was also the mother of science. From making pills of immortality to smelting steel, from astrology to astronomy . . . science descended from the swaddling clothes of superstition."[23]

Scientists were not the only ones to welcome the conquest of legend by science. Fiction writer Sòng Yōuxīng vehemently denied the accusation that his 1985 story "A *Yěrén* Seeks a Mate" was a "ghost story" from "ancient myths and primitive religions" of "ignorant times." He went on to lambast the critic: "That he took this objective scientific fact [of *yěrén*'s existence] and turned it into a 'spirit' or 'ghost' just completely reveals this 'science disseminator' as an impostor and a little copycat."[24] Several of the readers who wrote Sòng fan letters defended him on this point, and one specifically noted the story's usefulness in suppressing, rather than encouraging, superstition. "Making it into a science fiction story . . . ," the reader opined, "helps disseminate scientific knowledge about *yěrén* among the masses and extinguish the masses' prejudice, ignorance, and superstition."[25]

It is not easy to identify precisely what separates proponents and opponents of "marginal" sciences like the study of *yěrén*.[26] In a sociologi-

---

21. Yuán Zhènxīn and Huáng Wànbō, "'Yěrén' zhī mí," 3.

22. Liú Mínzhuàng, *Jiēkāi 'yěrén' zhī mí*, 100. Compare with Máo's "On Practice," in Mao Tse-tung, *Selected Works of Mao Tse-tung* (Peking: Foreign Languages Press, 1967–1071), 1:296.

23. Zāng Yǒngqīng, *Yěrén mí zōng* [Tracking the *Yěrén* Mystery] (Shěnyáng: Liáoníng chūbǎnshè, 1996), preface, 2.

24. The short story entitled "A *Yěrén* Seeks a Mate," written by popular science writer Sòng Yōuxīng and published in 1985 by the Guǎngxī periodical *Short Stories* (小説報), generated such a positive response that presses in Hong Kong and Běijīng soon republished it in short story collections on the subject. Sòng Yōuxīng, *Yěrén de chuánshuō* [*Yěrén* Legends] (Hong Kong: Xiānggǎng hǎiwān chūbǎnshè, 1986). Zǐ Fēng, ed. *Yěrén qiú'ǒu jì* [A *Yěrén* Seeks a Mate] (Běijīng: Zhōngguó mínjiān wényì chūbǎnshè, 1988). The quotation appears in Sòng Yōuxīng, *Yěrén de chuánshuō*, 3.

25. Sòng Yōuxīng, *Yěrén de chuánshuō*, 101.

26. I use the term "marginal" here because *yěrén* research has never gained complete acceptance as a valid form of scientific inquiry. Especially in the late 1970s and early 1980s, however, *yěrén* research was far more accepted among scientific circles in China than similar research has been in the United States and other places. Moreover, from a popular angle, *yěrén* can hardly be considered "marginal."

cal study of ufology (research into UFOs), Joseph Blake divides writers on UFOs into two categories. The first group seeks "to define them as natural phenomena, thus including them within the bounds of normal science," while the second group seeks "to present UFOs as something beyond the confines of normal science." Within the first group are both "debunkers" who "are convinced that UFOs are 'nothing more than' stars, birds, swamp gas, hoaxes, or 'mass hysteria'" and also "hopefuls" who "hope to demonstrate that UFOs are secret weapons, extra-terrestrial vehicles or something else subsumable under normal science."[27] The vast majority of nonfiction writers on *yěrén* and many fiction writers as well, at least during the 1970s and 1980s, can be classified as "hopefuls" within the "normal science" approach. (In this way, they are similar to Qián Xuésēn and others who have sought to make scientific sense of supernormal *qìgōng* abilities.) Few if any have taken an explicit stand that *yěrén* lie outside the bounds of science as we know it. Rather, they have tried to confirm the existence and evolutionary origins of *yěrén* using the same methodologies found in established sciences. Shàng Yùchāng's position could perhaps be classified as that of a "debunker" in that he suggested *yěrén* were probably nothing more than ordinary animals. However, since the "hopefuls" also consider *yěrén* to be animals, the division is a fuzzy one and perhaps best characterized as a difference in degrees of confidence in the reliability of the evidence and of enthusiasm about the potential of the research to make a significant contribution to science.

The Scientific Significance of *Yěrén*

What, then, constitutes "scientific knowledge about *yěrén*" to those seeking to disseminate it? While not the only issue, the question of *yěrén's* place in evolution has been important.[28] In answering this question, many writers on *yěrén* have relied on Engels's theory that labor created humanity and general Marxist theories on evolution and social development. This pattern appeared already in Chinese writings from the late 1950s and 1960s on yeti and *yěrén*. One theory in the Soviet Union, advanced by Boris Porshnev, posited that yeti were living descendants of Neanderthals.[29] Máo Guāngnián and the respected physical anthropologist Wú Dìngliáng doubted this hypothesis. In 1962, the *Xīnmín Evening*

---

27. Joseph Blake, "Ufology: The Intellectual Development and Social Context of the Study of Unidentified Flying Objects," in *On the Margins of Science: The Social Construction of Rejected Knowledge,* Sociological Review Monograph 27, ed. Roy Wallis (Keele, U.K.: University of Keele, 1979), 315.

28. Other aspects include *yěrén's* life cycle, mating habits, and ecology.

29. Others in the world of cryptozoology share this view. See especially Shackley, *Wildmen.*

News printed an interview with Wú in which he concluded that either yeti sightings were cases of mistaken identity or that yeti represented a primate group somewhere between apes and humans in evolutionary terms. For Wú, yeti could not be Neanderthals in part because eyewitness accounts indicated their inability to speak and engage in labor. Neanderthals, on the other hand, "belonged to the second stage of human development, had their own language, and could engage in labor."[30] In a treatise on the *yěrén* question written between 1964 and 1969 (not published until after the Cultural Revolution), Máo Guāngnián echoed these sentiments, discounting the Neanderthal theory on the basis of evidence for Neanderthal tools, language, and thought.[31]

Máo began his study by searching imperial-era texts for descriptions of similar creatures, focusing especially on *jué*. Máo then turned to the fossil record, and particularly to Chinese research on *Gigantopithecus*, a fossil ape (sometimes classified as a hominid) found only in China. Having rejected the Neanderthal theory, Máo postulated that *yěrén* were the same as *jué* and were living descendants of *Gigantopithecus*.[32] Engels's theories on human evolution—especially his discussion of labor, tool manufacture, and the liberation of the hands through bipedalism—dominated Máo's analysis.[33]

Work on *yěrén* conducted in the post-Máo period shared Máo Guāngnián's interest in human evolution and often recirculated his theories. Yuán Zhènxīn and Huáng Wànbō's 1979 article in *Fossils* quoted at length from materials published on *Gigantopithecus* by Wú Rǔkāng beginning in 1962. Noting the similarities between Wú's descriptions of *Gigantopithecus* and descriptions of *yěrén* supplied by eyewitnesses, they concluded optimistically, "Some people say *Gigantopithecus* is extinct. This is not true!"[34] In an article published the same year in *Scientific Experiment*, Zhōu Guóxīng also made positive reference to the *Gigantopithecus* theory.[35] Another book, published in 1983, even borrowed and adapted a phylogenetic tree from a recent popular book on paleoanthropology depicting human and ape evolution in order to demonstrate the relationships between humans, *Gigantopithecus*, and *yěrén*. *Gigantopithecus* has

30. Wú Dìngliáng, "'Xuěrén' de xíngxiàng yánjiū" [Research on the Form of 'Yeti'], *Xīnmín wǎnbào*, 29 April 1962, 1.

31. Máo Guāngnián in Wáng Bō, *Yěrén zhī mí*, 162–163.

32. Whether he knew it or not, a 1959 Soviet text had suggested the possibility that yeti were descended from *Gigantopithecus*. Yuán Zhènxīn and Huáng Wànbō, "'Yěrén' zhī mí," 9.

33. Wáng Bō, *Yěrén zhī mí*, 117–165.

34. Yuán Zhènxīn and Huáng Wànbō, "'Yěrén' zhī mí," 9.

35. Zhōu Guóxīng, "Shénnóngjià 'yěrén' kǎochá" [Investigation of Shénnóngjià "Wildman"], *Kēxué shíyàn*, no. 2 (1979): 30.

remained the most popular postulated fossil ancestor of *yěrén* in China and has attracted a following among cryptozoologists abroad as well.

"Professor of *Yěrén*" Liú Mínzhuàng appreciated Máo Guāngnián's insight with regard to *Gigantopithecus* but suggested another possibility, that *yěrén* were most closely related to *Australopithecus robustus*, a very early hominid not yet found in China but whose discovery could help support claims for human origins in China rather than Africa.[36] Liú's conclusion, which he claims to have made by 1976, was based once again on tool manufacture. In a 1979 article printed in *Encyclopedic Knowledge*, Liú noted that, according to eyewitness accounts, *yěrén* "walk erect, have their hands liberated, and can use natural tools and hunt small animals," but that they "do not have social division of labor and cannot construct even the crudest stone knives with which to labor."[37] Here, Liú was recalling Engels's "The Part Played by Labor in the Transition from Ape to Human" in which he asserted, "No simian hand has ever fashioned even the crudest stone knife"—a quotation that, not incidentally, also appeared in many popular materials on human evolution of the time.

Other researchers of the late 1970s and 1980s also invoked Engels in their discussions of *yěrén*'s evolutionary significance. For example, a 1983 book on *yěrén* quoted Engels several times with respect to physiology and life activities, the importance of the concept of the species, and the liberation of the hands in human evolution.[38] Engels's influence is also keenly felt in Liú Mínzhuàng's concern about depictions of *yěrén* living in monogamous pairs. In his 1988 book, Liú asserted that "*yěrén* families should be polygamous or promiscuous matrilineal families."[39] While he did not say so explicitly, this "should" clearly reflected Engels's understanding that early humans lived in a matriarchal system.

If Engels's theories provided helpful analytical tools for the study of *yěrén*, *yěrén* in turn offered the promise of additional evidence to support these theories and other Marxist principles. In 1979 Liú enthused that if a *yěrén* were captured, "Engels's theory that labor created humanity and the theory that humans originated in South Asia will have gained their strongest evidence."[40] The same year, Zhōu Guóxīng similarly asserted that data on *yěrén* "will be precious scientific material for research into

---

36. Liú Mínzhuàng, *Jiēkāi 'yěrén' zhī mí*, 72–73.

37. Liú Mínzhuàng, "Yánzhe qíyì de jiǎoyìn" [Following Strange Footprints], *Bǎikē zhīshí*, no. 6 (1979): 80.

38. Jiāng Tíngān and Yún Zhōnglóng, *"Yěrén" xún zōng jì* [Tracking the "Yěrén"] (Xī'ān: Shǎnxīshěng rénmín chūbǎnshè, 1983), 84–86.

39. Liú Mínzhuàng, *Jiēkāi 'yěrén' zhī mí*, 199.

40. Liú Mínzhuàng, "Yánzhe qíyì," 82.

biological evolution [and] will supply important scientific evidence for Marxist dialectical materialism and historical materialism."[41] Such claims were still alive a decade later when another author suggested that *yěrén* research "not only can enrich Marx's accurate thesis regarding the creation of humans, but can also add to people's clear understanding of human ancestors and their process of evolutionary development."[42]

In addition to distinguishing *yěrén* from early hominids, Engels's theory about tool manufacture in human evolution was at times extended to differentiate *yěrén* from "primitive" humans living today. This kind of comparison was almost inevitable, not least because the term *yěrén* (as in "savage") had long been used to refer to tribal peoples.[43] In a 1986 book on the "marvels" of Shénnóngjià, the authors interviewed Liú Mínzhuàng and asked him if *yěrén* could be either "a mutated group of primitive people or evolutionarily retarded modern people." Liú responded in the negative: tool manufacturing defined humans, and there was no evidence of tool manufacturing among *yěrén*.[44]

Chinese researchers traveling to the Tibetan plateau in 1980–1981 also expressed the need to determine "whether or not these '*yěrén*' were monkeys, apes, bears, or even more primitive nationalities."[45] Here, "even more primitive" suggests a comparison with the minority nationalities the authors interviewed in their research. Similar comparisons can be found in the work of other scientists. Yuán Zhènxīn and Huáng Wànbō used the example of the Kǔcōng people (a group discovered in the 1960s living in the forests of Yúnnán Province) as evidence against a local Húběi legend that *yěrén* were the descendants of Great Wall conscript evaders.

41. Zhōu Guóxīng, "Shénnóngjià 'yěrén' kǎochá," 30.

42. Wáng Bō, *Yěrén zhī mí*, preface, 1.

43. In 1908, the Chinese magazine *Science* used the term to describe indigenous Australian people. *Kēxué zázhì* 13 (11 August 1908): 409. The usage persisted after 1949, for example, in an appallingly racist passage in Wáng Shān, *Láodòng chuàngzào rénlèi* [Labor Created Humanity] (Shànghǎi: Shànghǎi qúnzhòng chūbǎnshè, 1951), 15–16. The many meanings of the term *yěrén* require that we read texts carefully and not simply rely on suggestive titles, as Dikötter appears to have done in erroneously citing a 1928 ethnology article on the "savages" of New Guinea and other places as evidence of discourse on half-ape, half-human monsters in the republican era. The article he cites is Cuī Dàiyáng's "Yěrén gètǐ de yuánsuǒ yǔ jièxiàn" [Characteristics and Boundaries of the Self among Savages], *Mínsú* [Ethnology] 23–24 (1928): 1–11. Dikötter translates the title as "Facts about the Wild Man." Cited in Frank Dikötter, "Hairy Barbarians, Furry Primates, and Wild Men: Medical Science and Cultural Representations of Hair in China," in *Hair: Its Power and Meaning in Asian Cultures*, ed. Alf Hiltebeitel and Barbara D. Miller (Albany: State University of New York Press, 1998), 67 and 71.

44. Máo Zhènhuá, Zhōu Hóngyóu, and Fù Wànměi, eds., *Shénnóngjià tàn qí* [Looking for Marvels in Shénnóngjià] (Běijīng: Gōngrén chūbǎnshè, 1986), 112.

45. Liú Mínzhuàng, Chén Nǎiwén, Zhāng Guóyīng, et al., *Yěrén, xuěrén, húguài* [Yěrén, Yeti, and Lake Monsters] (Běijīng: Zhōngguó jiànshè chūbǎnshè, 1988), 29. Chén Nǎiwén and Zhāng Guóyīng participated in the expedition.

Yuán and Huáng noted that the Kǔcōng people's life in the primeval forests had not caused them to grow hair all over their bodies, so why, they asked, should this have happened to conscript evaders?[46] If these authors saw primitive people and *yěrén* to be subject to the same evolutionary processes, at least they distinguished the physical appearance of Kǔcōng from their hairy counterparts. The contributions of nonscientists to the discourse could be less sensitive. In a *Fossils* article, a man from Yúnnán wrote that when he first saw a *yěrén* as "a woman with straggly hair coming out of the forest laughing," he "thought it was a local Wǎ-nationality woman climbing the mountains to collect pig food."[47]

Writers on *yěrén*, moreover, invoked a strongly progressivist understanding of evolution in their theories about *yěrén* and human evolution, in keeping with those found in popular paleoanthropology in general. Liú Mínzhuàng recalled that early investigations in Shénnóngjià gave rise to the theory that the area's hospitable climate and abundant food sources had prevented the competition necessary for *yěrén* to evolve into modern humans. The researchers involved in the 1980–1981 expedition to the Tibetan Plateau shared this theory, and they explicitly compared yeti to the local minority nationalities in this regard. "This environment richly endowed by nature has not only preserved many backward minority nationalities . . . but has also created excellent conditions for '*yěrén*' to escape the cold, seek food, and avoid extinction."[48]

Thus, the *yěrén* literature contributed to discourses on science, superstition, and human origins already very familiar in materials on human evolution the socialist state produced as part of its science dissemination mission. But as he has in popular paleoanthropology in general, Engels has gradually lost his prominence in materials on *yěrén* and fails to appear entirely in most recent books on the subject. Moreover, the scientific community in general is significantly less enthusiastic about *yěrén* today than it was in the late 1970s and early 1980s. *Fossils* magazine stopped publishing articles on *yěrén* in the late 1980s because some at IVPP felt that the failure of scientific expeditions to prove the existence of *yěrén* undermined its validity as a scientific topic.[49] *Fossils* readers today are divided: in a survey of fifty-four people, 32 percent stated that they did not believe *yěrén* existed, 23 percent stated that they believed

46. The authors further referred to the real-life "white-haired girl" who came out of hiding in 1959 after living in the Sìchuān wilderness for twelve years (see chapter 2, note 111), noting that she had not become a *yěrén*. Yuán Zhènxīn and Huáng Wànbó, "'Yěrén' zhī mí," 8.

47. Lǐ Míngzhí, "Wǒ hé 'yěrén' bódòu" [I Fought with a "Yěrén"], *Huàshí*, no. 4 (1984): 6.

48. Liú Mínzhuàng, Chén Nǎiwén, Zhāng Guóyīng, et al., *Yěrén, xuěrén, húguài*, 29.

49. My interview subjects were either unsure who specifically opposed it or unwilling to divulge their names.

yěrén probably existed, and 17 percent stated they thought *yěrén* might exist.[50] Yet even as many scientists have lost hope that the creature will be found, they do not usually scorn the original impetus to study it. And a few paleoanthropologists—most notably IVPP's Yuán Zhènxīn and to a lesser extent the Natural History Museum's Zhōu Guóxīng (formerly of IVPP)—remain committed to the research. In 2005, the anthropology wing of the Natural History Museum still displayed evidence collected from *yěrén* research. And scientists and professional writers alike continue to use the mantle of science dissemination in the materials they produce on *yěrén* for popular audiences.

From Mass Science to Scientific Heroism

If the paired goals of disseminating science and squelching superstition represented one legacy from the Máo era found in *yěrén* literature, mass science was another, though far weaker and shorter lived. As in previous decades, people in early post-Máo China were deeply conflicted about mass science. With the pressure to remain true to socialist principles fading, this conflict has given way to a remarkable lack of interest in the idea that scientists should "rely on the masses." In its place is an enthusiasm for heroic individuals alienated from society in pursuit of personal glory or spiritual fulfillment. Wrapped in these new values, *yěrén* research nonetheless continues to encourage lay participation and challenges to expert authority, both central goals of mass science.

At the heart of discussions about the relationship between *yěrén* and superstition lies the question of the trustworthiness of the masses, and particularly the rural masses. This is always a sticky problem, and in *yěrén* research an unavoidable one. *Yěrén* research, particularly in the early years, consisted primarily of systematic interviews with "eyewitnesses." "Professor of *Yěrén*" Liú Mínzhuàng and other investigators tracked down villagers willing to discuss their encounters with *yěrén*, photographed them, recorded their biographical information and narratives, asked them standard questions, and had them identify the creature they had seen from pictures of apes, bears, fossil hominid reconstructions, and legendary animals found in old encyclopedias.

This was precisely the problem, according to some commentators. A 1989 book on the subject noted that one of the main criticisms of

---

50. Nineteen percent did not answer the question; 4 percent stated they did not know; 2 percent gave unclear answers; 4 percent thought the term *yěrén* referred to primitive peoples; 4 percent said they exist but are humans gone wild; and 6 percent said they exist but are simply a kind of ape.

yěrén research was that eyewitnesses were all "local people," and never zoologists or anthropologists.[51] Nobel laureate Gāo Xíngjiàn gave voice to such sentiments in his play *Yěrén*, which, though fictional, insightfully and accurately identified prevailing sentiments. A reporter in the play explained, "You see, you cannot rely on what people say. The key is to get hold of some hard evidence."[52] The play's village secretary further questioned the veracity of eyewitness reports: "You can't believe a word these country bumpkins say. They may look honest, but deep down, forget it."[53]

This attitude was certainly well known to people living in Shénnóngjià and other areas rich in *yěrén* sightings. It helps explain the antagonism with which a young macho, Lí Guóhuá, proclaimed he would "kill a *yěrén*, bring its head to the scientific institute, and ask them if there was or wasn't such a thing as *yěrén*!"[54] A cultural troupe member from the nearby city of Yíchāng, Lí had abandoned his job in 1979 to hunt *yěrén* in Shénnóngjià. He had probably encountered and been offended by sentiments such as the one expressed by Gāo's fictional village secretary. Nonetheless, such derogatory claims about rural people's knowledge and trustworthiness were outweighed in the early post-Máo literature by those defending the masses' reliability. For example, in a 1977 speech, Qián Guózhēn—an East China Normal University biology professor—declared, "We must . . . respect the observations of the masses."[55]

"Professor of *Yěrén*" Liú Mínzhuàng took a particularly strong line on this question in his 1988 book. It was with almost audible anger that he recounted the story of the first scientist to respond to reports of *yěrén* in Shénnóngjià. In 1974, the biologist interviewed an eyewitness and found him to be "a peasant lacking knowledge of scientific culture and having backward consciousness." The biologist concluded that the man had fabricated the account to explain his truancy from his work cutting kudzu vines in the mountains. According to Liú, the biologist was not only prejudiced against the man but also very "rude" to him.[56] In contrast, Liú painted paleoanthropologist Wú Rǔkāng as a fair-minded scientist with a proper socialist attitude. Wú was in Húběi in 1975 on unrelated fieldwork when he heard reports of Lǐ Jiàn's investigations.

51. Wáng Bō, *Yěrén zhī mí*, preface, 2.

52. Gao Xingjian, "Wild Man: A Contemporary Chinese Spoken Drama," trans. Bruno Roubicek, *Asian Theatre Journal* 7, no. 2 (1990 [1985]): 217.

53. Gao Xingjian, "Wild Man," 237.

54. Stories about Lí Guóhuá are common in post-Máo *yěrén* literature. See, for example, Liú Mínzhuàng, *Jiēkāi 'yěrén' zhī mí*, 164–168.

55. Qián Guózhēn in Wáng Bō, *Yěrén zhī mí*, 115.

56. Liú Mínzhuàng, *Jiēkāi 'yěrén' zhī mí*, 89–90.

He was said to have concluded that the stories were "probably not fabricated" and that investigators "should believe the masses."[57]

If Liú and other *yěrén* researchers professed their trust in the masses as reliable eyewitnesses, they also saw them as necessary participants in *yěrén* investigations. For Liú, rural people were Shénnóngjià's "permanent residents," and therefore they alone could be relied upon to provide accurate information about the local area.[58] In 1980, another observer blamed the failure of early investigations to find *yěrén* on the lack of local residents' participation. Science worker Fán Jīngquán wrote a letter to the editor of *Fossils* in which he told of his encounter with a *yěrén* mother and child in 1954 while on a geological prospecting mission. Fán had not simply stumbled upon the *yěrén*. Rather, old people living deep in the forest had carefully instructed him on where to go to find them, and how to behave once he had seen them. Fán attributed the ease with which he subsequently discovered the *yěrén* to the forest dwellers' extensive knowledge of their life habits. He chided the investigation team for failing to make adequate use of local hunters and herb collectors as guides, instead employing a large number of soldiers who undoubtedly just "drove the fish into deeper waters."[59]

The notion that *yěrén* research required mass participation was a practical consideration: local people do make good guides. However, it was also rooted in a core vision of socialist science manifested in established practices of paleoanthropology, which depended on local people to report fossil finds, help on the digs, and protect the sites when scientists were away. It is ironic that Máo Guāngnián's Máo-era work on *yěrén* did not involve much if any mass participation, but rather relied extensively on imperial-era texts and more recent scientific reports. Significantly, mass participation in *yěrén* research only flourished in the early post-Máo period. And flourish it did. *Yěrén* proved to be a topic of scientific inquiry with which "the masses" could readily identify. People read fictional and nonfictional accounts of *yěrén* in popular magazines and connected what they read with the stories they had heard from friends and neighbors. For example, one of Sòng Yōuxīng's fans recognized elements of the story recounted in "A *Yěrén* Seeks a Mate" and noted, "I've heard it in my hometown, and it's true."[60]

57. Liú Mínzhuàng, *Jiēkāi 'yěrén' zhī mí*, 91. See note 14 above on the question of whether Wú supported *yěrén* research.

58. Liú Mínzhuàng, *Jiēkāi 'yěrén' zhī mí*, 135.

59. Fán Jīngquán, "Hé 'yěrén' mǔ zǐ xiāngyù zài lì shùlín" [A Meeting with Mother and Child "Yěrén" in a Chestnut Grove], *Huàshí*, no. 1 (1980), 30.

60. Sòng Yōuxīng, *Yěrén de chuánshuō*, 103.

Other Sòng admirers praised the story for "encouraging people to solve the *yěrén* mystery" or "encouraging the masses from relevant places to aid in searching for '*yěrén*' tracks."[61] Scientists Yuán Zhènxīn and Huáng Wànbō were particularly excited by the power of *yěrén* to bring many segments of society together for a single purpose: "All these scientists, cadres, soldiers, and masses . . . some researching the vast stores of ancient texts, some translating foreign materials, some asking to participate in investigation teams, some writing letters to scientific institutions to contribute their own materials, some using their vacation time and their own funds to investigate . . . are all dreaming the same dream of turning this heart-stirring dream into a living fact."[62] Yuán and Huáng could well have added "reading articles in *Fossils* magazine" to this list of activities, since *Fossils* provided an important means for the masses to participate vicariously in IVPP *yěrén* research.

Despite Fán Jǐngquán's concerns to the contrary, mobilizing the masses to participate in *yěrén* investigation arose again and again as a priority among *yěrén* researchers in the early post-Máo era. When "Minister of Yěrén" Lǐ Jiàn and scientists from IVPP and other institutions first formed an investigation team in 1976, nine peasants from the Shénnóngjià area participated, and subsequent missions also usually included local people. Investigators further encouraged mass participation by posting rewards. In 1978, IVPP scientists announced a reward they hoped would spur local Shénnóngjià people to organize their own investigations.[63] At the 1984 meeting of the Chinese Yěrén Investigative Research Association, held in Guǎngxī, members decided to begin posting rewards in several likely places throughout the country in order to "mobilize the masses."[64] The local government in Shénnóngjià offered ten thousand yuán, in those years a princely sum.[65] Occasionally, as was the case with Lí Guóhuá, people inspired by these mobilization efforts ended up dedicating years of their lives to the cause of "solving the *yěrén* mystery."

Like scientists in the early People's Republic, post-Máo *yěrén* researchers also strove to cultivate scientific personas that reduced the apparent distance between intellectual elite and worker and so enhanced the semblance of "mass science." On several occasions, Liú Mínzhuàng empha-

61. Sòng Yōuxīng, *Yěrén de chuánshuō*, 106 and 101–102, respectively.

62. Yuán Zhènxīn and Huáng Wànbō, "'Yěrén' zhī mí," 4.

63. Liú Mínzhuàng, *Jiēkāi 'yěrén' zhī mí*, 96–97, 139.

64. Jùn Gǔ, "'Yěrén' jìxù zài mǒuxiē dìqū chūmò" ["Yěrén" Continue to Appear in Certain Places], *Huàshí*, no. 4 (1984): 7.

65. Zhōu Guóxīng, interview with the author, July 2005.

sized the relative importance of physical experience over book learning in *yěrén* research. For example, he gently criticized Shàng Yùchāng—the scientist who participated in the 1959 survey of the Mount Everest region—for thinking that "all things in books could also be found in nature, and [that] the converse was also true."[66]

Yěrén investigation was not a walk in the park, at least not in the sense that it was easy. *Yěrén* researchers, in the spirit of socialist science, gloried in the struggles and hardships their quest required of them. Yuán Zhènxīn and Huáng Wànbō summarized this position neatly in their 1979 article: "In order to take a dreamlike legend and turn it into a real scientific story, the road is long and winding. . . . Marx has said, 'At the entrance to science, as at the entrance to hell, the demand must be made: 'Here must all distrust be left; All cowardice must here be dead.'"[67] A 1980 article published in *China Youth News* portrayed *yěrén* investigators in a similar light. The authors characterized them as having two chief characteristics: they were "poor," enduring harsh conditions with limited supplies; and they were "fanatical" (迷, *mí*), absolutely dedicated to their cause.[68] In the preface to his 1988 book, Liú Mínzhuàng waxed poetic, and self-congratulatory, in his description of the "hardships" he had endured to "solve this ancient riddle." "In the deep mountains and old forests of Shénnóngjià, in that limitless ocean of clouds, I welcomed wind and rain, scaled cliffs, ate wild fruits, drank spring water, and lived a *yěrén's* life. . . . I climbed one thousand mountains, crossed ten thousand *lǐ*, and wrote many hundreds of thousands of words."[69] The trope of the heroic scientist willing to sacrifice himself (or, occasionally, herself) in the name of science is by no means unique to socialist China.[70] Nonetheless, in the immediate post-Máo period, such depictions of struggle and hardship inevitably recalled Máo-era directives to brave any danger on the road to socialism.

Yěrén hunters continue to cultivate heroic personas, but as the association with "mass science" has faded, their individuality—and even isolation—has been highlighted instead. Dedicating one's life to *yěrén* has become not only sacrificing oneself to science, but also doing so at the expense of one's social obligations. When he was interviewed in the

66. Liú Mínzhuàng, *Jiēkāi 'yěrén' zhī mí*, 58–59.

67. Yuán Zhènxīn and Huáng Wànbō, "'Yěrén' zhī mí," 5. Translation borrowed from Karl Marx, *A Contribution to the Critique of Political Economy* (Moscow: Progress Publishers, 1977 [1859]), 23.

68. Zhāng Xiǎosōng and Bǎi Hóng, "'Yěrén' shì bàn shuì," 4. The suggestive relationship between fanatical (*mí*, 迷) and superstition (*míxìn*, 迷信) was apparently not intended.

69. Liú Mínzhuàng, *Jiēkāi 'yěrén' zhī mí*, preface, 4. One *lǐ* is equivalent to about one-third of a mile or half a kilometer.

70. See chapter 4, note 46.

late 1980s, Lí Guóhuá said he often went "for weeks or months without seeing people." He said, "I've lived my life in solitude. . . . Maybe I'll be without a family. . . . Maybe I'll die in the forest someday, but I don't regret any of it."[71] A 2000 magazine article described another *yěrén* tracker, Zhāng Jīnxīng, who has spent years in the Shénnóngjià forest, as "an environmentalist and wanderer who, some would say, has completely lost touch with society."[72] The flight to the forest to track *yěrén* has thus been a flight from society.

It has also been a flight from women and family. The men (and they are almost invariably men) who devote themselves to *yěrén* often portray this decision as a choice between *yěrén* and family.[73] Lí Guóhuá was said to have been a heartbreaker among his female coworkers before he took up *yěrén* hunting. According to Liú Mínzhuàng, Lí "gave his heart" instead to China's "newest scientific field."[74] Zhāng Jīnxīng's determination has perhaps been even greater: he left a wife and family when he began his pursuit of *yěrén* in 1995, and proclaimed shortly thereafter, "I won't return to them until I find [a *yěrén*]."[75] The ecologist in Gāo Xíngjiàn's play displays similar tendencies. His marriage is falling apart because of his increasing obsession with the wilderness and with *yěrén*. When he tells his wife he loves her, she replies, "I don't feel it. You love your exploring. The forest and your darling Wild Man."[76] In this respect, the post-Máo *yěrén* enthusiasts represent a sharp break from the Máo era. To transform oneself through struggle was a Maoist notion; to abandon society for personal glory or salvation was not.

A few people, notably Liú Mínzhuàng, continued to talk about the importance of the masses to science well into the 1990s. A 1993 book Liú wrote on Shénnóngjià's geological history and contemporary ecology likewise unabashedly recirculated earlier Maoist sentiments. Quoting from a paper he presented in 1982, he again echoed Máo: "We must have a spirit of smashing cooking pots and sinking boats [i.e., a determination

71. Gene Poirier and Richard Greenwell, *The Wildman of China* (New York: Mystic Fire Video, 1990).

72. Anne Loussouarn, "What's Out in the Woods?" *City Weekend* 1, no. 1 (2000, trial issue): 10.

73. Liú Mínzhuàng recounted the story of a woman doctor who sought to join the search for *yěrén*. He was reluctant to accept her at first, but because she was a basketball player and stronger than the average man, he eventually conceded. Liú called her the first and only woman tracker of *yěrén*. Liú Mínzhuàng, *Jiēkāi 'yěrén' zhī mí*, 244–245. While women scientists and trackers are almost nonexistent in *yěrén* literature, heroic writings on science in the Máo era did sometimes include women.

74. Liú Mínzhuàng, *Jiēkāi 'yěrén' zhī mí*, 164.

75. Andrew Marshall, "In Search of the Wild Man," *South China Morning Post*, 2 September 1995, 1.

76. Gao Xingjian, "Wild Man," 211.

to achieve victory at all costs], and we must study the local language and deeply rely on investigation by the masses."[77]

Liú, however, was unusual in his loyalty to Maoist notions of "trusting the masses." Such sentiments had quietly faded from most writings on *yěrén* by the mid- to late 1980s. And in the 1980s there were already signs from above that "the masses" were not to be trusted at all. In 1985, *People's Daily* printed a story about a monkey recently caught in Húnán Province and profitably displayed to large audiences as a possible living *yěrén* specimen. Rumors spread quickly before primatologists and paleoanthropologists were able to gather and determine that it was not in fact a *yěrén* at all. The author of the article concluded, "We sincerely hope that *yěrén* research will achieve results, and at the same time hope that it will not artificially create a *'yěrén'* fever.'"[78] Such concerns intensified in the 1990s, when escalating commercialism prompted some travel agencies to encourage tourism in Shénnóngjià by posting bounties for *yěrén*.[79] The impact of these bounty hunters on the protected forests has provoked official scrutiny. In 1998, another *People's Daily* article went so far as to denounce "unscientific, large-scale *'yěrén'* investigative activities of mass character."[80]

Part of the change in the official value of "mass character" in the 1990s resulted from the government's response to the mass protests at Tiānānmén Square in 1989. In the late 1980s, "people-run institutes" became increasingly important forms of organization for intellectuals seeking democratic reforms.[81] In the wake of 4 June 1989, the government clamped down on people-run organizations like the Yěrén Association and compelled them to submit to the supervision of the Ministry of Civil Affairs. After 1994, the name was changed to Strange and Rare Animals Exploration and Investigation Committee (奇異珍稀動物探險考察專業委員會), and the group officially became part of the Chinese Association for Scientific Expedition (中國科學探險協會), itself overseen by the Chinese Association for Science and Technology (an institution with strong ties to the party). It has remained a "people's" organization, however, in that it does not receive central government funding and

77. Liú Mínzhuàng, *Zhōngguó Shénnóngjià* [China's Shénnóngjià] (Shànghǎi: Wénhuì chūbǎnshè, 1993), 314.

78. Péng Zhùpíng, "Hóu shì zěnyàng 'biàn' rén de? Bǔhuò 'huó yěrén' zhēnxiàng" [How Do Monkeys "Become" Human? Capturing the Truth about the "Living Yěrén"], *Rénmín rìbào*, 14 November 1985, 5.

79. Loussouarn, "What's Out in the Woods?" 10.

80. Niè Xiǎoyáng, "Kǎochá 'yěrén' yào zūnzhòng kēxué hé fǎlù" [In Investigating "Yěrén," People must Respect Science and the Law], *Rénmín rìbào*, 24 December 1998, 11.

81. Gu, "Cultural Intellectuals," 408.

receives little even in the way of funding from the local government in Shénnóngjià. While it has some authority to issue letters of introduction for its members when they wish to enter a remote area for research, the members themselves have to pay any related expenses. Occasionally a generous business owner contributes alcohol or food to its meetings or helps offset printing costs, but otherwise the committee tends to be on its own.[82]

The post-Máo period began with the same contradiction between trust in the masses and doubts about their reliability that characterized China under Máo. "The masses" as an abstracted class category had political capital, while the thoughts and customs of the actual people were politically suspect and often branded "superstition." Moreover, beginning in the 1980s and especially after 1989, the very notion of "following the masses" itself became politically dubious and even dangerous. Despite these changes, however, *yěrén* research has continued to encourage popular participation and even to foster challenges to "expert" authority. Wáng Fāngchén, for example, participates actively in *yěrén* research alongside scientists like IVPP's Yuán Zhènxīn and serves as secretary of the Strange and Rare Animals Exploration and Investigation Committee despite his lack of science education—his background is in media. Wáng was recently quoted in an epigraph to a report on *yěrén* investigations: "So-called experts are only regular people with a little more specialized knowledge. In actuality, experts can also make mistakes, and they can even use scientific methods to turn an entire subject into mistake upon mistake."[83]

Like other "marginal" sciences, the study of *yěrén* necessarily continues to involve contested claims to the authority to speak on questions of science. There are other questions, however, that Wáng and others like him have not explicitly raised. Is there a limit to science's authority to speak? Are there areas of experience outside the scope of what science can explain? Such questions were never broached by mass science, and indeed would have been labeled "superstitious" in the Máo era. Yet, increasingly in the post-Máo period, such questions and the answers generated from popular culture have begun to form a new kind of bottom-up challenge to the top-down model of science dissemination.

---

82. Interview data, 2002 and 2005. The exhibitions described above (in the sixth paragraph under "*Yěrén* Fever") and other such activities provide some revenue.

83. Zhāng Jīnxīng, *Zhēng zuò Gǔdàoěr, yǒng tàn 'yěrén' mí* [Striving to Be like Goodall, Bravely Exploring the "Yěrén" Mystery] (Běijīng: Zhōngguó kēxué tànxiǎn xiéhuì, 2002), no page number.

Popular Culture Goes Wild

No matter how far the political rhetoric swings away from a celebration of "the masses," what we know about *yěrén* remains deeply rooted in stories told by rural people. Eyewitness accounts and local legends make up the meat and potatoes, the heart and soul, and the bread and butter of *yěrén* literature. The stories inevitably embody ideas about people and their relationships to society and nature. In them, humanity is no longer unambiguously defined by labor. Rather, a host of other categories has come to the fore, generated from the rich troves of popular culture, both rural and urban. These myriad and intertwined themes have included sex, emotion, culture, environment, and even such scientifically dubious concepts as legend and soul. Together, they may be thought of as an exploration of mystery and wild(er)ness. Fiction writers, poets, and other artists have been best positioned to explore these ideas. However, they are by no means absent even in the literature most self-consciously identified as science dissemination.

Receptivity to such ideas has grown stronger as doubts have begun to loom, even among officials in charge of science dissemination, about the infallibility of science in the face of environmental destruction and other problems of the industrial age.[84] Increasingly, people have begun to look away from science for answers about the human condition. Such receptitivity has also been an outcome of the market economy, in which popular science writers have faced new pressures to make their work salable. It is not just sex and sensationalism, however, that the market has favored. *Yěrén*, as a symbol of wildness, have served as a mirror for humanity. In the aftermath of the Cultural Revolution, they have helped people reflect on the perceived restrictiveness and inhumanity of the Máo era. And *yěrén* have continued to offer ways to talk about a range of social and cultural issues pressing in post-Máo China.

Yěrén have been exciting to post-Máo Chinese people in large part because they are mysterious. The concept of "mystery" is common in popular writings on science. A mystery is generally something that is not understood. Within a scientific worldview, a mystery is more accurately not *yet* understood—it is solvable through the application of scientific methodology provided sufficient material evidence is available. In science dissemination literature, mysteries typically form the starting points for scientific investigation; they are the reason for beginning a scientific investigation in the first place. Writings on *yěrén*, like other types

---

84. Léi Qíhóng and Shí Shùnkē, interview with the author, 26 July 2002.

of science dissemination literature, have emphasized mystery. Many use the word "mystery" (謎, *mí*) in the title and express an interest in "solving" or "cracking" the "*yěrén* mystery." Nonetheless, I would argue that much of the popular appeal of the *yěrén* mystery lies in its resistance to being solved, in its open-endedness. The public's interest in such phenomena is greatly aroused both by "the sensational and mysterious aspects of anomaly sightings" and by a resentment of science's "posture of infallibility."[85] The mystery that science cannot—or at the very least, has been persistently unable to—solve is thus an alluring subject.

Yěrén have also been exciting to post-Máo Chinese people because they are wild. Wildness may often bear an ugly face and a violent temper, but the very horror of these traits provides the basis for a kind of liberation and even healing.[86] The wildness of *yěrén* and their ancestors in imperial-era Chinese texts is expressed most vividly in their pronounced sexuality. The *jué* of old were said to be "purely male and without females . . . and good at stealing human women to serve as their mates and bear their children" (figure 21).[87] Another creature, *zhǒu* (貁), was a *jué* "with the males and females reversed," in other words, the species consisting only of females who stole human men for their mates.[88] "Wild women" or "wild wives" (野婆), also known as *xīngxing* (猩猩), were similarly "groups of females without males who dash up mountains and down valleys . . . [and] whenever they encounter men, they carry them off on their backs to seek a union."[89] Encounters with such creatures could be quite grisly, particularly when the abductee was female. The famous Qīng-dynasty writer and folklorist Yuán Méi recorded a story from Shǎnxī Province in which a "hairy person" abducted a woman who had ventured outside at night to urinate. When her husband and village neighbors found her

85. Ron Westrum, "Knowledge about Sea Serpents," in Wallis, *On the Margins of Science*, 308.

86. Michael Taussig, *Shamanism, Colonialism, and the Wild Man: A Study in Terror and Healing* (Chicago: The University of Chicago Press, 1987), 220.

87. The *jué* appear in Chinese writings at least as early as the *Records of the Search for Spirits* (搜神記, ca. A.D. 350), where they are already said to steal human women. This text further locates *jué* in the southwest of Shǔ (modern day Sìchuān) and suggests that many people who live in the area and are surnamed Yáng are the descendants of such unions. The account is in the twelfth chapter of the twenty-chapter edition and is the 308th story. See Gān Bǎo, *Sōu shén jì quán yì* [Records of the Search for Spirits, Complete and Interpreted], ed. Huáng Dímíng (Guìyáng: Guìzhōu rénmín chūbǎnshè, 1991 [ca. 350]), 350–351. For an English translation, see Kenneth J. DeWoskin and J. I. Crump, Jr., *In Search of the Supernatural: The Written Record* (Stanford, Calif.: Stanford University Press, 1996), 148–149.

88. Chén Mèngléi, ed. *Gǔjīn túshū jíchéng, qínchóng diǎn* [Synthesis of Books and Illustrations Past and Present, Volume on Animals] (Táiběi: Dingwen shuju, 1971 [1726–1728]), 51:852–853. This encyclopedia gathered materials from diverse sources and listed them together under each entry. The quotations cited here originated in Lǐ Shízhēn's 1602 *Classified Materia Medica*.

89. Also from Lǐ Shízhēn, who cited a Táng-dynasty text as his source. In modern usage, *xīngxing* means orangutan. Chén Mèngléi, *Gǔjīn túshū jíchéng*, 51:856.

the next day, she had been tied up, raped, and killed. "Her lips had giant bite marks, the area around her genitals was broken open and torn apart [to the point that] all her bones could be seen, and there was more than a pint of blood mixed with white semen on the ground."[90]

This story, and others like it, have been recirculated in post-Máo writings on *yěrén*. Similar accounts, though none so gruesome, have continued to be reported in China, particularly in Shénnóngjià and in the southwestern province of Guǎngxī. In Shénnóngjià, *yěrén* researchers have investigated accounts of "monkey babies" (猴娃) resulting from rapes of human women by monkeys and/or *yěrén*. Cautious even from the beginning, serious researchers now appear to agree that the "monkey babies" are results of birth defects or atavism rather than interspecies crossing.[91] Nonetheless, abduction stories have provided the juiciest fodder for creative writers of *yěrén* fiction. And particularly when the *yěrén* abductor is female, the encounter is not always portrayed in an entirely negative light.

In Sòng Yōuxīng's "A *Yěrén* Seeks a Mate," published in 1985, a voluptuous and kind-hearted female *yěrén* abducts a man in Guǎngxī Province to sire her child. The *yěrén*'s large breasts and long, attractive hair (figure 3) do not reflect merely this author's imagination. Female *yěrén* are often portrayed this way in fiction, and such descriptions frequently make their way into scientists' prose via eyewitness testimonies.[92] In biologist Wáng Zélín's recollection of a female *yěrén* he had seen in 1941, "The two breasts were very big, and the nipples very red." A witness in 1982 reported a female *yěrén* with big breasts and long hair who was "prettier than a monkey."[93] Gāo Xíngjiàn took up this breast obsession in his play *Yěrén* when one of his characters, an eyewitness, described the *yěrén* he saw. "It had two breasts. Obviously a female. Those breasts . . ."[94] Unfor-

---

90. "A Big Hairy Person Snatches a Woman" (大毛人攫女) is reprinted in Jiāng Tíngān and Yún Zhōnglóng, "*Yěrén*" *xún zōng jì*, 107–108. It originated in Yuán Méi's collection *What Confucius Did Not Discuss* (子不語), which contained several accounts of "hairy people" and other apelike creatures. Yuán Méi, *Zhèng xù Zǐ bù yǔ* [What Confucius Did Not Discuss, Corrected and Continued] (Táiběi: Xīnxìng shūjú, 1978), 5149, 5241, 5272.

91. In 1980, Liú Mínzhuàng rejected the theory that "monkey babies" were monkey-human crosses because of the differences between the two species. He held out for the possibility of *yěrén*-human hybridization, but suggested an alternative theory that "monkey babies" were instances of atavism (返祖). Liú Mínzhuàng, "Hóu wá zhī mí" [The Mystery of Monkey Babies], *Kēxué huàbào*, no. 4 (1980): 32–33. On the birth defect theory, see Loussouarn, "What's Out in the Woods?" 11; J. Richard Greenwell and Frank E. Poirier, "Is There a Large, Unknown Primate in China? The Chinese Yeren or Wildman," *Cryptozoology* 11 (1992): 72.

92. For examples in fiction, see Chén Yùlóng, "'Yěrén' yìshì," 41; Liú Yǒuhuá, "Yěrén zhài," 58, 61; Zǐ Fēng, *Yěrén qiú'ǒu jì*, 11, 141.

93. Liú Mínzhuàng, *Jiēkāi 'yěrén' zhī mí*, 52, 241.

94. Gao Xingjian, "Wild Man," 214.

tunately for us, his interviewer impatiently interrupted him at this point in his narrative.

Sòng Yōuxīng's story provoked quite a stir with its titillating account of the intimate and erotic feelings shared by the female *yěrén* and her human mate. Yet many readers appeared more moved by the *yěrén*'s dispositional virtues than her physical assets. The Hong Kong reprinting of the story included an appendix of fan letters received by the author. A "woman worker" wrote, "As a woman, I was holding back tears as I finished [the story]. I really empathized with this warm-hearted '*yěrén*' girl." A "single male" reader was struck specifically by the *yěrén*'s concern for her family. "She loves her husband, much surpassing some of the women in our own lives. She has feelings and righteousness. In the end she cannot find her husband and child so she stops eating and dies. How can this not cause people to be moved?"[95]

Sòng Yōuxīng and his fans would seem to have answered the question posed by a later nonfiction writer on the subject: "Do *yěrén* have the emotion of love? One would never dream that love resides in the primeval savage wilderness."[96] The notion that *yěrén* indeed experience and express love for their families is found not only in fictional accounts, but in eyewitness narratives and investigators' analyses. In 1976, villagers in Shénnóngjià reported a pregnant female *yěrén* traveling through the area to meet up with her "husband." "Professor of *Yěrén*" Liú Mínzhuàng theorized that all the *yěrén* sighted in Fāngxiàn County near Shénnóngjià belonged to the same "family," and that one particularly belligerent *yěrén*'s behavior could be explained as a wild rage after his mate had been killed.[97]

These portraits of *yěrén* love present an interesting mixture of wildness, eroticism, and passion on the one hand, and human compassion, loyalty, and traditional values on the other. Both of these elements can be traced to a reaction against Cultural Revolution–era gender neutrality and attacks on traditional family values. The 1980s saw a dramatic change in ideals of womanhood. "The fervor and enthusiasm with which women beautified themselves, the widespread support for moving women back into 'suitable' lines of work, the discussions of womanly virtues in the press, must all be understood in part as a reaction to Cultural Revolution norms."[98]

95. Sòng Yōuxīng, *Yěrén de chuánshuō*, 104, 105.
96. Zāng Yǒngqīng, *Yěrén mí zōng*, 91.
97. Liú Mínzhuàng, *Jiēkāi 'yěrén' zhī mí*, 95, 180.
98. Emily Honig and Gail Hershatter, *Personal Voices: Chinese Women in the 1980s* (Stanford, Calif.: Stanford University Press, 1988), 7.

The resulting tension between a renewed interest in feminine sexuality and pressures on couples to return to more traditional relationships is clear in Sòng's short story, published not long after the campaigns against pornography and other forms of "spiritual pollution" that began in 1983. Just as Sòng had defended his choice to write about *yěrén* by saying that he was promoting science and not superstition, Sòng and several of his readers defended the story against charges of obscenity by saying that it promoted good human values. They pointed especially to the female *yěrén*'s traditional womanly virtues—her warm-heartedness and her loyalty to husband and family. Sòng's statements about his own marriage further established his position on this subject. In an interview, he praised his wife for her attention to household chores and called her by the appelation popular in the nineteenth century, "dutiful wife and loving mother" (賢妻良母). Sòng's female *yěrén* symbolized a break not only from the sexual restrictiveness of the Cultural Revolution, but also from its ideals of gender neutrality, and moreover from its privileging of politics over loyalty to one's mate and family.

The use of a "wild person" as a symbol of human sentiment was striking to contemporary readers. The "single male" reader of Sòng's story found the female *yěrén* to surpass the human women of his experience in womanly virtue. With "feelings and righteousness," Sòng's *yěrén* embodied a humanity that resonated strongly within Confucian tradition—and with the writings of Máo-era intellectuals like Wáng Rènshū, whose theory of a "human nature" based on "universal human feeling" had suffered attacks in the Cultural Revolution. Indeed, one review criticized Sòng's humanistic portrayal of the *yěrén*. The reviewers found Sòng's style "too simple," with the result that "the 'human nature' [人性] of the *yěrén* seems greater than that of the human."[99] Yet this is precisely the strength of *yěrén* symbolism. Granting human nature to a literal "savage" makes possible a damning criticism of less-than-civilized humans. When the protagonist of Gāo Xíngjiàn's novel *Soul Mountain* arrives in Shénnóngjià in the early 1980s, he wonders, "When the world is becoming increasingly incomprehensible, where man and mankind's behavior is so strange that humans don't know how humans should behave, why are they looking for the Wild Man?"[100] An answer may be found in Zhōu Liángpèi's 1986 poem "I Am *Yěrén*" (我是野人), in which the "I" who is *yěrén* tells the reader: "If we look at one another,

99. Wáng Píng and Lǚ Xuě, "Guǎngxī 'tōngsú wénxué rè' diàochá jì" [An Investigation of Guǎngxī's "Popular Literature Fever"], *Wényì bào*, no. 2 (1985): 40.

100. Gao Xingjian, *Soul Mountain*, trans. Mabel Lee (New York: Harper Collins, 2000 [1990]), 364.

we'll see who still retains a tail / You are my bright mirror, and I am yours."[101]

For both Gāo and Zhōu, *yěrén* became a "bright mirror" held up to reflect the barbarity of civilization, the inhumanity of humanity. The suitability of *yěrén* for this task lay first of all in their own primitiveness and thus their immunity from the evils associated with "civilization." Here Gāo and Zhōu evoked a theme made long familiar in China by the Daoist classics of Zhuāngzǐ and Lǎozǐ and in the West by Rousseau's "natural man."[102] *Yěrén* were also connected to the "traditional" cultures that had preserved them in legend. Zhōu Liángpèi's poem frequently refers to *yěrén* through such images of the distant past as "the history book," "the primeval soul not stained by the dirt of custom," "the life fixed in prehistory yet still living today," and "the not-yet-civilized wildness." In a particularly telling stanza, he explains, "I only fear that people's hearts are not ancient; it is a modern anxiety / The search for searching often resides in returning to the origin / The art of art often resides in going back to truth."[103] Gāo Xíngjiàn was engaged in a similar root-searching endeavor when he wrote *Yěrén*. In his preface to the play, he notes that his research included visits to the Yángzǐ River valley, where he witnessed priests and shamans performing rituals he associates with the "the origins of Chinese theatre." A character in the play gives voice to Gāo's concern for these people and the ancient practices they have preserved: "These days sorcerers are a rare commodity. If we want to understand the past, we must respect them and try to understand their ways. Like Wild Men, they are dying out fast." Another character suggests: "We could link [*yěrén*] with mankind's evolution and countless folktales and legends and help to explain these myths. Perhaps we can give a new lease on life to some of those old stories, and . . ." At this point, we are told, the speaker's "excitement turns to sorrow."[104] As Gāo rightly ascertained, legends themselves were threatened with extinction. If rehabilitating them through science was their best hope for survival, this was a sad state of affairs. Gāo and Zhōu's views on legend thus

101. Zhōu Liángpèi, *Yěrén jí* [*Yěrén* Collection] (Běijīng: Huáxià chūbǎnshè, 1992), 97. The "tail" here is symbolic only. In legends and eyewitness accounts, *yěrén* do not have tails.

102. On Rousseau's influence on his contemporaries' approaches to feral children, see Julia V. Douthwaite, *The Wild Girl, Natural Man, and the Monster: Dangerous Experiments in the Age of Enlightenment* (Chicago: University of Chicago Press, 2002); Michael Newton, *Savage Girls and Wild Boys: A History of Feral Children* (London: Faber and Faber, 2002), 106–108, 112–113.

103. Zhōu Liángpèi, *Yěrén jí*, 97–98.

104. A Chinese literary movement that began in the 1980s, root searching (尋根) is identified with the search for the origins of Chinese culture, especially among rural and ethnic minority people. Gao Xingjian, "Wild Man," 192, 239, 244.

differed from those of science disseminators and some fiction writers who stood firmly on the side of science and against superstition.

Despite the continued vigilant attitude toward superstition on the part of officials and many intellectuals, the interest Zhōu and Gāo expressed in primitivism was widely shared among intellectuals in the post-Máo period and was actually encouraged by changes in official policy regarding minority nationalities. While still pursuing integration, officials began once again to encourage ethnic minorities to identify themselves as such by wearing traditional dress and speaking their mother tongues (although only in addition to the national language).[105] The late 1970s and early 1980s saw an artistic renaissance inspired by the arts of minority nationalities, as in the Yúnnán School of painting that emerged during this period.[106] Filmmakers and novelists of the 1980s also eagerly tapped the subject of minority nationalities. These primarily Hàn Chinese intellectuals were drawn to "the minority other" because they "enjoyed freedoms that had been denied to the Hàn subject, who was seen to be doubly tormented by the strictures of Confucianism and the aesthetic practices of Maoist socialism."[107] For writers like Zhōu and Gāo, *yěrén*'s wildness and association with "primitive" cultures evoked a similar sense of freedom.

Zhōu and Gāo thus shared with Sòng Yōuxīng an attraction to the freedom symbolized by *yěrén*, but the types of freedom were different. While Sòng focused on a kind of sexual freedom and rejected "legend," Zhōu and Gāo celebrated legend and moreover expressed concerns that were explicitly political. Both Zhōu and Gāo used *yěrén* as a mirror to reflect on the viciousness of Máo-era political campaigns. Zhōu's introduction to his 1992 book of poetry, entitled *Yěrén Collection*, explained that his motivation for writing "I Am *Yěrén*" in 1986 lay in his experience during the Cultural Revolution. Zhōu's foot was hurt, and his inability to obtain medical care made the injury permanent.[108] Gāo Xíngjiàn likewise linked *yěrén* to the inhumanity of the Cultural Revolution and other Máo-era campaigns. In *Yěrén*, a villager informs a visiting scientist, "There was a professor here once who criticized someone for shooting a songbird. . . . They tortured him to death during the Cultural

---

105. June Teufel Dreyer, *China's Forty Millions: Minority Nationalities and National Integration in the People's Republic of China* (Cambridge, Mass.: Harvard University Press, 1976), 237–247.

106. Joan Cohen, *Yunnan School: A Renaissance in Chinese Painting* (Minneapolis: Fingerhut Group Publishers, 1988), 25.

107. Ralph Litzinger, *Other Chinas: The Yao and the Politics of National Belonging* (Durham, N.C.: Duke University Press, 2000), 231.

108. Zhōu Liángpèi, *Yěrén jí*, 1–7.

Revolution. If a man tortures another, he's worse than a wild animal."[109] In *Soul Mountain*, the protagonist hears of a disturbing encounter with a creature thought at first to be a *yěrén*. It turns out instead to be a man who has escaped from a labor camp where he had been sent during an antirightist campaign. Too frightened to return to society, he has been living in the mountains ever since.[110] This story and others like it recall the much older story of the Great Wall conscript evaders, who lost their humanity when the first emperor's cruelty forced them to flee into the wilderness. Such stories accomplish sophisticated political critiques by suggesting that inhumane treatment can result in a loss of humanity, a passing back into the realm of beasts.

The use of *yěrén* to question civilization from the perspective of "wild people" cannot be separated from the emergence of an environmental movement in China in the 1980s and 1990s. Zhōu Liángpèi's poem portrayed Shénnóngjià as "trees full of ancient mosses" threatened by the "felling circle pressing closer." He warned, "Cutting down [trees], capturing [wild animals], the great mountain falls into a terror from which it has no power to save itself."[111] In Gāo Xíngjiàn's play, *yěrén* served as spokespeople for the forests. In one telling moment, an ecologist responds to news of a logging road under construction by asking, "What would our Wild Man think about that?" Bruno Roubicek, who translated the play into English, credited Gāo for "presag[ing] the warnings against deforestation that have appeared in the Chinese press since its premiere" in 1985.[112]

Gāo's play certainly stands as a beautiful and strong defense of nature, and especially forests, against wanton destruction in the name of socialism and private profit alike. Nonetheless, Gāo was not the first to introduce this theme, nor was he original in linking deforestation with the question of *yěrén*. Already in the late 1950s, the socialist state had become attuned to the consequences of deforestation for endangered animals, most notably the giant panda.[113] These priorities have had enormous influence on *yěrén* research from the beginning. As early as 1964, Máo Guāngnián stated emphatically, "Because of their small numbers, their [evolutionary] closeness to humans, and some other characteristics, I

109. Gao Xingjian, "Wild Man," 225.
110. Gao Xingjian, *Soul Mountain*, 385–388.
111. Zhōu Liángpèi, *Yěrén jí*, 97–98.
112. Gao Xingjian, "Wild Man," 201, 186.
113. Elena Songster, "A Natural Place for Nationalism: The Wanglang Nature Reserve and the Emergence of the Giant Panda as a National Icon" (Ph.D. diss.: University of California, San Diego, 2004).

believe that the *jué* species [his preferred term for *"yĕrén"*] are worth earnestly researching, protecting, and breeding."[114]

Post-Máo writings on *yĕrén* have increasingly focused on environmental protection and have frequently tied the question of *yĕrén* directly to the question of pandas. In a 1977 speech at a meeting of the *yĕrén* investigation team in Shénnóngjià, Qián Guózhēn noted that fossils of pandas and of *Gigantopithecus* had been discovered in the same areas, what he termed an "ancient 'panda-*Gigantopithecus* belt.'" He professed his belief that pandas were not alone in their ability to change their food habits and survive the ravages of time, but rather shared this characteristic with *Gigantopithecus*. Like pandas, *Gigantopithecus* (or *"yĕrén"*) were now on the verge of extinction. Protecting them was "an issue of great international significance in terms of politics, philosophy, and science."[115] Yuán Zhènxīn and Huáng Wànbō's 1979 article in *Fossils* made the same point. They worried that with widespread deforestation *yĕrén* would become extinct before their existence could be proven, and they explicitly linked their own research and interest in protecting *yĕrén* with panda research and preservation.[116] Similarly, Zhōu Guóxīng cited the possibility that Shénnóngjià harbored *yĕrén* as a compelling reason to create a nature reserve there.[117] While there were many other good reasons, it is likely that the excitement over *yĕrén* helped make the reserve a reality in 1983.[118]

More recently, *yĕrén* trackers like Lí Guóhuá, Yuán Yùháo, and Zhāng Jīnxīng have become directly involved in environmental protection. Lí and Yuán both participated as lay members of the 1970s *yĕrén* investigations; both now work for the Shénnóngjià nature reserve (Lí as a photographer and disseminator, and Yuán as a ranger). Zhāng has modeled himself on Jane Goodall, both in his immersion style of research and his commitment to educating the public about the importance of saving the wilderness.[119]

What was new to environmental discourse beginning in the 1980s was the criticism of modernity, progress, and human civilization itself.

114. Máo Guāngnián in Wáng Bō, *Yĕrén zhī mí*, 122.
115. Máo Guāngnián in Wáng Bō, *Yĕrén zhī mí*, 115–116.
116. Yuán Zhènxīn and Huáng Wànbō, "'Yĕrén' zhī mí," 3–4.
117. Zhōu Guóxīng, "Shénnóngjià 'yĕrén' kǎochá," 30.
118. Dù Yǒnglín attributes the inspiration for the Húbĕi Forestry Department's initial 1978 proposal of a nature reserve in Shénnóngjià to a 1977 report from a *yĕrén* investigation team. See Dù Yǒnglín, *Yĕrén: Lái zì Shénnóngjià de bàogào* [*Yĕrén*: Report from Shénnóngjià] (Bĕijīng: Sānxiá chūbǎnshè, 1995), 101.
119. Lí Guóhuá, interview with the author, 17 April 2002; Yuán Yùháo, interview with the author, 16 April 2002; Zhāng Jīnxīng, interview with the author, 18 April 2002. Zhāng Jīnxīng, *Zhĕng zuò Gǔdàoĕr*, 4.

In the historical context of a growing awareness of environmental degradation, the dominance of humans over all other creatures has been questioned. The signs have been reversed: the all-too-human capacity for destruction is labeled "savage," and people are called upon to reshape themselves in the image of nature—as caretakers of the forest that once sheltered our tree-dwelling ancestors. To quote from the 1986 poem "I am *Yěrén*," "Heaven and earth are inverted and spinning in karmic retribution."[120]

Also new in the post-Máo era is an appreciation for nature as a source of spiritual comfort and enlightenment. For many of the most devoted *yěrén* hunters and popularizers, saving the environment has come to mean saving themselves, and searching for *yěrén* has meant searching for their own souls. For the poet Zhōu Liángpèi, what was needed was a "primeval soul not stained by the dirt of custom." Such a soul could be found, for Zhōu and others, in a return to nature. Lí Guóhuá, who in the late 1970s claimed to want to kill a *yěrén* and bring its head to the scientific institute, told American film makers a decade after he began his search: "My life came from nature, and I might end my life going back to nature. It's as if my soul has been possessed by nature, possessed by *yěrén*."[121] Similarly, a newspaper reporter commented in 1988 that "Professor of *Yěrén*" Liú Mínzhuàng "long ago committed his life to the search for nature's mysteries. His spirit resides in Shénnóngjià."[122] Nature offered a space away from the troubles of civilization to find oneself again. "In the pathless forest, to be lost does not count as being lost," said the poet Zhōu Liángpèi.[123] "I seek your protection, you seek my shelter," says Zhōu's *yěrén* narrator, speaking for the forests.

It is just this shelter, this finding oneself by losing oneself, that has drawn some people deeper and deeper into the forests in their search for wild people, wilderness, or wildness. It is not uncommon to find them consciously drawing on the rich resources of Daoist philosophy to express this impulse. When the narrator of *Soul Mountain* arrives in

---

120. Zhōu Liángpèi, *Yěrén jí*, 97. This is where my analysis of the cultural meaning of wildness in post-Máo China departs from Taussig's analysis of it in other times and places. In the cases Taussig examines, the healing power of wildness resides in the way it "tears through the tired dichotomies of good and evil, order and chaos" and inevitably "emerge[s] on the side of the grotesque and the destructive" (*Shamanism, Colonialism, and the Wild Man*, 220).

121. Poirier and Greenwell, *The Wildman of China*. Lí currently works as a wildlife photographer in the propaganda and education department of the Shénnóngjià nature reserve. Lí Guóhuá, interview with the author, 17 April 2002.

122. Yǐn Sǔnjūn, "Hún xì Shénnóngjià" [His Spirit Is Tied to Shénnóngjià], *Chángjiāng rìbào*, 19 January 1988, 3.

123. Zhōu Liángpèi, *Yěrén jí*, 98.

Shénnóngjià, he feels ambivalent about the idea of looking for *yěrén*, and says, "Not having a goal is a goal, the act of searching itself turns into a sort of goal, and the object of the search is irrelevant."[124] Zhōu Liángpèi's "pathless forest" and his line "The search for searching often resides in returning to the origin" similarly evoke a Daoist rejection of ambition in favor of the freedom that can be found only in nature.

The most enthusiastic of the *yěrén* enthusiasts, and the most desperate of the socially persecuted, do not simply search for *yěrén*, they become *yěrén*. When Lí Guóhuá returned from tracking *yěrén*, his coworkers exclaimed that he had "turned into a *yěrén*." Liú Mínzhuàng said that he "lived a *yěrén*'s life." The narrator of *Soul Mountain* imagined himself fleeing the inhumanity of society and becoming a *yěrén*.[125] Some *yěrén* trackers have intentionally altered their appearance to resemble *yěrén* with the hope that this will facilitate encounters with the real thing.[126] Zhāng Jīnxīng, who has forsworn shaving until he finds a *yěrén*, reportedly has "detractors" who "joke that he's searching for himself" (figure 22).[127] Clearest of all is Zhōu Liángpèi, the poet who survived the ravages of the Cultural Revolution to become a *yěrén*, spokesperson for the trees and for all threatened by inhumanity and carelessness.

I am the *yěrén* you yearn for, you fear, you hate,
and because of this don't understand, the *yěrén* your wide open eyes are searching for!

. . .

I, Shénnóngjià's *yěrén*!
I am the being in which you believe,
I am the outcome of your thoughts and searches.[128]

While such metamorphoses are undoubtedly quite rare in practice, they are common enough in the popular imaginary that Liú Mínzhuàng included a warning in his 1988 book. Just as Liú distinguished *yěrén* from members of "primitive tribes," he also explained the difference between *yěrén* and "people gone wild" (野化人), who are "modern humans who run away to the deep mountains and eat wild fruits, do not wear clothes, and do not make tools."[129]

124. Gao Xingjian, *Soul Mountain*, 342.
125. Gao Xingjian, *Soul Mountain*, 389.
126. Liú Mínzhuàng, *Jiēkāi 'yěrén' zhī mí*, 175.
127. Zhāng Jīnxīng, *Zhēng zuò Gǔdàoěr*, 11; Loussouarn, "What's Out in the Woods?" 9.
128. Zhōu Liángpèi, *Yěrén jí*, 98–99.
129. Liú Mínzhuàng, *Jiēkāi 'yěrén' zhī mí*, 265. A 1951 book on human evolution described "feral"

22 Zhāng Jīnxīng, who has sworn not to shave his beard until he catches a *yěrén*. This photograph nicely captures Zhāng's deliberate efforts to blur the lines between human and animal,
scientist and subject, hunter and quarry. The photograph, by Sahid Maher, appears in Anne
Loussouarn's "What's Out in the Woods?" the cover article for a 2000 trial issue of an expa-
triate English-language magazine, *City Weekend*, published in Běijīng.

It is this constant need to distinguish *yěrén* from humans, to explain
in what ways they are and are not similar to us, that has made them such
a rich subject in popular culture. Their slippery status requires constant
attention to definitions, and the process of defining inspires thoughtful
reflection on what humanity is, what it is not, and what it should be.
Creatures that are almost, but not quite, human have made their way
into stories all over the world in ancient as well as modern times. The

---

people who grew up among wild animals as *yěrén*. The author explained that although they had the
physical ability to speak, their lack of social labor precluded it. Lín Yàohuá, *Cóng yuán dào rén de
yánjiū* [Research on from Ape to Human] (Běijīng: Gēngyún chūbǎnshè, 1951), 59.

opening of the publishing market in the 1980s and 1990s helped such stories circulate in China more freely than before. However, the emergence of *yěrén* fever in post-Máo China is not simply the sudden satisfaction of long-held interest. Changes in post-Máo China have raised deep questions that call out for cultural icons like *yěrén*. As a symbol of the wild, *yěrén* have been ideally suited to aid in the exploration of sex and human values, tradition and modernity, development and environment, society and self.

As a "mystery," *yěrén* have also helped raise questions about the omnipotence of science. While many continue to see *yěrén* along with other "mysteries" as scientific questions bound to be answered conclusively, others have been reluctant to cede all to science. Appeals to legend, spirit, and wildness have allowed a space for ideas generated from popular culture to take on fundamental questions about humanity and society. This space is *not* outside science. Science is not sheltered from popular culture. Rather, popular culture has mingled with science, and thus made popular science truly popular.

Conclusion

Yěrén literature reflects enormous uncertainties about science and culture in post-Máo China. The crusade against superstition in the name of science is by no means over—as evidenced in recent years by the official response against the Fǎlún Gōng movement and the increased attention science dissemination has received as a result. Much of *yěrén* writing is still justified in terms of an attack on superstition via the time-honored means of science dissemination. Moreover, this top-down model of scientific knowledge production and dissemination no longer meets any officially sanctioned challenge from a bottom-up model of "mass science," either in rhetoric or in practical programs.

Nonetheless, the boundaries between science and popular culture have become more, not less, porous in the post-Máo period. Paleoanthropologists have authored popular accounts of *yěrén* research, which in turn is based on stories told by rural people. Fiction writers have used the *yěrén* research as inspiration for sensational stories and sensitive explorations into what it means to be human. And laypeople have used the marginal position of *yěrén* research as a way to enter the scientific arena and challenge the reigning authorities therein.

This is not to say that the old prejudices against the masses as superstitious are disappearing—if anything, they may be more virulent. The market is perfectly capable of exploiting the riches of popular culture without giving credit where credit is due. The heroes of *yěrén* literature are rarely local people; only urbanites can credibly "return to nature," and only moderns can search for a "primeval soul." Whether acknowledged or not, however, popular culture—even in its most allegedly superstitious forms—has unquestionably gained greater influence in science dissemination writing and even scientific research.

This chapter opened with an account of a scientist who "recognized" a statue of Peking Man. He had "seen it before" during an encounter with a creature he learned to call "*yěrén*." *Yěrén* and Peking Man have become intertwined in the post-Máo period, an association that has enhanced the popularity of Peking Man and the science she represents. People drawn by the mystery of *yěrén* to purchase books on the subject are often exposed to established theories of and evidence for human evolution. *Yěrén* literature has similarly glamorized the scientific investigation of human origins and the profession of science as a whole. The rugged, heroic personas that scientists in the early People's Republic worked to cultivate are well disseminated in recent writings on *yěrén*. And the *yěrén* does something for popular paleoanthropology that its slightly less popular "twin," Peking Man, does somewhat less well. As a paradoxical "living legend," it has the power simultaneously to seem real and to inspire the imagination. *Yěrén* have encouraged people to reflect on humanity and society and to imagine "a different and better condition."[130]

Not everyone will agree that *yěrén* research—or Bigfoot research, for that matter—is a legitimate scientific enterprise. Some question the scientific rigor of its assumptions and methodology; others worry what harm is done to legend when it is shoe-horned into science. Yet I believe it has had cultural value in post-Máo China. It has offered a way for science, literature, and legend to come together in an exploration of humanity and inhumanity, wild(er)ness and civilization, so necessary in China's rapidly changing social and natural worlds.

---

130. Constance Penley, *NASA/TREK: Popular Science and Sex in America* (London: Verso, 1997), 15–16. My thoughts on the relationship between Peking Man and *yěrén* are strikingly similar to Penley's analysis of NASA and *Star Trek* as popular science "twins."

EIGHT

"Have We Dug at Our Ancestral Shrine?" Post-Máo Ethnic Nationalism and Its Limits

The Scope and Limitations of Chinese Ethnic Nationalism

On 28 December 2001, I attended the opening of a new Paleolithic site museum in the basement of a shopping mall at Wángfǔjǐng in downtown Běijīng.[1] I went to the opening as an observer, but as the only foreigner there, I quickly became the focus of attention myself. The reporters were especially excited to photograph me next to one of the exhibit's statues, saying that it made for an interesting study in contrast: a Western, modern woman next to a Chinese, ancient man (figure 23). When I sent the picture as it appeared in the *Běijīng Evening News* to friends and family in the United States, one person wrote back suggesting that the reporters had actually seen a resemblance between me and the "caveman." It was a joke, but a perceptive one. A scientist who helped oversee the excavations and the design of the exhibit is disappointed with the statue. The tall bridge of the nose and the round eyes make the man

1. The site was discovered in 1996 during the construction of Oriental Plaza (東方廣場), an enormous mall and business center in the historically commercial Wángfǔjǐng district. The museum is in a small area in the basement of the mall where it connects with the subway station.

23 The author attending the opening of the Wángfǔjǐng Paleolithic Exhibit in the basement of the Oriental Plaza shopping mall, 2001. The photograph appeared in Yáng Wēi, Zuǒ Yíng, and Cài Wénqīng, "Jīntiān yuēhuì gǔrénlèi" [A Date Today with Ancient Humans], Běijīng wǎnbào, 28 December 2001, 2.

look too Western to be the ancestor of the Chinese people he is supposed to represent.

In China today, as in previous years, questions of ethnic and national identity are central to the story of human evolution as told in both professional and popular arenas.[2] Exhibits on human evolution at the Běijīng and Shànghǎi natural history museums, the Institute of Vertebrate

2. On the relationship between national and ethnic identities, see Etienne Balibar, "Racism and Nationalism," in Race, Nation, Class: Ambiguous Identities, ed. Etienne Balibar and Immanuel Wallerstein (London: Verso, 1991 [1988]).

Paleontology and Paleoanthropology, and the shopping mall at Wángfǔ-jǐng seek especially to educate visitors about human evolution in *China* and so to arouse their patriotic sentiment. The government also encourages and expects history teachers to use discussions of Chinese human fossils to "expound on the origins of the Chinese people and cause students to realize the long history of the ancestral country and [its status as] one of the origin places of the world's humans."[3]

Nationalist influence on Chinese paleoanthropology has been thoroughly exposed and denounced in Barry Sautman's article "Peking Man and the Politics of Paleoanthropological Nationalism in China," published in the *Journal of Asian Studies* in 2002.[4] Sautman examines a large number of newspaper articles and other sources in order to demonstrate the consistent preference given in China to theories of human evolution that preserve a Chinese origin for humanity as a whole and Chinese people in particular. His discussion focuses especially on Chinese scientific and media support for the "multiregional theory" of modern human origins in contrast with the "recent out-of-Africa theory" that the majority of Western paleoanthropologists have come to support.

The latter theory—also known as the "replacement theory" or "Eve theory"—suggests that all modern humans are descended from a common ancestor who lived in Africa perhaps less than 100,000 ago and whose descendants migrated to Asia and Europe, replacing existing human populations. Peking Man and other early Chinese human fossils lose their status as human ancestors in this model. In contrast, the multiregional theory, or "continuity theory," maintains that humans evolved from *Homo erectus* to anatomically modern *Homo sapiens* in Africa, Asia, and Europe. Modern populations have retained a detectable degree of morphological continuity with the fossil humans in their areas. Some critics of the multiregional theory brand it racist: while the replacement model, or recent out-of-Africa theory, emphasizes the recency of racial differences and thus works to undermine race as a significant category, the multiregional theory proposes a far longer history for racial differences.[5]

---

3. Zhōu Fāzēng and Gōng Qízhù, *Lìshǐ jiàoxué yǔ àiguózhǔyì jiàoyù* [History Pedagogy and Patriotic Education] (Jǐ'nán: Shāndōng jiàoyù chūbǎnshè, 1984), 40.

4. Barry Sautman, "Peking Man and the Politics of Paleoanthropological Nationalism in China," *Journal of Asian Studies* 60, no. 1 (2001): 95–124. On nationalist influence in archaeology (in China and elsewhere), see Philip L. Kohl and Clare Fawcett, eds., *Nationalism, Politics, and the Practice of Archaeology* (Cambridge: Cambridge University Press, 1995).

5. Robert N. Proctor, "Three Roots of Human Recency: Molecular Anthropology, the Refigured Acheulean, and the UNESCO Response to Auschwitz," *Current Anthropology* 44, no. 2 (2003): 215, 224–225.

Sautman makes a persuasive argument that a nationalistic state agenda has worked to privilege scientific theories that root Chinese ethnic identity in the remote past. He offers a very compelling analysis of contemporary Chinese nationalism and the manipulation of discourse on human evolution to its advantage. He is also correct, I think, in seeing this trend to have accelerated beginning in the mid-1980s. Nonetheless, I cannot agree that the mid-1980s marked a watershed before which Peking Man stood for the origins of humanity as a whole and after which it stood for Chinese origins.[6] Rhetoric consistently linking Peking Man with China's "glorious" past dates to the early 1950s, and there is plenty of room left for humanity in contemporary Chinese discourse on Peking Man.

In the process of proving his points, Sautman also paints a picture of paleoanthropology that is somewhat unbalanced. This results from the question driving his research. He seeks to show that state-sponsored Chinese nationalism has encouraged and drawn strength from scientific theories that emphasize the longevity of the Chinese as a biological race and the connection of this race to the Chinese land. With this I am in complete agreement. My purpose, however, is to understand the significance of Chinese discourse on human origins as a whole, including but not limited to issues of ethnic nationalism. The question then becomes, is ethnic nationalism a sufficient explanation of Chinese preferences for theories that place human evolution on a Chinese stage?

I answer no. First, nationalism alone does not account for the ongoing debates in China and elsewhere. The jury is still out on many questions about human evolution, including the question of continuity versus replacement with respect to modern human origins. In addition to political issues, there are empirical and methodological disagreements that drive the debates. Moreover, Chinese scientists are hardly alone in arguing for the continued significance of Chinese fossils (meaning fossils unearthed in China) as human ancestors. Rather, Western scientists who have built their careers on Chinese fossils share with Chinese scientists a professional interest in preserving their ancestral status. Conversely, Chinese geneticists, whose careers are not tied to fossils, are less likely to be wedded to multiregionalism. Second, where nationalism does play a role (and I agree that the role is significant), it is not only or even primarily an issue of race or ethnicity. Chinese scientists, along with the Chinese government, are proud of China's place in the history of paleoanthropology and hope that China will continue to be a "center

6. Sautman, "Peking Man," 109.

for research on human origins." This is a question more of prestige in international science than of ethnic identity.

Third, while the Chinese state does frequently mine theories of human origins to construct a concept of the Chinese nation rooted in a biological concept of race, this construction is simultaneously destabilized by other meanings produced by scientists, laypeople, and the state itself. Discourse on human evolution works to construct identities for smaller units than the Chinese nation, including for "southern Chinese" versus "northern Chinese" and for "minority nationalities." It also works to demonstrate connections between Chinese people and people in other parts of the world—for example, other East Asians, Native Americans, and Australians. Furthermore, the Máo-era commitment to seeing "all the world as one human family" continues to shape some important science dissemination materials on human evolution.

Finally, there is the question of how people on the receiving end of science dissemination view the fossils. I suggest that many people have embraced Peking Man, Yuánmóu Man, and other human fossils not simply as early representatives of their nation or race, but in much more personal ways. Published materials, survey responses, and interview data provide hints about how some Chinese people from diverse backgrounds have come to understand their relationships to the human fossils unearthed in China. Paleoanthropology has worked to support not only a Chinese national identity, but also family, community, regional, professional, and human identities.

Earliest Origins of Humanity

The search for the site of human origins is a very ambiguous project. The answer necessarily depends on what one considers the critical juncture in human evolutionary history. For example, some scientists have suggested that one branch of fossil apes in southern China remained there and evolved into the great apes, while another branch migrated to Africa and then evolved first into *Australopithecus* and then into early humans before migrating back into Asia. The Chinese press has seen this as supporting the claim that "China is one of the birthplaces of early humans," even though such a theory suggests that the first humans emerged in Africa.[7] Still more strangely, in 1995 the Chinese press heralded the discovery of "dawn ape" fossils, said to be the earliest primates

---

7. Gōng Dáfā, "Nányuán huàshí chūtǔ jì" [On the Excavation of *Australopithecus*], *Rénmín rìbào*, 25 November 1989, 3.

yet unearthed, as evidence supporting China's claim to be the cradle of humanity. Chinese scientists, however, criticized this interpretation, noting that humans evolved from just one branch of the primate order.[8] If we celebrate the site of each link in the chain extending back to earliest life, humans may be said to be "from" virtually everywhere on the globe.

More commonly, however, the question, Where did humans originate? has come to center on just two junctures. Ernst Mayr explained the issue in a paper he presented in 1950 at the Cold Spring Harbor symposia on Quantitative Biology, the seminal meetings in which what is known as the "modern synthesis" in evolutionary science was brought to bear on paleoanthropology: "The analysis of this problem [of the 'missing link'] will be facilitated by the realization that it is an oversimplification to use in this case the uninomial alternative 'ape' versus 'man.' . . . Classifying man binomially as *Homo sapiens*, it at once becomes apparent that we must look for two missing links, namely that which connects *sapiens* with his ancestor and that which connects *Homo* with his ancestor."[9] The recent out-of-Africa theory and the multiregional theory introduced above both address the question of where fully modern humans (anatomically modern *Homo sapiens*) emerged. Before turning to that debate, however, we shall first address the question of where the earliest creatures we may consider human (*Homo*) arose—the same question that occupied scientists of the republican era and the early People's Republic.

When writers in the nineteenth and early twentieth centuries speculated on the likely place of human origins, they meant the site where the earliest humans had split away from the apes. In his influential *Man's Place in Nature*, first published in 1863, Thomas Huxley considered "either the Chimpanzee or the Gorilla" to be "the Ape which most nearly approaches man, in the totality of its organization."[10] Charles Darwin agreed, and for this reason believed it "somewhat more probable that our early progenitors lived on the African continent than elsewhere."[11] Ernst Haeckel, on the other hand, considered the Asian apes to be more similar to humans and so suggested Asia as the probable

---

8. Wú Rǔkāng, "Gāoděng língzhǎnglèi zǔxiān bù děngtóng yú rénlèi zǔxiān" [The Ancestors of Higher Primates and Humans Are Not Equivalent], *Zhōngguó kēxué bào*, 29 April 1996; Zhōu Guóxīng, "Rénlèi qǐyuán dà sōusuǒ" [The Great Search for Human Origins], *Zhōngguó guójiā dìlǐ zázhì*, no. 10 (2000): 60.

9. Ernst Mayr, "Taxonomic Categories in Fossil Hominids," *Cold Spring Harbor Symposia on Quantitative Biology* 15 (1950): 115.

10. Thomas H. Huxley, *Man's Place in Nature* (New York: Random House, 2001 [1863]), 72.

11. Charles Darwin, *The Descent of Man, and Selection in Relation to Sex*, 2nd ed. (New York: Clarke, Given, and Hooper, 1874), 177.

cradle of humanity.[12] Eugene Dubois's 1891 discovery of Java Man reinforced Haeckel's position, following which the work of William Diller Matthew and Henry Fairfield Osborn in the second and third decades of the twentieth century convinced many of the likelihood that central Asia in particular was the birthplace of humanity. In the 1940s, Franz Weidenreich promoted a human lineage beginning with *Gigantopithecus* (ancient ape fossils recently discovered in southern China) through Java Man and Peking Man to modern humans.[13]

Not everyone was convinced by Asian origins. Some continued to argue for Europe.[14] More important, Raymond Dart's 1924 discovery of *Australopithecus*, together with supporting evidence found by Robert Broom in the late 1930s and then again in 1947, built confidence in African origins.[15] Broad changes in the dominant theoretical approach to questions of human evolution served to bolster support for the *Australopithecus* fossils. Known as the "modern synthesis," the integration of evolutionary theory and modern genetics accomplished from the 1920s to the 1950s required evolutionary change to be explained with respect to adaptation to the natural environment rather than in terms of innate trends. This helped resolve the old question of which came first, an advanced brain or bipedal locomotion. (The latter had been Engels's position, though this apparently went unnoticed by participants in the modern synthesis.) The small-brained, erect-walking *Australopithecus* was easier to accept as a human ancestor in this light.[16] Nonetheless, support for an Asian origin of humanity did not fully die out even among Western paleoanthropologists. *Dryopithecus*, *Gigantopithecus*, *Ramapithecus*, and *Sivapithecus* fossils from Eurasia continued at least through the 1970s and early 1980s to keep the possibility of Asian origins alive for some.[17]

12. He specifically suggested that a postulated, now sunken continent in the Indian Ocean named Lemuria was the most likely place of origin. Ernst Haeckel, *The History of Creation*, trans. E. Ray Lankester, 5th ed. (New York: D. Appleton Company, 1910 [1876]), 2:437.

13. Franz Weidenreich, *Apes, Giants, and Man* (Chicago: University of Chicago Press, 1946).

14. Bowler cites Ales Hrdlička as defending European origins in 1926. Peter Bowler, *Theories of Human Evolution: A Century of Debate, 1844–1944* (Baltimore: Johns Hopkins University Press, 1986), 106. Until they were proved fraudulent, the *Eoanthropus* (Piltdown Man) fossils, unearthed in 1912, also helped defend European origins.

15. B. G. Campbell and R. L. Bernor, "The Origin of the Hominidae: Africa or Asia?" *Journal of Human Evolution* 5, no. 5 (1976): 441–454.

16. Bowler argues that the *Australopithecus* fossils themselves were not the key factor in moving the consensus from a brain-first to a bipedalism-driven theory of evolution. Rather, the transformation occurred as a result of the theoretical revolution of the evolutionary synethesis. Proponents of the brain-first model were able to integrate the *Australopithecus* fossils into their paradigm. Bowler, *Theories of Human Evolution*, 185.

17. Campbell and Bernor, "The Origin of the Hominidae." The authors cite von Koenigswald, who found the first *Gigantopithecus* tooth in a Hong Kong apothecary in 1935, as one scientist who continued to favor Asian origins for hominids.

Chinese popular science materials on human evolution from the 1950s and 1960s continued to favor Asia as the most likely origin of earliest humanity while often presenting the subject as an open question in which Africa and Asia were both candidates.[18] This pattern has continued in many books published in the post-Máo period, only now it is southern Asia (including southwestern China), rather than central Asia, that receives the most attention.[19] Books published in the 1990s and beyond have offered diverse opinions on the subject, including strong support for Asian origins, strong support for African origins, and a "wait and see" approach.[20]

Support for Asian origins in the post-Máo period has been sustained in part through a series of suggestive fossil discoveries. The first evidence came in the form of the *Lufengpithecus* fossils of Lùfēng, Yúnnán, discovered in 1975 by an employee of the local cultural center who was also "a famous 'fossil fan' in the county."[21] In 1979, Chinese scientists attributed the fossils to *Ramapithecus* and *Sivapithecus*, the former at that point having been widely and internationally considered the earliest hominid representative.[22] The significance of this research was widely celebrated in the Chinese press as offering new support for the Asian origins of hominids. The IVPP scientist Xú Qìnghuá, who led the excavation team in

18. Zhū Xǐ, *Wǒmén de zǔxiān* [Our Ancestors] (Shànghǎi: Wénhuà shēnghuó chūbǎnshè, 1950 [1940]), 146; Jiǎ Lánpō, *Zhōngguó yuánrén* [Peking Man] (Shànghǎi: Lóngmén liánhé shūjú, 1950), 11–16; Huáng Wéiróng, *Zhōngguó yuánrén* [Peking Man] (Shànghǎi: Shàonián értóng chūbǎnshè, 1954), 29–32; Fāng Shàoqīng [pseud. for Fāng Zōngxī], *Gǔ yuán zěnyàng biànchéng rén* [How Ancient Apes Became Human] (Běijīng: Zhōngguó qīngnián chūbǎnshè, 1958), 91–92; Fāng Shàoqīng, *Gǔ yuán zěnyàng biànchéng rén* [How Ancient Apes Became Human] (Běijīng: Zhōngguó qīngnián chūbǎnshè, 1965), 109.

19. Several of Wú Rǔkāng's books are good examples. See his *Rénlèi de qǐyuán hé fāzhǎn* [Human Origins and Development] (Běijīng: Kēxué chūbǎnshè, 1976), 58–59; Wú Rǔkāng, Wú Xīnzhì, Qiū Zhōngláng, and Lín Shènglóng, *Rénlèi fāzhǎn shǐ* [The History of Human Development] (Běijīng: Kēxué chūbǎnshè, 1978), 94–96; and, largely authored by Wú Rǔkāng and his associates, Shànghǎi rénmín chūbǎnshè, ed., *Shíwàn ge wèishénme (19): Rénlèi shǐ* [100,000 Whys, Vol. 19: Human History] (Shànghǎi: Shànghǎi rénmín chūbǎnshè, 1976), 41–42. See also Shèng Lì, Zhāng Lóng, and Shèng Lín, *Mànhuà cóng yuán dào rén* [Conversations on from Ape to Human] (Níngxià: Níngxià rénmín chūbǎnshè, 1983), 106–107.

20. For respective examples, see Jiǎ Lánpō, *Rénlèi qǐyuán zhī wǒ jiàn* [My View on Human Origins] (Shànghǎi: Shànghǎi kējì jiàoyù chūbǎnshè, 2000), 49–61; Yuǎn Jìn, *Yuǎngǔ yǔ wèilái* [The Past and the Future] (Chéngdū: Sìchuān kēxué jìshù chūbǎnshè, 1998), 11–15; Wèi Qí [魏奇, not to be confused with 衛奇] and Wèi Shìfēng, *Zǒuchū Yīdiàn yuán: Rénlèi de qǐyuán hé yǎnhuà* [Out of Eden: Human Origins and Evolution] (Xī'ān: Shǎnxī rénmín chūbǎnshè, 1994), 151–155.

21. Zhāng Xīngyǒng, "Lùfēng gǔyuán fājué sǎnjì" [Notes on the Excavation of *Lufengpithecus*], *Kēxué zhī chuāng*, no. 3 (1981): 2–4.

22. Xinzhi Wu and Frank E. Poirier, *Human Evolution in China: A Metric Description of the Fossils and a Review of the Sites* (New York: Oxford University Press, 1995), 241. D. A. Etler, T. L. Crummett, and M. H. Wolpoff, "Longgupo: Early *Homo* Colonizer or Late Pliocene *Lufengpithecus* Survivor in South China?" *Human Evolution* 16, no. 1 (2001): 2. *Ramapithecus* fossils have since been reinterpreted as female specimens of *Sivapithecus*.

Lùfēng, wrote in *Science Times* that the discovery was "a step forward in proving that southwest China was one of the important areas in human origins."[23] In 1981, after doubts had been raised about the ancestral status of *Ramapithecus*, Wú Rǔkāng wrote more cautiously in the *Guāngmíng Daily*: "If *Lufengpithecus* is on the human line, then the birthplace of humanity may well be in Asia. On the other hand, if *Lufengpithecus* is not on the human line, then the earliest fossil representative of the human line is *Australopithecus*, and since this ancient ape has mainly been found in Africa, then the birthplace of humanity is very likely in Africa and not in Asia."[24] The state press, however, was not always so restrained. A *Běijīng Evening News* article from 1981 boldly proclaimed that the fossils "proved . . . that humans originated in Asia, not in Africa."[25]

The subsequent dismissal of *Ramapithecus* as a human ancestor by the majority of paleoanthropologists internationally was accepted by Xú Qìnghuá, among other Chinese scientists; Xú even explained the new interpretation and its consequences in a 1985 article in *Fossils*.[26] And in 1987 Wú Rǔkāng reassigned the fossils to a separate genus and species, *Lufengpithecus lufengensis*, on the basis of significant differences between the fossils and other *Ramapithecus* and *Sivapithecus* specimens. Nonetheless, excitement about the possibility that the ape fossils from Yúnnán might serve as evidence for an Asian origin of humanity has remained strong among some scientists and especially in the popular media. In 1998, the popular science magazine *Mysteries Illustrated* suggested that the fossils could support a human lineage from *Sivapithecus* to *Homo erectus* to archaic *Homo sapiens* to modern *Homo sapiens*.[27] Jiǎ Lánpō also remained until his death in 2001 a strong supporter of southwest China as the cradle of humanity.[28]

23. Xú Qìnghuá, "Rénlèi qǐyuán de xīn zhèngjù" [New Evidence of Human Origins], *Kēxué bào*, 4 December 1979.

24. Wú Rǔkāng, "Lùfēng Làmǎ gǔyuán tóugǔ fāxiàn de yìyì" [The Significance of the Discovery of the *Lufengpithecus* Skull Fossil], *Guāngmíng rìbào*, 6 March 1981, 4.

25. Běijīng wǎnbào, "Yúnnán Lùfēng fāxiàn de Làmǎ gǔyuán tóugǔ huàshí zhèngmíng rénlèi qǐyuán zài yīqiānsìbǎiwàn nián qián, qǐyuán dìdiǎn zài Yàzhōu bù zài Fēizhōu" [The Discovery of *Lufengpithecus* Skull Fossil in Lùfēng, Yúnnán Proves That Humans Originated 14 Million Years Ago in Asia, Not in Africa], *Běijīng wǎnbào*, 12 January 1981 (IVPP clippings file).

26. Xú Qìnghuá, "Làmǎ gǔyuán shì rénlèi de zǔxiān ma?" [Is *Ramapithecus* a Human Ancestor?], *Huàshí*, no. 4 (1985): 1–3. For a history of the "rise and fall" of *Ramapithecus*, see Roger Lewin, *Bones of Contention: Controversies in the Search for Human Origins* (Chicago: University of Chicago Press, 1987).

27. Kūn Shēng, "Xīwǎ gǔyuán shì rénlèi de zǔxiān ma?" [Is *Sivapithecus* a Human Ancestor?], *Àomì huàbào*, no. 4 (1998): 46–47.

28. See, among many examples, Jiǎ Lánpō, *Rénlèi qǐyuán zhī wǒ jiàn*, 49–61; and Tāng Hǎifān, "Jiǎ Lánpō chídào de fāxiàn" [Jiǎ Lánpō's Late-Coming Discoveries], *Běijīng qīngnián bào*, 6 August 1999, 13 (IVPP clippings file).

For those who continued to see Chinese fossil apes as potential human ancestors, the new goal became to bridge the gap between the *Lufengpithecus* fossils (the most recent of which are the Yuánmóu variety, at 4 million years) and the oldest Chinese *Homo erectus* fossils, which may be the Yuánmóu Man specimens if the date of 1.7 million years is correct. In 1985–1986, the IVPP scientist Huáng Wànbō claimed to have taken a step in that direction with the discovery of fossils said to represent humans more primitive than *Homo erectus* and dating from 2 million years ago. Huáng narrated the events leading to the discovery in a three-part article published in *Fossils* magazine in 1988. He began with a lengthy discussion of his participation in the state-sponsored scientific search for *yěrén* in the late 1970s in the wilderness area of Shénnóngjià. It was through this experience that he gained a deep appreciation for the flora, fauna, and geology of these wild mountains, and in the mid-1980s he traveled to the nearby Three Gorges area to search for fossil evidence of early humans.[29] His team made the key find in a cave on Dragon Bone Hill (龍骨坡), on Sorcerer's Mountain (巫山 Wūshān), high above the Three Gorges.

The cultural potency of the site and its significance in Chinese history undoubtedly contributed to the excitement the discovery caused in the press. Fossil gibbons discovered by the team were celebrated as evidence that China's most famous poet, the Táng-dynasty Lǐ Bái (Li Bo), had been zoologically correct when he wrote of the Gorges, "Gibbons' voices cry unceasingly from both banks."[30] Huáng himself frequently waxes poetic when describing the site. In the last paragraph of the final installment of his 1988 article, he wrote, "The high places of the Three Gorges with their seductive power have finally begun to reveal the hidden secrets of the past 2 million years [of human history]."[31]

For Huáng, as for the authors of many articles in mainstream newspapers and popular science magazines, the fossils point to a Chinese origin of *Homo*. The apparent presence of late fossil apes and early fossil humans in the same time and place suggests to Huáng that "maybe the high reaches of the Three Gorges is the place where ancient apes crossed over into humanity."[32] *Mysteries Illustrated* echoed this optimistic

29. Huáng Wànbō, "Sānxiá gāodì mì zōng: Jì Yàzhōu dōngbù zǎoqī rénlèi huàshí de fāxiàn" [Searching for Traces in the High Places of the Three Gorges: Notes on the Discovery of East Asian Human Fossils], *Huàshí*, no. 1 (1988): 20–21; no. 3 (1988): 6–7; and no. 4 (1988): 20–21 .

30. Yuán Shùhuá and Hú Yǒuquán, "Fùyánjiūyuán Huáng Wànbō jiēkāi qiān gǔ zhī mí" [Assistant Researcher Huáng Wànbō Cracks an Ancient Mystery], *Guāngmíng rìbào*, 23 February 1988, 2 (IVPP clippings file).

31. Huáng Wànbō, "Sānxiá gāodì," *Huàshí*, no. 4 (1988): 21.

32. Huáng Wànbō, "Sānxiá gāodì," *Huàshí*, no. 3 (1988): 7.

perspective still more loudly: "At this point, the Three Gorges is the world's single best place for researching the split between apes and humans; even Africa's Olduvai Gorge cannot compare with it. We have reason to believe that the ancestors of the Chinese people, the ancestors of the yellow race, and even the ancestors of all the other races, came from the high places of the Three Gorges."[33]

Not everyone shares this interpretation of "Wūshān Man." Some Western scientists have instead welcomed the fossils as evidence supporting the theory that the Asian fossils classified as *Homo erectus* represent a unique side branch of human evolution. In this variation of the recent out-of-Africa theory, *Homo erectus* was an exclusively Asian species, and African specimens formerly attributed to the species should thus be reclassified as *Homo ergaster*. An earlier species of African *Homo* migrated to Asia, evolved into *Homo erectus*, and then became extinct. Russell Ciochon and Roy Larick distinguished this view, which they themselves hold, from that of "Chinese paleoanthropologists [who] tend to see [the] primitive features [of the Wūshān fossils] as deriving from Asian apes and suggest a local Asian origin for *Homo erectus*."[34] Yet in actuality many Chinese paleoanthropologists have come to support a third interpretation. Wú Xīnzhì quoted Ciochon and Larick's characterization of Chinese paleoanthropologists' views and rejected it, noting that he and Zhōu Guóxīng both consider the fossils to be apes closely related to *Lufengpithecus* and not humans (*Homo*) at all.[35] Several Western scientists have also come to this conclusion.[36]

For Huáng, the next step is clear. The obstacle in the way of widespread acceptance for China as the cradle of humanity is the "gap" in the Chinese human fossil record between 4 million years ago (the date of the *Lufengpithecus* specimens from Yuánmóu) and 2 million years ago (the date of "Wūshān Man").[37] He has thus set his sights on discovering fossils to fill this gap, which he imagines as a Chinese form of *Australopithecus*.[38] In a newspaper interview on the significance of the project,

33. Fàn Bǐngzhōng, "Hōngdòng shìjiè de Wūshān rén" [Wūshān Man Rocks the World], *Àomì huàbào*, no. 8 (1993): 31.

34. Russell Ciochon and Roy Larick, "Early *Homo erectus* Tools in China," *Archaeology* 53, no. 1 (2000): 15.

35. Wu Xinzhi, "Longgupo Hominoid Mandible Belongs to Ape," *Rénlèixué xuébào* 21 (supplement, 2002), 23–24.

36. Etler, Crummett, and Wolpoff, "Longgupo." The authors note that the humanlike tooth that most strongly suggested the fossils were human probably became mixed with the other fossils at a later date. The article gives a good overview of the debate over these fossils.

37. Huáng Wànbō, "Sānxiá gāodì," *Huàshí*, no. 4 (1988): 20.

38. In the meantime, artifacts discovered in Fánchāng, Ānhuī; and Wèixiàn, Héběi, have been

Huáng has even resurrected the old phrase "western origins theory," which originally referred to the origins of Chinese *civilization* (not the origins of humanity) and was hotly debated in the republican era and resurrected previously in the 1950s. "Some people," says Huáng, "use Asia's *Homo erectus* and *Homo sapiens* fossils to support a 'western origins theory,' namely, that they came from the migration or diffusion of early African *Homo erectus*. . . . We do not oppose this kind of inference, but we want to convince people through reason, so we must find human fossils older than *Homo erectus*, such as *Australopithecus*."[39] The Chinese government has been very supportive of Huáng's cause, in 1998 pledging 5 million yuán to fund it.[40]

The Chinese state press clearly favors theories and discoveries that root human origins in China, and the Chinese government funds projects for nationalist reasons. Several Chinese scientists expressed their frustration with this kind of bias in interviews with me in 2001 and 2002. Among other problems, the bias masks the very real diversity of opinion among Chinese paleoanthropologists with respect to earliest human origins. Nonetheless, many scientists and other producers of science dissemination materials in China have continued to be willing and able to present the question as an open one awaiting further research.

The Origin of Modern Humans

While Chinese paleoanthropologists are divided on the question of African or Asian origins for the *Homo* genus, there are few if any who argue for the *recent* out-of-Africa theory, namely, the notion that all *modern* humans are descended from a common ancestor who lived in Africa about 100,000 years ago and subsequently replaced existing human populations in other parts of the world. In order to understand the contemporary debate and the position of Chinese scientists in it, it is necessary to review the history both internationally and in China itself. The question of modern human origins is especially complicated because it has involved recurrent competing hypotheses that have combined and diverged over decades of debate. These complexities preclude any

---

dated to 2–2.4 and 3 million years, respectively, but these finds have been controversial and have in any case not been associated with reliable fossil remains.

39. Táng Tànfēng, "'Wūshān rén' fāxiàn shǐmò" [The Complete Story of the Discovery of "Wūshān Rén"], *Guāngmíng rìbào*, 16 August 1997 (IVPP clippings file).

40. Sautman, "Peking Man," 105.

attempt to construct a lineage of allegedly racist approaches to human evolution from contemporary multiregionalists back through earlier proponents of theoretically similar hypotheses.[41]

The geographic origins of modern humans was not a central issue until fairly recently. Nonetheless, we can find the roots of the current debate in earlier theories. While at the turn of the twentieth century, a linear model of human evolution was widely accepted, from around 1910 to the 1940s, the dominant model placed fossil hominids like Java Man, Peking Man, and the Neanderthals on side branches of the family tree. These "cousins" were understood to have become extinct, replaced by our unknown direct ancestors.[42] As one critic complained in 1943, the human genealogical tree was "all branches and terminal twigs which have no fibers running through the trunk specifically connecting with definite ancestral forms. So men, those of the present and of lower levels, stand ancestorless."[43]

Franz Weidenreich, the scientist who replaced Davidson Black at Zhōukǒudiàn after Black's death in 1934, was in this area as in many others in the minority position. Weidenreich anticipated the "single-species" model of the modern synthesis, which held that "never more than one species of man existed on the earth at any one time."[44] At the root of his theoretical understanding of human evolution was an explicit commitment to orthogenesis, the notion that evolution has proceeded according to certain internal trends—in the case of human evolution, erect posture and expansion of the brain—and not primarily through natural selection.[45] Weidenreich's work on the Chinese and Indonesian fossils, together with his knowledge of the work others had done elsewhere, convinced him that "just as mankind of today represents a morphological and generic unity in spite of its being divided into manifold races, so has it been during the entire time of evolution." For Weidenreich, the "old theory, claiming that man evolved exclusively from *one* center whence he spread over the Old World each time afresh after having entered a new phase of evolution, no longer tallies with the paleontological facts."[46] Weidenreich thus strongly advocated that

---

41. Milford Wolpoff and Rachel Caspari explicitly defend multiregionalism against charges of racism in their book *Race and Human Evolution* (New York: Simon and Schuster, 1997).

42. Bowler, *Theories of Human Evolution*, 75.

43. J. M. Gillette, "Ancestorless Man: The Anthropological Dilemma," *Scientific Monthly* 57, no. 6 (1943): 533.

44. Mayr, "Taxonomic Categories," 112. Quoted also in Proctor, "Three Roots," 221.

45. Franz Weidenreich, "The Trend of Human Evolution," *Evolution* 1, no. 4 (1947): 221–236.

46. Franz Weidenreich, "Some Problems Dealing with Ancient Man," *American Anthropologist* 42, no. 3 (1940): 380 and 381–382, respectively.

Peking Man and the other early human fossils be understood as legitimate ancestral predecessors of modern humans. All of this was almost precisely what Ernst Mayr concluded at the Cold Spring Harbor symposia. The new human tree became a straightforward lineage from *Homo transvaalensis* (which included *Australopithecus*) through *Homo erectus* to *Homo sapiens*.[47]

Yet part and parcel of Weidenreich's view was a theory on the origin of the races that for both epistemological and political reasons did not sit as well among participants in the modern synthesis. Weidenreich identified key apparent morphological similarities between Peking Man and modern Mongoloid peoples, and between other fossils and the modern peoples among whom they were found. Each of these populations retained specific racial characteristics while following the basic, shared "trends" of human evolution. In order to make sense of apparent differences in the development of hominids in different parts of the world at the same time, Weidenreich suggested that "development was not going on simultaneously everywhere but was accelerated in one place and retarded in another."[48] Although Weidenreich was notable for his relative lack of racism compared with other anthropologists of his generation, ideas like these were nonetheless uncomfortably reminiscent of notions of "superior" and "inferior" races that had undergirded Nazi policies.[49] Robert Proctor has suggested that the acceptance of the "single-species" model in the modern synthesis was related to the general response to Nazi racism in the scientific community after the Second World War.[50] Thus, while Weidenreich's "single-species" model and his consequent acceptance of Peking Man and others as ancestors became part of the modern synthesis, his orthogenism was rejected and his perspective on race sidestepped.

Proctor is probably right in associating the commitment to "one species at one time" with the antiracist politics of the period. Nonetheless, the new paradigm helped support not only antiracist theories of modern human origins but also theories founded on racist prejudice and used to buttress discriminatory policies. One of the most egregious, and certainly the most influential, of these was Carleton S. Coon's *The Origin of Races*,

47. Mayr, "Taxonomic Categories."

48. Weidenreich, "Some Problems," 381.

49. Weidenreich emphasized diversity within races and the overall similarities among all members of the human family. Weidenreich, *Apes, Giants, and Man*, 67–91. See also Robin Dennell, "From Sangiran to Olduvai, 1937–1960: The Quest for 'Centres' of Hominid Origins in Asia and Africa," in *Studying Human Origins: Disciplinary History and Epistemology*, ed. Raymond Corbey and Wil Roebroeks (Amsterdam: Amsterdam University Press, 2001), 54–55.

50. Proctor, "Three Roots."

first published in 1962 and dedicated to the late Franz Weidenreich. Coon argued for the great antiquity and inequality of human races, and he even worked behind the scenes to support segregationist forces in *Brown v. Board of Education.*[51]

At the same time, the single-species model of the modern synthesis also served as the foundation for a strident critique of Coon and anthropological racism in general. Ashley Montagu and Theodosius Dobzhansky were both firmly committed to the single-species model and also fierce critics of racist science and politics. Both Montagu and Dobzhansky wrote scathing reviews of Coon's *Origin of Races* for *Current Anthropology* in 1963. The key difference between Coon's understanding of the process of human evolution on one hand, and that of Dobzhansky and Montagu on the other, lay in the role of gene flow. While Coon emphasized the reproductive isolation of racial populations, Dobzhansky saw gene flow among populations to have prevented any significant divergence among the races. He was in this respect closer to Weidenreich, whose model involved considerable genetic exchange across racial groups.[52]

In 1976, W. W. Howells pointed to the emerging crystallization of two positions on human evolution, which he characterized as "evolutionists vs. migrationists" or "candelabra" versus "Noah's Ark." As Howells characterized the positions (in reverse order), "either there was a single local centre of origin for modern man with subsequent outward migration in all directions, or else there were many centre [*sic*] of origin and no dispersal."[53] These may be seen as the seeds of the debate that was to emerge in the 1980s, and that is still with us today, between supporters of the multiregional theory and those of the recent out-of-Africa theory. Over the years, Milford Wolpoff, in particular, pursued the "candelabra" theory, and in 1984 he teamed up with Wú Xīnzhì and Alan Thorne on a paper that served on the one hand as a corrective to theories of human evolution that had for too long neglected to consider the evidence from Asia, and on the other hand as an argument for "local regional continuity" or "multiregional evolution" based on this evidence. Consciously modeling themselves on Weidenreich, the authors assembled evidence to demonstrate that there had been no localized speciation event (that

51. Proctor, "Three Roots," 223.

52. Theodosius Dobzhansky, Ashley Montagu, and C. S. Coon, "Two Views of Coon's *Origin of Races* with Comments by Coon and Replies," *Current Anthropology* 4, no. 4 (1963): 366–367.

53. W. W. Howells, "Explaining Modern Man: Evolutionists *versus* Migrationists," *Journal of Human Evolution* 5, no. 5 (1976): 478. Howells cited Brose and Wolpoff as representatives of the former. D. S. Brose and M. H. Wolpoff, "Early Upper Paleolithic Man and Late Middle Paleolithic Tools," *American Anthropologist* 73 (1971): 1156–1194.

is, no emergence of a new species of modern humans in a single locality) followed by migration and replacement of existing forms. Rather, human evolution was a gradual process that occurred simultaneously across the old world, so gradually that (again, following Weidenreich) the distinction between *Homo erectus* and *Homo sapiens* itself was suspect.[54] As had Dobzhansky and Montagu earlier, the authors explicitly distanced themselves from Coon in part on the basis of the greater importance of gene flow in their model of human evolution.[55]

The strongest evidence for the recent out-of-Africa theory of modern human origins has been genetic, as was the case for the similar but less place-specific "Noah's Ark" model that preceded it. Results from a genetic study conducted at the University of California, Berkeley in the late 1980s suggested that all modern humans derive from a shared ancestor who lived in Africa between 100,000 and 200,000 years ago.[56] Emphasizing this common ancestor, some proponents have called it the "Eve" theory. Paleoanthropologists Chris Stringer and Peter Andrews were then quick to integrate these findings with the fossil record from Africa and Europe in an explicit challenge to the multiregional theory.[57] Supporters of multiregionalism, however, responded in force, and they highlighted in their rebuttal Stringer and Andrews's failure to do justice in their analysis to the evidence from Asia.

Despite Weidenreich's influence and prestige at Zhōukǒudiàn, few Chinese scientists and writers on science in the republican era had followed his interpretations of Peking Man and the other key fossils. Chinese writings on human evolution in those years rarely claimed Peking Man as a direct ancestor of modern Chinese people or of modern humans as a whole. By about 1952, however, Peking Man was universally regarded in Chinese science dissemination materials as such an ancestor. This shift was in agreement with the radical simplification of the human family tree produced by the modern synthesis, and it also supported, and was supported by, the early socialist state's emphasis on national identity and scientific education about human evolution.

---

54. Milford H. Wolpoff, Wu Xin Zhi, and Alan G. Thorne, "Modern *Homo sapiens* Origins: A General Theory of Hominid Evolution Involving the Fossil Evidence from East Asia," in *The Origins of Modern Humans*, ed. Fred H. Smith and Frank Spencer (New York: Alan R. Liss, 1984), 465–467.

55. Wolpoff, Wu, and Thorne, "Modern *Homo sapiens* Origins," 457.

56. Rebecca L. Cann, Mark Stoneking, and Allan C. Wilson, "Mitochondrial DNA and Human Evolution," *Nature* 325 (1987): 31–36.

57. C. B. Stringer and P. Andrews, "Genetic and Fossil Evidence for the Origin of Modern Humans," *Science* 239, no. 4835 (1988): 1264–1268. See also the critical response from eight scientists (led by Wolpoff) in *Science* 241, no. 4867 (1988): 772–773.

Yet embracing Peking Man as an ancestor did not entail embracing Weidenreich. The specific brand of anti-imperialism waged in the name of international socialism compelled even Weidenreich's protegé Jiǎ Lánpō to criticize Weidenreich's theory on racial continuity in the early years after the revolution. Soviet and Chinese scientists and science writers emphasized the racist and imperialist strains in Western anthropology and distinguished themselves as recognizing that "all the world is one human family." In 1963, the Shànghǎi Natural History Museum even prepared an exhibit specifically on racial discrimination against black people in the United States.[58] This political perspective did not, however, preclude attention to racial differences in science dissemination materials. Rather, Soviet and Chinese books on human origins devoted space to analysis of the physical differences among human races.[59] As political discourses, emphasis on "one family" and analyzing racial differences appear to run counter to each other, but they were not necessarily contradictory. In analyzing race to demonstrate difference without inequality, Soviet and Chinese writers reified race as a natural category.[60]

The existence of races was thus not a problem for Máo-era Chinese scientists, as it was for Ashley Montagu. Weidenreich, however, was a problem, both because he was associated with an allegedly racist perspective on human evolution and, more fundamentally, because he had participated in republican-era Chinese science as an "imperialist" and had then returned to live in the United States. In the post-Máo era, these considerations are far less weighty. Considered more significant are Weidenreich's contributions to Chinese paleoanthropology through his work at Zhōukǒudiàn and his invaluable descriptions of the Peking Man fossils. And with the new threats to Peking Man's ancestral status, Chinese scientists have recognized in Weidenreich an invaluable resource.

Already in 1979, Zhōu Guóxīng was alert to the early stages of the debate on modern human origins and wrote an article on the subject for the popular magazine *Scientific Experiment*. He provided a summary of the points in favor of each position and concluded by lending his sup-

58. Rénlèi zǔ, "Shànghǎi zìrán bówùguǎn chóuwěihuì jǔbàn 'Fǎnduì Měiguó dìguózhǔyì zhǒngzú qíshì, zhīchí Měiguó hēirén fǎnduì zhǒngzú qíshì de dòuzhēng'" [Shànghǎi Natural History Museum Planning Committee Prepares [the Exhibit] "Oppose American Imperialism and Racial Prejudice, Support American Blacks in Their Struggle against Racial Prejudice"], *Zìrán bówùguǎn qíngkuàng jiāoliú* 1 (1963): 30–31.

59. See chapter 3, note 64.

60. This pattern has also existed elsewhere. Alice L. Conklin, "Skulls on Display: The Science of Race in Paris' Museum of Man, 1920–1950," in *Museums and Difference*, ed. Daniel J. Sherman (Bloomington: Indiana University Press, 2007).

port to the "multiple centers" theory. He rejected, however, any notion that these centers were genetically isolated from one another. He particularly noted that the Liǔjiāng fossils—acclaimed as "the Far East's first modern humans"—bore not only characteristics of the "yellow race" but also of the south Asian and Australian races.[61]

Gene flow has been less significant in Wú Xīnzhì's writings on multiregional theory, and he may be contrasted in this respect with Milford Wolpoff. In 1999, he summarized his model as "evolutionary continuity and incidental hybridization" (連續進化併附帶雜交).[62] As he consistently characterizes it, the fossil evidence testifies to the existence of "a small amount of gene flow."[63] Far more central to his and others' writings is the morphological continuity that Weidenreich first identified and that subsequent fossil finds are said to support. According to proponents of the multiregional theory, *Homo sapiens* fossils from many Chinese sites display features linking Peking Man and other Chinese *Homo erectus* with contemporary people in China and East Asia more generally.[64]

Race and racism have thus played complex and changing roles in both Western and Chinese paleoanthropology. To some extent, these roles have reflected political changes. In the Chinese case, international socialism competed more effectively with nationalism in the Máo era than it has in the post-Máo. On the other hand, science has not been a perfect mirror for politics. If it were, we might expect Peking Man to have been embraced far more enthusiastically as a Chinese ancestor in the republican period. Both sides of the contemporary debate between multiregionalists and supporters of the out-of-Africa theory can claim antiracist predecessors and antiracist motivations. At the same time, neither side is completely divorced from their shared history of racist science.

61. Zhōu Guóxīng, "Huàshí zhìrén" [Fossil *Homo sapiens*], *Kēxué shíyàn*, no. 6 (1979): 41–43, 43.

62. Xīnhuá shè Běijīng, "Liánxù jìnhuà bìng fùdài zájiāo" [Evolutionary Continuity and Incidental Hybridization], *Běijīng qīngnián bào*, 15 November 1999 (IVPP clippings file).

63. Wú Xīnzhì, *Rénlèi jìnhuà zújī* [The Tracks of Human Evolution] (Běijīng: Běijīng shàonián értóng chūbǎnshè, 2002), 132.

64. Wú Xīnzhì, *Rénlèi jìnhuà zújī*, 132; Mǎ Róngquán, "Xiàndài rén zǔxiān dànshēng zài nǎlǐ?" [Where Were the Ancestors of Modern Humans Born?], *Huàshí*, no. 1 (1994): 7–8; Xīnhuá shè, "'Nánjīng yuánrén' bèi què rènwéi yǒu wǔshí duō wàn nián lìshǐ" [Nánjīng Man's Age Firmly Considered Greater Than 500,000 Years], *Běijīng wǎnbào*, 25 February 2001, 12 (IVPP clippings file). Marta Mirazón Lahr has challenged the empirical basis for claims of morphological continuity in "The Evolution of Modern Human Cranial Diversity: Interpreting the Patterns and Processes," in *Conceptual Issues in Modern Human Origins Research*, ed. Geoffrey A. Clark and Catherine M. Willermet (New York: Aldine de Gruyter, 1997). More recently, Noel T. Boaz and Russell L. Ciochon have proposed "clinal replacement" as a compromise theory that satisfies both the genetic evidence behind out-of-Africa and the continuities in Asian fossil morphologies that support multiregionalism. See their *Dragon Bone Hill: An Ice-Age Saga of Homo Erectus* (New York: Oxford University Press, 2004).

Ethnic Nationalism, Defensive and Assertive

Theories of human evolution were important to Chinese intellectuals in the early twentieth century because they helped make sense of China's position in the world. Race and nation became fused as people began to see China's resistance to imperialism as a struggle for survival akin to that found in nature. Chinese scientists and nonscientists alike remain sensitive to experiences that suggest imperialism. The title of a particularly telling article published in a popular science magazine in 1998 warned, "Chinese People, Look After Your Genes." The authors noted the important research that American and other foreign scientists were conducting using Chinese subjects, often in collaboration with Chinese institutions, to learn more about the human genome and the diseases that afflict us. While proud that China should contribute to science in this important way, the authors cautioned that China was becoming a "hunting ground" and Chinese genes a commodity. Instead of allowing "outsiders" to "steal" Chinese genes, the authors proposed that Chinese people grasp their precious genetic resources in their "own hands" and make their "own contribution to research on the human genome."[65] This kind of fear, arising when a foreign power (often specifically represented by scientists or other modernizers) is seen to be stealing bodies or bodily essences, is common in countries experiencing political or economic colonization.[66] And Chinese people are not always in the victim's position: rumors have recently circulated in Fiji that Chinese scientists have been acquiring Polynesian fetuses for use in genetic experiments to create a race of superhumans.[67]

Such ongoing fears help explain why Huáng Wànbō and others publicly dredge up the republican-era controversy over the western origins theory when they promote Chinese origins for *Homo* or *sapiens*.[68] They

65. Méi Zǐ and Lì Tāo, "Zhōngguó rén kànhǎo nǐ de jīyīn" [Chinese People, Look after Your Genes], *Kēxué dàzhòng*, no. 11 (1998): 9. See also Laurence Schneider, *Biology and Revolution in Twentieth-Century China* (Lanham, Md.: Rowman and Littlefield, 2003), 258–263.

66. Christian missionaries in China were often suspected of such atrocities as gouging out and stealing Chinese people's eyes and kidnapping children for nefarious purposes resulting in the abnormal decomposition of their corpses. Paul A. Cohen, *China and Christianity: The Missionary Movement and the Growth of Chinese Antiforeignism, 1860–1870* (Cambridge, Mass.: Harvard University Press, 1963), 31 and 230.

67. I learned of this rumor from a Fijian friend.

68. For an example of someone other than Huáng Wànbō invoking the "western origins theory," see "Wǒguó gǔrénlèi fāzhǎn shǐ shàng de zhòngdà fāxiàn: Dīngcūn wénhuà yízhǐ" [A Great Discovery in China's Ancient Human Developmental History: The Dīngcūn Culture Site], in *Xiāngfén wénshǐ zīliào 9*, ed. Zhèngxié Xiāngfén xiàn wěiyuánhuì wénshǐ zīliào yánjiū wěiyuánhuì (Xiāngfén: Zhèngxié Xiāngfén xiàn wěiyuánhuì wénshǐ zīliào yánjiū wěiyuánhuì, 1997), 18.

also shed light on the defensive character of Chinese scientific and media representations of the Chinese archaeological record. Ironically, one argument against the recent out-of-Africa theory rests on the persistence in Asia of lithic technologies considered "crude" in comparison with African and European examples. If more advanced humans with more sophisticated technologies had replaced local populations, one would expect to see a change in the archaeological record. Thus, some have argued that the lack of such change throws doubt on the replacement hypothesis.[69] This is strikingly reminiscent of the uncomfortable alternative the western origins theory once appeared to offer: either you are Chinese and inferior or you are equal but not Chinese. As Fa-ti Fan has summarized the debate over the western origins theory, "What was at stake was the genius or creativity of the Chinese people."[70] It comes as no surprise, then, to see articles in scientific journals and newspapers celebrating a new archaeological discovery associated with *Homo erectus* in southern China as evidence that "Eastern and Western humans [are/were] equally smart."[71]

Warning Chinese people about foreigners out to "steal" their genes, invoking the western origins theory, and applauding the finding that Asians are of equal intelligence are all examples of racial identity or ethnic nationalism constructed defensively in reaction to perceived threats. A more assertive kind of ethnic nationalism, however, is also at play in scientific research and writing on human evolution in China. Scientists and media reporters have used paleontology and paleoanthropology to make ties between China and two of its controversial "regions"—Táiwān and Tibet—appear historically or even naturally valid.[72]

As early as 1980, Jiǎ Lánpō quoted an article printed in the Táiwān

69. Geoffrey G. Pope, "Replacement versus Regionally Continuous Models: The Paleobehavioral and Fossil Evidence from East Asia," in *The Evolution and Dispersal of Modern Humans in Asia*, ed. Takeru Akazawa, Kenichi Aoki, and Tasuku Kimura (Tokyo: Hokusen-sha Publishing Company, 1992), 4–6.

70. Fa-ti Fan, "How Did the Chinese Become Native? Science and the Search for National Origins in the May Fourth Era," in *In Search of Modernity: Re-examining the May Fourth Movement*, ed. Kai-wing Chow and Tze-ki Hon (Lanham, Md.: Lexington Books, forthcoming).

71. The Chinese original does not indicate the tense. See Yú Zhēng, "Dōng xī fāng rénlèi yīyàng cōnghuì" [Eastern and Western Humans Equally Smart], *Qiánjiāng wǎnbào*, 26 March 2000, 6 (IVPP clippings file). See also Yán Hóng, "Bǎisè shǒufǔ: Zhāidiào Yàzhōu zhílìrén 'wénhuà zhìhòu' de màozǐ" [The Bǎisè Handaxes: Casting Off the Asian *Homo erectus*'s "Stagnant and Backwards" Label], *Kēji rìbào*, 24 March 2000, 4 (IVPP clippings file); and "Zǎoqī dōngfāng rén bìng bù 'dīzhì'" [Early Eastern People Were Definitely Not of "Low Intelligence"], *Wénhuì bào*, 18 March 2000 (IVPP clippings file). The research was first reported in Hou Yamei et al., "Mid-Pleistocene Acheulean-Like Stone Technology of the Bose Basin, South China," *Science* 287, no. 5458 (March 2000): 1622–1626.

72. Sautman makes a similar argument with respect to Táiwān and Tibet in his "Peking Man," 107.

newspaper, *Central Daily* (中央日報): "Everyone is deeply interested in exploring the question of the blood and land relationships [血緣、地緣] between Táiwān and the mainland." Jiǎ agreed enthusiastically, and he proposed that scientific colleagues on both sides should work together to complete the important task of establishing the archaeological and geological connections between the "province" of Táiwān (implying its inclusion in a single "China") and the southeast provinces of Zhèjiāng, Fújiàn, Guǎngdōng, and Jiāngxī.[73] In the late 1980s, *Fossils* magazine began devoting serious attention to the subject. An article in 1987 discussed a discovery by Fújiàn fishers the previous year of a fossil deer off the coast of the island of Dōngshān, demonstrating the existence of a land bridge between Táiwān and the mainland during the Tertiary period (66 million to 1.6 million years ago). The author highlighted such geological connections in the remote past to imply that the two states were naturally a single unit. The article concluded, "Now we hope that the two shores of the strait will soon be unified, so that flesh-and-blood kindred [骨肉親人] will be able to come and go freely and together make a contribution to constructing our prosperous and strong ancestral country [祖國]."[74]

Here we see a fascinating inversion of a familiar pattern: while it is common for nationalists to invoke the land to represent the people, in these writings the blood ties of the people are also being used to knit together two pieces of land representing two political entities. An article written two years later on recent discoveries of human fossils on Dōngshān made this point even more clearly: "The geological connection between Fújiàn and Táiwān cannot be sundered. One can say that the Dōngshān land bridge was the fossil Dōngshān Man's mother, and Dōngshān Man's two hands pulled the two shores tightly together."[75] In 1995, newspapers celebrated research by mainland scientists on "Táiwān's first water buffalo," whose fossilized remains in the Pénghú archipelago demonstrated that "10,000 years ago Táiwān and the Chinese mainland were connected."[76]

73. Jiǎ Lánpō, "Táiwān yǔ dàlù" [Táiwān and the Mainland], 5 March 1980, unpublished draft, courtesy of Jiǎ Yǔzhāng.

74. Sūn Yīnglóng, "Yuánshǐ shíqī de Mín Tái guānxī" [Fújiàn-Táiwān Relations in Prehistoric Times], *Huàshí*, no. 4 (1987): 15.

75. Zēng Wǔyuè, "'Dōngfāngrén' huàshí de niándài yǔ jiàzhí" [The Age and Value of the Dōngfāng Man Fossils], *Huàshí*, no. 2 (1989): 1.

76. Zōng Xìng, "Táiwān fāxiàn gǔjǐzhuī dòngwù huàshí yīwàn nián yǐqián liǎng àn xiānglián" [Discovery of Paleontological Fossils in Táiwān Proves 10,000 Years Ago the Two Shores Were Connected], *Guāngmíng rìbào*, 6 November 1995, 7; "Wànnián qián Táiwān yǔ dàlù xiānglián" [10,000 Years Ago Táiwān and the Mainland Were Connected], *Zhōngguó kēxué bào*, 17 January 1996.

Fossil discoveries have similarly buttressed Chinese nationalistic claims over another disputed territory, that of Tibet. In 1989, the *People's Daily* reported that morphological studies on Neolithic remains discovered in Tibet and on modern Tibetans put to rest long-standing debates over the origin of the Tibetan ethnicity. Different theories had posited western Sìchuān, India, and Tibet itself as likely places of origin. The new research suggested that "Tibetans, together with Hàn and minority nationalities from the other regions, all originated from northern Chinese late *Homo sapiens*, as represented by the Upper Cave Man of Zhōukǒudiàn." These findings, moreover, pointed to a more general conclusion: "Tibetans without a doubt are members of [the group of] modern Chinese people—the East Asian type of the yellow race."[77] Similarly, in 1994 *China Science News* reported that artifacts discovered in 1990 on the Qīnghǎi-Tibetan Plateau shared key characteristics with the "northern Chinese Paleolithic cultural system of the Běijīng Upper Cave Man."[78]

It is significant that these articles consistently refer to Upper Cave Man, the late Paleolithic humans unearthed in the 1930s at Zhōukǒudiàn near the older Peking Man remains. Weidenreich's initial analysis of the Upper Cave Man fossils suggested that they represented more than one race living in the area at that time. After decades of discussion, however, in the 1960s IVPP scientists concluded that the fossils were ancestral to the "Mongoloid" race. Upper Cave Man has thus served as an ancestral link between modern Chinese people and Peking Man. Jiǎ Lánpō and Huáng Wèiwén, in the several editions of their book on the history of Zhōukǒudiàn, dramatically emphasized this ancestral relationship. Noting the presence of burial plots at the site, they titled the chapter on Upper Cave Man in the English edition of the book "Have We Dug at Our Ancestral Shrine?"[79] Jiǎ then used the same language in a later

---

77. "藏族無疑屬於現代中國人——黃種人的東亞類型." Lǐ Xīguāng, "Shāndǐngdòngrén shì Zàngzú Hànzú gòngtóng zǔxiān" [Upper Cave Man Is the Common Ancestor of the Hàn and Tibetan Ethnicities], *Rénmín rìbào*, 18 April 1989, 4.

78. Yáng Cúnruì, "Wūlánwūlā hú fāxiàn liǎngwàn nián qián jiùshíqì" [Paleoliths from 20,000 Years Ago Discovered at Wūlánwūlā Lake], *Zhōngguó kēxué bào*, 18 July 1994, 2.

79. Jia Lanpo and Huang Weiwen, *The Story of Peking Man: From Archaeology to Mystery*, trans. Yin Zhiqi (Beijing: Foreign Languages Press, 1990), 100, 92. In the Chinese-language editions, this sounds less pretty. Either from an archaeologist's sense of guilt or, more likely, from an earthy sense of humor, the authors described the excavation of Upper Cave Man by invoking one of the most serious punishments the imperial state inflicted on criminals—digging up the ancestral graves. The authors use the phrase "拔祖墳," a colloquial alternative for the more formal "掘墳." Jiǎ Lánpō and Huáng Wèiwén, *Zhōukǒudiàn fājué jì* [The Excavation of Zhōukǒudiàn] (Tiānjīn: Tiānjīn kēxué jìshù chūbǎnshè, 1984), 62 and 67; Jiǎ Lánpō and Huáng Wèiwén, *Fāxiàn Běijīngrén* [The Discovery of Peking Man] (Táiběi: Yòu shī, 1996), 82 and 96.

book to call Zhōukǒudiàn as a whole "our ancestral shrine" based on the identification of Peking Man and Upper Cave Man as "close to" and a member of the "yellow race," respectively.[80]

Jiǎ Lánpō and Huáng Wèiwén are not the only ones to treat Zhōukǒudiàn as a shrine for ancestors of an ethnically and nationally defined group. Responses to public concern over environmental threats to Peking Man's "home" and the continued interest in the whereabouts of the missing Peking Man fossils testify to the strong feelings many people share about Peking Man. The authors of the book *The Search for "Peking Man,"* published in 2000, titled their first chapter "We Lost Our Ancestor!" specifically referring to Peking Man as "our bloodline ancestor."[81] While this book is remarkable for its international approach and its concern for the impact of the loss on people of all races, it is nonetheless clear who the "we" are who lost "our" ancestor. In their afterward, entitled "Why Search for 'Peking Man'?" the authors summarize their position: "In the past, we did not have the power to resist foreigners' plundering or the ability to protect 'Peking Man' . . . But the China of today is not the China of yesterday. China is no longer the 'sick man of Asia'; the Chinese people have stood up [i.e., liberated ourselves] from the dirge of history. . . . Although Peking Man belongs to the world, belongs to humanity, it first and foremost belongs to China. . . . By searching for 'Peking Man' we will be making history grant justice to the Chinese people and we will be sculpting for our great and proud nation an image of strength to show the world."[82] Such ethnic nationalist sentiment about Peking Man, frequently described as "our ancestor," certainly helps explain the widespread rejection of scientific theories that throw Peking Man's ancestral status into question. Similarly, investment in Zhōukǒudiàn as "our ancestral shrine" undoubtedly contributes to vehement reaction against not only environmental threats to the site, but also scientific theories that threaten Zhōukǒudiàn's status as a "home" by questioning whether Peking Man actually lived there and made the hearths that long stood as the earliest evidence of human use of fire.[83]

80. Jiǎ Lánpō, *Zhōukǒudiàn jìshì* [Chronicle of Zhōukǒudiàn] (Shànghǎi: Shànghǎi kēxué jìshù chūbǎnshè, 1999), preface (no page numbers).

81. Lǐ Míngshēng and Yuè Nán, *Xúnzhǎo "Běijīngrén"* [The Search for "Peking Man"] (Běijīng: Huáxià chūbǎnshè, 2000), 1.

82. Lǐ Míngshēng and Yuè Nán, *Xúnzhǎo "Běijīngrén,"* 449. For an example of their more humanist take, see page 2.

83. Lù Shí, "Zhōukǒudiàn shì 'Běijīngrén' zhī jiǎ ma?" [Was Zhōukǒudiàn Peking Man's Home?], *Huàshí,* no. 4 (1987): 16–17; Kē Jī, "Běijīng yuánrén yòng huǒ wúyōng zhìyí" [No Reason to Doubt That Peking Man Used Fire], *Huàshí,* no. 4 (1998): 9. On the other side, see Máo Zhènqí, "Běijīng

Making a Contribution: China as a Research Center

One of the core principles in science studies, as put forward in the Strong Program, is symmetry.[84] If we are to examine social, cultural, and political influences on the production of scientific theories, we must do so evenly and not only for theories we know or suspect to be flawed. However, some critics of multiregionalism accept the recent out-of-Africa explanation of modern human origins and use the empirical evidence marshaled in its favor to explain its popularity in Western countries, while emphasizing political motivations, almost to the exclusion of empirical ones, for the Chinese preference for the multiregional theory.

This approach is just as effective on the other side of the argument. For example, when I showed a member of IVPP a Chinese article from 1952 that criticized Franz Weidenreich's theory of racial continuity (a precursor to the multiregional theory) for its alleged imperialist roots, she dismissed it as a predictable consequence of the political context of the time. In her eyes, the morphological continuity between Peking Man and present-day Chinese people is empirically grounded, while challenges are likely to arise from political motivations. At the same time, there are important political motivations in the West for supporting the recent out-of-Africa theory; it helps counter racism by insisting on the recency, and thus the negligible importance, of racial differences.[85]

Ethnic nationalism is clearly a potent force in many Chinese scientists' writings. It is not, however, the only force. There are also empirical questions, and along with them disputes that are disciplinary in character. Moreover, nationalism comes in many forms. The nationalism of contemporary Chinese discourse on human evolution is not just about naturalizing Chinese ethnic identity, but also about the contributions of Chinese scientists and Chinese fossils to international science, and about how such contributions are or are not recognized in foreign countries.

The Chinese press has generally given more favorable attention to multiregionalism than to the recent out-of-Africa theory. But the situation is not quite as unbalanced as Sautman suggests, and media even

---

yuánrén bìng fēi qúnjū dòngxué" [Peking Man Did Not Live Together in the Cave], *Zìrán yǔ rén*, no. 5 (1986): 9.

84. David Bloor, *Knowledge and Social Imagery* (London: Routledge and Kegan Paul, 1976), 10–13. The Strong Program is an approach to the sociological study of science developed in the 1970s at the University of Edinburgh. The approach recognizes science as a set of social activities requiring analysis that takes into account forces generally considered "external" to scientific research.

85. See Stephen Jay Gould's convincing and moving rebuttal of Carleton Coon and others in "Human Equality Is a Contingent Fact of History," in *The Flamingo's Smile: Reflections in Natural History* (New York: Norton, 1985), 185–198.

appear to have warmed up to the out-of-Africa theory in recent years. A Chinese proponent of the theory, Jīn Lì (L. Jin in English-language publications), finds the media increasingly interested in his research and conclusions. He even suggests that some scientists and political figures consider the out-of-Africa theory's potential for reducing prejudice to be consistent with the current political agenda of increasing Chinese participation in international affairs.[86]

The prospects for diverse viewpoints are perhaps better outside the party-directed Xīnhuá News Agency articles that make up the vast majority of Sautman's sources. Media run by government departments and scientific institutes—from youth newspapers to science magazines—are still accountable to the state but appear to have much more flexibility with respect to content and viewpoint. While the multiregional theory still receives greater support across the board, many materials present the issue as an open question awaiting resolution on the basis of future scientific research, and some even favor the recent out-of-Africa theory.[87] Survey responses from readers of *Fossils* magazine were likewise divided on the question of where humans originated. Enthusiasm for specific theories that glorified China's role in human evolution was often tempered by a careful agnosticism that served in part to convey the importance to the scientific worldview of impartiality and the willingness to postpone judgment until the evidence becomes clear.

Moreover, while Chinese paleoanthropologists appear united in favor of multiregionalism, the debate in China—as in the West—has fractured to some extent along disciplinary lines.[88] Some geneticists in China have been outspoken supporters of recent African origins. The most influential is Jīn Lì. Jīn received his doctoral degree at the University of Texas and a tenured position at the University of Cincinnati before returning to China in 1997, where he became the dean of Fùdàn University's Department of Life Sciences. It is his vision, to the dismay of his critics, that is shaping the anthropology section of the new and spectacular Shànghǎi Science and Technology Museum.[89]

86. Jīn Lì, interview with the author, 21 July 2005.

87. For examples that are either agnostic or support the recent out-of-Africa theory, see "Fēizhōu Bǐgémǐ rén shì rénlèi shǐzǔ" [African Pygmies Are Humanity's Ancestors], *Zhīshí jiùshì lìliàng*, no. 4 (1992): 42; "Zhōngguó rén lái zì Fēizhōu?" [Did Chinese People Come from Africa?], *Kēxué shíbào*, 29 July 2000 (IVPP clippings file); Wú Rǔkāng, *Rénlèi de guòqù, xiàndài hé wèilái* [The Past, Present, and Future of Humanity] (Shànghǎi: Shànghǎi kēxué jiàoyù chūbǎnshè, 2000), 64–69; and Duàn Jīn, "Tā shì Zhōngguórén zǔxiān ma" [Is He the Ancestor of Chinese People?], *Běijīng qīngnián bào*, 29 July 2002, 32.

88. For a mainstream Chinese newspaper article characterizing the debate as divided between geneticists and paleoanthropologists, see Duàn Jīn, "Tā shì Zhōngguórén," 32.

89. Jīn Lì, interview with the author, 21 July 2005.

Jīn's work on the Y chromosome has added important evidence to support claims for a recent, common, African origin of modern humans. After inviting Jīn Lì to visit IVPP, however, Wú Xīnzhì remains unconvinced that geneticists have the tools necessary to interpret the data.[90] Jīn is attempting to overcome these disciplinary divides through the creation of the Center for Anthropological Studies, bringing geneticists together with physical anthropologists, linguists, historians, ethnologists, and others.[91] The hard work required to enable geneticists and paleoanthropologists to hear one another serves as a reminder that not just political considerations but methodological questions as well are at work in the debate over modern human origins in China today. Chinese paleoanthropologists are fighting to defend their discipline's primary data set—fossils—as the key to resolving questions about human evolution.

Chinese paleoanthropologists are not just defending the role of fossils. They are specifically defending the role of Chinese fossils, and Chinese researchers, in international science. Western ignorance about Chinese fossils has played an important role in contemporary debates over human origins. The concerns of Chinese paleoanthropologists thus take on a nationalist meaning, but it is a nationalism based more on professional than on ethnic identity.

The African turn in paleoanthropology and the extremely limited access Westerners enjoyed in China during the 1950s and 1960s combined to remove China as a central stage in international research on human evolution. Chinese paleoanthropologists insist that Chinese fossils represent crucial evidence for testing theories about human evolution, but decades of political barriers, followed by lingering linguistic ones, have inhibited the full consideration of Chinese fossils in much Western work on the subject. Western paleoanthropologists who work extensively on Chinese fossils often share their Chinese colleagues' frustrations with Western ignorance of the Chinese materials, and some have made important efforts to remedy the situation.[92] Many such "Asianists" support multiregionalism based on their interpretations of the Chinese fossils on which they have built their careers. Geoffrey Pope, for example, has suggested that it is largely ignorance of Chinese archaeological

90. Wú Xīnzhì, interview with the author, 7 July 2005.

91. Jīn Lì, interview with the author, 21 July 2005.

92. John W. Olsen and Wu Rukang, eds., *Paleoanthropology and Paleolithic Archaeology in the People's Republic of China* (Orlando: Academic Press, 1985); Wu and Poirier, *Human Evolution in China*. Also note long-time Asianist Ciochon's recent efforts to reconcile the genetic evidence for out-of-Africa with Asian fossils (see note 64).

and fossil materials that has prevented proponents of the recent out-of-Africa theory from recognizing the flaws in their arguments.[93] Pope has also criticized Western researchers of the past and present for shoehorning Chinese materials into preconceived categories based on European evidence, a charge that should be familiar to historians. Not coincidentally, Pope has accepted and recirculated the Chinese account of Western "colonial" attitudes at Zhōukǒudiàn and has pointed out the continued lack of respect some Western scientists show for their Chinese colleagues' abilities.[94]

A careful look at Chinese writings on human evolution for lay audiences shows that the ancestry of modern Chinese people is not all that is at stake in debates over the "birthplace" of humanity. One theme that emerges again and again is the "contribution" that Chinese fossils and Chinese research are making to science writ large. Newspaper articles celebrate the "role of our country in human origins research" and the "worldwide recognition of China as one of the countries with bounteous discoveries of fossil hominids."[95]

While this bounty has been achieved through research at sites all across the country, the Peking Man site at Zhōukǒudiàn remains the symbol of China's status in international paleoanthropology. A flurry of articles in 1984, marking the fifty-fifth anniversary of the discovery of the first Peking Man skullcap, expressed the hope that China would again become a "world center for paleoanthropology."[96] On the seventieth anniversary, an article in *Běijīng Daily* entitled "Peking Man, Everlasting Scientific Value" concluded that people continued to pay so much attention to Zhōukǒudiàn because it held the answers to key research questions about human evolution.[97] Throughout these years, reports consistently referred to Peking Man, and sometimes other fossil discov-

93. Pope, "Replacement versus Regionally Continuous Models." See also Wolpoff, Wu, and Thorne, "Modern *Homo sapiens* Origins," 411–415.

94. Geoffrey G. Pope, "Paleoanthropological Research Traditions in the Far East," in Clark and Willermet, *Conceptual Issues*, 269–282.

95. Wú Rǔkāng, "Rénlèi qǐyuán yánjiū de xiànzhuàng hé zhǎnwàng" [The Current Situation and Future Prospects for Research on Human Origins], *Zhōngguó kēxué bào*, 23 December 1994, 2 (IVPP clippings file); Zhāng Shūzhèng, "Wǒguó Zhōukǒudiàn jí gǔrénlèixué yánjiū shuòguǒ lěilěi" [China's Zhōukǒudiàn and Innumerable Great Achievements in Paleoanthropological Research], *Rénmín rìbào*, 30 November 1989, 3.

96. Kēxué bào, "Jìniàn Běijīng yuánrén dìyī ge tóugàigǔ fāxiàn 55 zhōu nián" [Commemorating the Fifty-fifth Anniversary of the Discovery of the First Peking Man Skullcap], *Kēxué bào*, 27 December 1984 (IVPP clippings file). See also Chén Zǔjiǎ and Lǐ Bǐngqīng, "Wǒguó zhèng chéngwéi rénlèi qǐyuán yánjiū de xīn zhōngxīn" [China Is Becoming a New Center for Research into Human Origins], *Rénmín rìbào*, 21 December 1984, 3.

97. Qiū Zhùdǐng, "'Běijīngrén': Kēxué jiàzhí yǒng bù shuāi" [Peking Man, Everlasting Scientific Value], *Běijīng rìbào*, 16 October 1999, 11 (IVPP clippings file).

eries as well, as "gold medals" in scientific research. One such article called the original Peking Man skullcap "the first gold medal the Chinese scientific community took in the world scientific arena, just like a gold medal in the Olympics."[98]

In fossil-poor 1929, Peking Man could attract worldwide attention whether or not she was understood to be an ancestor of modern humans. In recent decades, as African fossils threaten to make China tangential to the story of human evolution, Chinese scientists are passionate about preserving China's status as "one of the important regions for human origins."[99] Western scientists who have built their careers on Chinese fossils, while perhaps less passionate, are also emphatic that the evidence from China not be ignored. Without denying other factors—both nationalist and empirical—in the support for the multiregional theory of modern human origins, we must recognize the professional stakes Chinese and some Western scientists have in Chinese fossils as an important motivation. This is not wholly unrelated to nationalism: scientists, like athletes at the Olympics, win respect for their nations through their successes.[100] It is, nonetheless, different in important ways from ethnic nationalism.

Making Connections:
China as a Center for the Human Family

Not only does nationalism come in many forms, but ethnic nationalism itself is a very unstable quantity in discourse on human origins and evolution. What defines ethnic nationalism also pulls apart into smaller identities. Post-Máo science dissemination has reflected a trend toward regionalism found in post-Máo society as a whole. Articles in popular science magazines have highlighted research demonstrating the morphological differences between northern and southern Chinese people.[101] Meanwhile, museums dedicated to the display of local fossils have

98. Sū Wényáng, "Shǒudū jìniàn Běijīng yuánrén wǎnzhěng tóugàigǔ fāxiàn 55 zhōu nián" [The Capital Commemorates the 55th Anniversary of the Discovery of a Complete Peking Man Skullcap], *Běijīng rìbào*, 20 December 1984 (IVPP clippings file). See also Kē Xuě, "'Běijīngrén': Kēxué yánjiū de yī kuài 'jīnpái'" [Peking Man, a Gold Medal in Scientific Research], *Běijīng rìbào*, 16 October 1999, 11 (IVPP clippings file); and Táng Tànfēng, "'Wūshān rén' fāxiàn shǐmò," 7.

99. Note the conservatism of the phrase "one of." Xú Qìnghuá, "Rénlèi qǐyuán." See also "Xúnzhǎo dìyī bǎ shí dāo: Zhōngguó shì rénlèi zuì zǎo qǐyuán dì zhī yī" [Searching for the First Stone Knife: Is China One of the Sites of Earliest Human Origins?], *Běijīng wǎnbào*, 18 January 1999, B20 (IVPP clippings file).

100. On nationalistic competitiveness in post-Máo science, see Schneider, *Biology and Revolution*, 277.

101. Note that Zhāng Zhènbiāo, author of numerous articles linking the Tibetan and Hàn

sprung up across the country, and vocal members of minority nationalities like Lǐ Xùwén, a Nàxī man from Lìjiāng, actively interpret such fossils in ways supporting distinct ethnic histories. What defines ethnic nationalism also contributes to broader identities. Discourse on human origins and evolution has emphasized commonalities among Chinese people on the one hand, and other "East Asians," or members of the "yellow" or "Mongoloid" race, on the other. It has also gone further to show the kinship among East Asian people, Native Americans, and Australians, creating a more inclusive Chinese identity as a geographic center for the human family.

Chinese and Japanese paleoanthropologists have enthusiastically collaborated on research projects designed to illustrate the evolutionary connections between the Chinese and Japanese peoples. Such research has received significant attention in science dissemination materials in both countries. In addition to the Peking Man exhibit that traveled to Japan in 1980, in 1988 IVPP assisted the Japanese National Science Museum in the creation of an exhibit to celebrate the museum's 110th anniversary. The title was "Exhibit on Japanese Origins: Where Did Japanese People Come From?" Focusing on the fossil evidence from China, the exhibit described successive waves of migration into Japan from southern China and Polynesia, northern Asia, and the Korean peninsula.[102] Also in 1988, the new anthropology exhibit at the Běijīng Natural History Museum devoted considerable space to the origins of the Japanese people, prominently including Péi Wénzhōng's research connecting early Japanese humans with northern China. In 1994, scientists from Japan and China participated in a conference held in the southern Chinese city of Liǔzhōu, Guǎngxī on the connections between Japanese fossils and Liǔjiāng Man. The conference resulted in renovations to the local site museum to focus more on the significance of the site for the origins of the Japanese people.[103] Throughout the post-Máo period, science dissemination print media have highlighted both the research collaboration and "friendship" of modern Chinese and Japanese, and the prehistoric relationship of the two peoples.[104] Such discourse has served

---

ethnicities (see above), has also worked on research demonstrating morphological differences between northern and southern Chinese people.

102. Kokuritsu kagaku hakubutsukan, *Nihonjin no kigen ten: Nihonjin wa doko kara kimashita ka* [Exhibit on Japanese People's Origins: Where Did Japanese People Come From?] (Tokyo: Yomiuri shinbun, 1988), 95.

103. Liú Wén, interview with the author, 13 May 2002.

104. Gài Péi and Liú Qiūshēng, "Zhōngguó hé Rìběn de wénhuà jiāoliú, kěyǐ zhuīsù dào jiùshíqì shídài" [Chinese and Japanese Cultural Exchange Can Be Traced to the Paleolithic Period], in *Kēxuéjiā tán kēxué* [Scientists Discuss Science], vol. 2, ed. Zhōngyāng rénmín guǎngbō diàntái kējì zǔ and Kēxué

as an important countercurrent to continued tensions between the two nations.

Migration from northern and southern China to Japan forms just one part of the reconfiguration of China as a center of humanity. In the conclusion to their seminal 1984 article on multiregionalism and East Asia, Wolpoff, Wú, and Thorne note the region's changing evolutionary role: "Continental east Asia does not remain a periphery. Instead it increasingly comes to act as a center in the Late Pleistocene, with gene flow extending from it to the east (the New World), west, and south (although this undoubtedly remained multidirectional, as the Late Pleistocene specimens from Indonesia and south China show)."[105] This construct of China as a center has itself become central to popular Chinese portrayals of human evolution. A 2000 *Fossils* article cited a French participant in a conference at Zhōukǒudiàn: "The Peking Man site makes clear that China indeed took on a pivotal role as a 'central country' [中央之國] in the process of human origins, evolution, and dispersal: after humans originated in Africa, they came to China and then migrated east to America and south to the Asian-Pacific Islands and Oceania."[106] Research linking the histories of Australians and especially Native Americans to China has received great attention in post-Máo science dissemination materials. Emphasizing Chinese origins for Australians sometimes explicitly arises from support for multiregionalism, demonstrating that this theory has the potential to construct more inclusive identities than just ethnic Chinese nationalism.[107] More generally, and particularly in the case of Native Americans, such discourse serves to construct Chinese roots for peoples on three continents. For example, an article printed in a 1986 issue of *Nature and Man*, entitled "The Stamp of China on American Indian Bodies," reported on serological research linking Native Americans and Chinese.[108]

---

pǔjí chūbǎnshè biānjíbù (Běijīng: Kēxué pǔjí chūbǎnshè, 1982), 110–114; Gǔ Zǔgāng, "Rìběn rén lái zì héfāng?" [Where Are Japanese People From?], *Huàshí*, no. 2 (1991): 15–17; Zhōngguó kēxué bào, "Rìběn guó shǒuxiàng fūrén cānguān Zhōngguó kēxuéyuàn gǔjǐzhuīsuǒ" [First Lady of Japan Visits the Chinese Academy of Science's IVPP], *Zhōngguó kēxué bào*, 25 March 1994 (IVPP clippings file).

105. Wolpoff, Wu, and Thorne, "Modern *Homo sapiens* Origins," 471.

106. Kē Pǔ, "Zhōukǒudiàn Běijīng yuánrén yízhǐ de 'tòushì'" [The Perspective from the Zhōukǒudiàn Peking Man Site], *Huàshí*, no. 1 (2000): 5. See also Zhōngguó kējì bào, "Zhōukǒudiàn Běijīng yuánrén yízhǐ shàng yǒu hěn dà de fājué qiánlì" [The Zhōukǒudiàn Peking Man Site Still Has Great Excavation Potential], *Zhōngguó kējì bào*, 25 November 1998, 4 (IVPP clippings file).

107. With respect to opposition to African origins for Australians, see Zhāng Zhènbiāo, "Àodàlìyà rén lái zì hé chù" [Where Are Australians From?], *Huàshí*, no. 1 (1985): 9–10. More generally on Chinese and Australian connections, see Zhāng Zhènbiāo, "Zhōngguó rén yǔ Dàyángzhōu rén yǒu tóngyuán guānxī ma?" [Do the Chinese and Oceanic Peoples Have a Common Origin?], *Kējì rìbào*, 19 April 1995, 2 (IVPP clippings file).

108. Yīng Qílóng, "Yìndiānrén shēn shàng de 'Zhōngguó' yìnjì" [The Stamp of China on American

Such texts may be read in multiple ways. Some see "the imputation of such connections" as "an almost mystical obsession of official promoters of paleoanthropological nationalism in China."[109] Positing Chinese ancestry for other peoples indeed serves to glorify China; of that there can be little question. Claiming Asian origins for the first inhabitants of Australia and the Americas—now dominated by Europeans—may also carry more aggressive overtones. At the very least, contemporary depictions of China as a "center" resonate with a cultural chauvinism that dates back many centuries. It is perhaps telling that two respondents to my survey of *Fossils* readers considered China's "central location" as evidence for believing that humans originated in China.[110]

It is less clear, however, that discourse on connections among Chinese, Australians, and Americans primarily buttresses ethnic nationalism. Rather, repeated emphasis on the hereditary links between Chinese and foreign peoples—especially when combined with discussions of differences among regional and ethnic groups within China—cannot but undermine the construction of a biologically based Chinese national identity. Moreover, Chinese writers on human evolution continue to take the opportunity in their science dissemination materials to emphasize the fallacy of racial prejudice. Although post-Máo paleoanthropological discourse on race lacks the political clout found in Máo-era battle cries against imperialism, one may still read that racial differences are superficial, that they are outweighed by similarities, and that racist tendencies in politics and in science must be exposed and challenged.[111]

In this case, as in many others, discourse is multivalent and thus flexible. Whether the construction of China as a center of human diaspora supports nationalist and even imperialist trends or encourages an inclusive identity celebrating "all the world as one human family" depends on social and political factors. Both paths remain possibilities.

The choice and its consequences are well expressed in a touchingly honest article published in *Běijīng Youth News* in 2002. Taking an agnos-

---

Indian Bodies], *Zìrán yǔ rén*, no. 1 (1986): 25–27. See also Chén Chún, "Zhuīzōng zuì zǎo de Měizhōurén" [Searching for the Earliest Americans], *Huàshí*, no. 4 (1983): 9–10; Zhāng Zhènbiāo, "Àisījīmórén de láilóng qùmài" [The History of the Eskimos], *Huàshí*, no. 1 (1986): 19–21.

109. Sautman, "Peking Man," 110.

110. The respondents were an entrepreneur (個體戶) from Jiāngsū, fifty-four years old and male, and a peasant from Shǎnxī, sixty-five years old and male.

111. Wú Xīnzhì, *Rénlèi jìnhuà zújī*, 143–144; Zhōu Guóxīng and Liú Lìlì, *Rén zhī yóulái* [Human Origins] (Běijīng: Zhōngguó guójì guǎngbō chūbǎnshè, 1991), 65. Interestingly, Wú Xīnzhì highlights American racism in these pages in a way that recalls 1950s socialist Chinese discourse. Here it further serves to defend Wú against charges that his own preference for multiregionalism stems from nationalist pride.

tic position, the author presents the implications of the recent out-of-Africa theory.

There exists this possibility: that 50,000 years ago you and a black man [literally a black "older brother," 黑人老兄] were part of one family. . . . Maybe it's because we have such a long history on this piece of land that with respect to our feelings we seem to be more willing to believe that the Peking Man and Upper Cave Man we've known since grade school are our ancestors. Conversely, we absolutely can't clap American white people and African black people of our own era on the shoulder and say, "Ah, brother, 500 years ago we were one family." . . . But what if you multiply it by 100? . . . According to [the recent out-of-Africa theory], African black people and American white people are actually our close relatives, while the fossil humans dating from 600,000 years ago within China's borders—Lántián Man, Yuánmóu Man, Peking Man, and so on—have no genetic relationship with us.[112]

The influence of state-sponsored Chinese nationalism is tangible here. There is also something worrisome in the opening sentence's implication that the reader should experience shock or at least discomfort at the thought of being related to black people. The racism that African students like Emmanuel Hevi experienced in the 1960s and 1970s has by no means lessened in subsequent decades—indeed, I heard a disturbing echo of Hevi's experience with the doctor (recounted in chapter 3) when an anthropologist told me in 2002 that black people are the "ugliest race" because their skin "looks dirty." (Again, part of what makes this disturbing is the expectation that doctors and anthropologists of all people should "know better.") And while racist attitudes have remained strong, the post-Máo era lacks the strong propaganda of earlier years that kept them in check. Where American racism once stood as the most shameful manifestation of the evils of capitalism and imperialism, Chinese students angry at the behavior of African exchange students in 1986 wrote to African embassies that they would base their response "on the experience of Americans, who know very well what to do to curb the Negroes in their country."[113]

Nonetheless, it is possible to understand and even sympathize with the dilemma the author of the *Běijīng Youth News* article faces. Chinese people today are bombarded with cultural images and products from foreign countries and encouraged to see themselves as part of a globalizing

112. Duàn Jīn, "Tā shì Zhōngguórén," 32.
113. Michael Sullivan, "The 1988–89 Nanjing Anti-African Protests: Racial Nationalism or National Racism?" *China Quarterly* 138 (1994), 446. See also Barry Sautman, "Anti-black Racism in Post-Mao China," *China Quarterly* 138 (1994): 413–437.

world. They remain, however, largely isolated from foreigners themselves—and especially from dark-skinned foreigners. The most cosmopolitan will meet foreign students and businesspeople, but few even of these will have the same opportunity to travel abroad. It is no wonder that some look upon the debate between multiregionalism and the recent out-of-Africa theory as a hard choice: to embrace "Chinese" humans from half a million years ago or people of different colors and countries living today.

Ancestors, National and Personal

The article just quoted strikes a very personal note that raises questions about the character of the connection Chinese people feel for the human fossils said to be their "ancestors." To what extent is this relationship framed nationalistically or ethnically, and to what extent is it a more specifically personal bond or a more inclusively human one? These are by no means mutually exclusive categories. It would be difficult to imagine nationalism succeeding without inspiring personal attachment, and as Frank Dikötter has suggested, in the mid- to late Qīng dynasty, "folk notions of patrilineal descent became widespread in the creation and maintenance of group boundaries," of which ethnic and national boundaries became increasingly important.[114] Ethnic differences, moreover, make the most sense when understood as subsets of a larger category of "human." Nonetheless, the significance of Chinese fossils for Chinese people cannot be reduced to nationalism, and in some cases the specific meanings people invest in the fossils undermine nationalistic and other kinds of state agendas. While it is impossible to speak for all Chinese people, I will share five cases that put a "human" face on Chinese passion for Chinese fossils. These cases lead me to question Dikötter's assertion that "multiple identities . . . and ambiguity in group membership are not likely to appear as viable alternatives to more essentialist models of group definition."[115] This may be the case for state rhetoric, but not for the everyday beliefs and practices of many Chinese people.

The *Běijīng Youth News* article cited above explained for its readers the big debate over whether modern humans emerged recently in Africa. Its specific focus, however, was the implications of this debate for one "person" in particular: Dàlì Man, who lived perhaps 290,000 years ago

114. Frank Dikötter, "Racial Discourse in China: Continuities and Permutations," in *The Construction of Racial Identities in China and Japan: Historical and Contemporary Perspectives*, ed. Frank Dikötter (Honolulu: University of Hawai'i Press, 1997), 15.

115. Frank Dikötter, "Introduction," in Dikötter, *The Construction of Racial Identities*, 10–11.

in what is now Shǎnxī Province. After briefly introducing Dàlì Man, the article began bluntly: "The first question we want to ask readers is this: Are you willing to believe that he was an ancestor of Chinese people, or are you willing to believe that he was an aboriginal of this place and afterwards was exterminated along with his relatives by our ancestors?" The consequences of the recent out-of-Africa theory are thus not just that the Chinese people may have lived in China for only 100,000 instead of 1 or 2 million years, or that a "black man" may be the reader's close relative. The humaneness or barbarity of one's ancestors is also at stake. If Dàlì Man was an "enemy of our ancestor" rather than the ancestor himself, "our ancestors" must have exterminated "the aboriginal people of this land under our feet," the ones "we've known since grade school."[116]

The author of the *Běijīng Youth News* article appeared to feel equally strongly about Dàlì Man, Peking Man, Yuánmóu Man, Lántián Man, and presumably any of the other Chinese fossil humans that have become part of the narrative of Chinese history. For those who live near a specific fossil human site, however, local fossils often take on a special significance, sometimes further invoking an ethnic minority identity. This is certainly the case with Lǐ Xùwén, the Nàxī tax collector and fossil hobbyist from Yúnnán. It is also true of others, for example, Zhāng Xīzhōu, a former head clerk in a local party office in Chángyáng Tǔjiā-Nationality Autonomous County of Húběi Province. Zhāng's story appeared in a 1995 issue of the popular science magazine *Nature and Man*. During the excavation of Chángyáng Man in 1956–1957, Zhāng ran errands and provided logistical support for Jiǎ Lánpō and other scientists. In the meantime, he developed a deep interest in the work and took the opportunity to ask questions. From Jiǎ Lánpō he learned about paleoanthropology, and from a group of ethnologists he learned about the possible connections between Chángyáng Man and an ancient people called the Bā (巴人). "This historical mystery pulled him in like quicksand from which he could not escape." Zhāng then traveled to the county seat, where he secretly visited a blacklisted scholar (demonstrating that his commitment to learning left no room for political fears) and applied himself to the study of "archaeology, anthropology, paleontology, Classical Chinese, geology, astronomy, intelligence gathering, and library science." Over the next several decades, Zhāng pored through the ancient text *Classic of Mountains and Seas* (山海經) and tried to identify the geography

---

116. The impetus for the article was recent research by Chinese and French scientists that extended the dates for Dàlì Man back from 150,000–200,000 years ago to 280,000–290,000 years ago. The author noted that the new dates did not make a difference with respect to multiregionalism versus the recent out-of-Africa theory. Duàn Jīn, "Tā shì Zhōngguórén," 32.

it described with the places surrounding the Chángyáng Man site. The theory he developed, that humans originated on Zhūlì Mountain in Chángyáng, "leaves people wide-eyed and agape."[117]

On Zhūlì Mountain, Zhāng found artifacts from the Paleolithic period, and in a cave he found evidence that humans had taken shelter. He interpreted this site as the "nest" of Nest Builder, the first of the legendary figures from the classics who taught people to be human. His most striking conclusion, however, arose from his reinterpretation of the phrase "Pángǔ opens the heavens" (盤古開天), typically understood to mean the first step that the legendary figure Pángǔ took in creating the world. Associating the characters 盤 and 古 with local place names, Zhāng identified Pángǔ as Zhūlì Mountain. "Opening the heavens" he then interpreted as opening up the highlands to human habitation. The artifacts he discovered on top of the mountain dated from about 180,000 years later than those discovered at the bottom. This matched well with records from the classics, which apparently recorded that Pángǔ opened heaven 180,000 years previously. Zhāng's startling conclusion that humans originated at Zhūlì Mountain thus came about because "the *Classic of Mountains and Seas*, which people have called an inexplicable book of myths, led him to find material evidence."[118]

Zhāng Xīzhōu is not the only one to believe that humanity began in his backyard, nor is he alone in mixing science with myth and religion. In the first pages of the introduction, we saw a very similar perspective voiced in the letter I received from Fāng Lì, an old peasant man who reads *Fossils* magazine (although he can no longer afford a subscription) and listens to the teachings of banned *qìgōng* sects. Echoing decades of science dissemination materials and turning them on their heads, he explained, "Humans came from nature, not from monkeys. Today's monkeys can't become human. They haven't created religion, and have no human consciousness; they can't transform nature or transform the ecological environment." The difficulty with Chinese scientists today, he said, is that they treat the Book of Changes (易經, Yìjīng) as superstition, and they do not realize that an understanding of paleoanthropology is dependent on the study of the Buddhist martial art of *qìgōng*.

Fāng is convinced that humans and all other living things originated on Yáo Mountain near his home in Shǎnxī Province, where the breasts of the Queen Mother flow in two springs. The people of Shǎnxī came from Yáo Mountain, the people of China came from Shǎnxī, and the

117. Liú Hóngjìn, "Shìjiè Bārén zhī mí" [The World Mystery of the Bā People], *Zìrán yǔ rén*, no. 1 (1995): 22–25.

118. Liú Hóngjìn, "Shìjiè Bārén," 25.

people of the world came from China. "China sits in the middle of the world, like the heart in the human body. She is . . . the origin of humanity and the ten thousand primitive things." All the people of the world "are from one mother," so "why can't we live in harmony?" Blending his religious and political educations, he speculated, "We humans will in the end realize communism—international communism. . . . One hundred families will unite to become one family." Though separated by many miles, in fulfilling our "fated" relationship he and I were moving closer to this future.

Is this nationalism? "China" certainly looms large in Fāng's view of the world, and the locale, province, and even the globe can be seen in his narrative as nested identities that support rather than contest the nation. Yet I am still struck more by the smallest and largest of these identities—by his spiritual ties to the mountain near his home and by his vision of the world as "one family." And, despite his obvious and demonstrated familiarity with officially disseminated scientific knowledge, his narrative of the human past and future is one with which the Chinese state would be very uncomfortable: it smacks of superstition.

In Yuánmóu, Yúnnán, home of China's oldest *Homo erectus* fossils, local people likewise mix science and other kinds of belief in ways the state would find problematic. The Yuánmóu Man site museum focuses on local fossils and encourages visitors to see themselves as having descended, over millions of years, from ancient apes and fossil humans. It appears that many local people accept this basic story. As one man I interviewed explained, "We are all Yuánmóu Man's descendants [子子孫孫的後代]."[119] This perspective delights science disseminators like the guide I interviewed at the Yuánmóu Man museum. Less satisfying to them, however, is what some people do with the knowledge that Yuánmóu Man is their ancestor: they pray to Yuánmóu Man for protection. They do not stop at accepting Yuánmóu Man as an ancestor; they worship him. People come to the site especially on the Chinese New Year to "pay their respects to their ancestors" (拜祖宗) and ask for protection and blessings (保佑).[120]

For the Chinese state, ancestor worship is superstition. In the Máo era, the practice of visiting ancestral tombs was further problematic since it emphasized clan relations instead of class differences.[121] The socialist state attempted to combat this form of "feudal superstition" that strengthened traditional family identities. Hence, the government

119. Local man in Yuánmóu, interview with the author, 16 May 2002.
120. Interview with the author, 16 May 2002.
121. Elizabeth Perry, "Rural Violence in Socialist China," *China Quarterly* 103 (1985): 428.

encouraged people to spend the Qīngmíng festival, a day for sweeping graves and ancestor worship, honoring the "revolutionary martyrs" who had died to create the new China.[122] National figures responsible for the birth of the "new China" thus replaced personal ancestors. Just as the state was largely unsuccessful in eradicating traditional practices in the observance of Qīngmíng, so it appears to have failed to convince some people to see human fossils as national, rather than personal, ancestors. The way state-supported science dissemination efforts portray the subject probably contributes to this phenomenon rather than discouraging it. The stone that marks the site of the first discovery of a Yuánmóu Man fossil tooth resembles a gravestone (figure 24), and a great many books, exhibits, and films refer reverently to Peking Man, Yuánmóu Man, and other fossil humans as "our ancestors."

While scientists typically distance themselves from explicit ancestor worship, among other "superstitions," their own practices are sometimes not entirely different. For the scientific community at IVPP, Zhōukǒudiàn has taken on a very personal meaning as an ancestral shrine. The bodies of Yáng Zhōngjiàn, Péi Wénzhōng, Jiǎ Lánpō, and a few others with deep connections to Zhōukǒudiàn are buried on the hill overlooking the excavation sites and exhibit hall. Through three scenes shown in quick succession, the 1986 movie that accompanies the exhibit makes an implicit connection between the fossil ancestors and the first generation to bring them to light. First comes a close-up of the statue that stands at the entrance to the exhibit hall, announced as "our ancestor, Peking Man" (我們的老祖宗, 北京猿人). Next the viewer sees the scientists' graves, where young students accompanying Jiǎ Lánpō (not yet among his colleagues in the ground) place flowers and bow respectfully. The narrator bids the old scientists "rest in peace" while the next generation prepares to continue their work. The students then follow Jiǎ down to the original excavation site, "returning to their ancestors' home to have class."[123] The professional lineage of paleoanthropologists is thus woven into the evolutionary lineage that ties all the figures in the film together.

As it is in the other examples, the Chinese nation is easily found at the Zhōukǒudiàn site and the film that memorializes it. But it is not the

122. A. P. Cheater notes that when one hundred thousand people mourned Zhōu Ēnlái's death on Qīngmíng in 1977, the festival was "*not* conceptualized as its official replacement, 'Revolutionary Martyrs Day,' even by the official press." See his "Death Ritual as Political Trickster in the People's Republic of China," *Australian Journal of Chinese Affairs* 26 (1991): 77.

123. Zhōngguó rénmín dàxué, *"Běijīngrén" yízhǐ de fāxiàn yǔ yánjiū* [The Peking Man Site: Discovery and Research], 1986.

24 Gravestonelike marker at the Yuánmóu Man excavation site in Yuánmóu, Yúnnán. The characters on the stone read, "Site of the discovery of the tooth." Photograph by the author.

only category that frames the way people identify with human fossils in China. Members of the scientific community invest the fossils and the sites of their burial and discovery with meanings specific to their profession and its distinguished history. People living near human fossil sites sometimes embrace and even worship the fossils as personal or community ancestors, rejecting attempts by the state and scientists to do away with such superstitious and clannishly divisive behaviors. Others, like the *qìgōng* practitioner in Shǎnxī, may interpret human evolution through multiple lenses of Daoism, Buddhism, Marxism, and popular science in ways that highlight both local and global identities. The Chinese classics, combined with archaeological and ethnological sources, offer Zhāng Xīzhōu and anyone who will listen a way of reviving an ancient origin story centered around a peripheral region inhabited by ethnic minorities. Finally, in a manner that recalls discourse on *yěrén* (in addition, perhaps, to ancient Chinese agricultural, as opposed to nomadic, identity), the author of the article in *Běijīng Youth News* interprets the recent out-of-Africa theory as threatening not only to Chinese ethnic identification with Dàlì Man and other Chinese fossils, but also to the notion of humans as settled and peaceful creatures rather than hostile and brutal invaders.

Choices and Interpretations

Interest in popular forms of resistance to hegemony has made a mark in the study of science in society. Scholars have shown that, far from passively accepting disseminated knowledge and simultaneously accepting authorities' characterization of their own ignorance, people reinterpret, or "deflect and redefine," the science they encounter in museums, books, classrooms, and clinics.[124] The five cases just shared, along with others presented in previous chapters, confirm this finding for post-Máo China. Seldom, however, do we find outright rejection of officially disseminated scientific knowledge. Rather, people selectively integrate elements of such materials into their own larger and more diverse sets of knowledge, which may include religious ideas, notions of family relations, local geographies, and a wide range of social and cultural identities.

---

124. For "deflect and redefine," see Sharon McDonald, "Authorising Science: Public Understandings of Science in Museums," in *Misunderstanding Science? The Public Reconstruction of Science and Technology*, ed. Alan Irwin and Brian Wynne (Cambridge: Cambridge University Press, 2003 [1996]), 167. Emily Martin's work also consistently explores such themes. See *The Woman in the Body: A Cultural Analysis of Reproduction* (Boston: Beacon Press, 1992 [1987]); and *Flexible Bodies: Tracking Immunity in American Culture from the Days of Polio to the Age of AIDS* (Boston: Beacon Press, 1994).

Moreover, this applies not only to laypeople, but to scientists as well. Scientists are not just professionals. At one time, they too were laypeople exposed to the same forms of science dissemination as others in their social status group. Even after entering the academy, they remain members of the larger society and retain many of the values and investments of their nonprofessional lives. There is thus no sharp dividing line between "scientific" and "nonscientific" understandings of the world. Paleoanthropologists bring their values to their work, and their scientific knowledge shapes their perspectives on their larger worlds. They lay their bones alongside those of the "ancestors" they study.

Scientists writing on human evolution have often been highly aware of the cultural and political significance of their work. Ashley Montagu, Carleton Coon, and Huáng Wànbō are just three of the most directly and self-consciously active in this regard. It would be a mistake, however, to see a one-to-one relationship between their political agendas and their scientific theories. Any one given theory can give rise to multiple meanings about humanity. For example, the "single-species model" that undergirds multiregionalism has supported both racist and antiracist interpretations of human difference (or nondifference). Even Weidenreich's orthogenetic theory of racial continuity could be turned to progressive purpose: the parallel evolution of distinct racial groups could be taken to indicate a fundamental shared humanity that overpowers any regional differences in propelling the entire human diaspora toward a single end. And as Robert Proctor has suggested, "out of Africa" can be read not only as "we are all Africans," but "Out of Africa: Thank God!"— that is, "hominids became human *in the process of leaving Africa*."[125] On the other side, the author of the *Běijīng Youth News* article on Dàlì Man interpreted multiregionalism to mean identifying with aboriginal people instead of with aggressive invaders—an understanding with much progressive political potential. Those who hope that humanism will overpower nationalism in China have less to fear from multiregionalism than they do from the social and political circumstances that lead to its nationalistic interpretation—for example, Chinese people's exposure to derogatory images of black people and their lack of exposure to living black people.

Scientific theories on human evolution can be, and have been, interpreted in so many different ways that it matters less which one triumphs than what is made of it. And as we have seen, the significant influence of ethnic nationalism in writings on human evolution has not prevented

---

125. Proctor, "Three Roots," 225. Emphasis in original.

Conclusion

A History of Human Identity

On 7 January 2007, a person calling himself "The *Yěrén* from Qiántáng River" (錢塘野人) posted an essay by the Marxist humanist Wáng Ruòshuǐ to an online document library, www.360doc.com. Based in the People's Republic, the Web site provides mainland Chinese people a space to share and discuss their favorite writings. Censors monitor the site and block users who attempt to post on the most sensitive subjects—for example, Fǎlún Gōng or the June Fourth Incident. But they generally allow more mildly provocative materials such as Wáng's "My Marxist Outlook" (我的馬克思主義觀), which he wrote in 1995 as a criticism of the Chinese socialist state's failure to embrace humanism as a core principle of Marxist theory. Wáng's analysis depended on the conception of human identity intensively disseminated in Máo-era popular science literature: "Marx's ideal was to liberate all humanity. What separates humans from animals is labor. Thus, to liberate humans, we must first liberate labor. Originally, labor created humanity. But in a class society, labor destroys the human mind and body and dehumanizes the human [使人不成其爲人]. This is the alienation of labor."[1] This was not the first time Wáng had

---

1. See http://www.360doc.com/showWeb/0/0/348689.aspx (viewed 6 May 2007). See also http://www.wangruoshui.net/CHINESE/MAYIGUAN.HTM (viewed 6 May 2007). In 1996, the overseas Chinese dissident journal *Běijīng Spring* published a revised version of the essay, which stated instead: "Labor is what first separated humans from animals, and labor is the realization of freedom." Wáng Ruòshuǐ, "Wǒ de Mǎkèsīzhǔyì guān" [My Marxist View], *Běijīng zhī chūn*, no. 1 (1996): 6–17.

evoked Engels's famous thesis that labor created humanity. In a 1986 article entitled "On the Marxist Philosophy of Humanity" (關於馬克思主義的人的哲學), Wáng had argued, "Labor created humanity. It not only transformed anthropoid apes into primitive humans, but also transformed primitive humans into civilized humans and [finally] modern humans."[2] Then too, Wáng's purpose was to establish humanism as the essential feature of Marxist philosophy and to argue that the Chinese socialist state had in fact contributed to the alienation of labor and thus prevented Chinese people from achieving freedom and realizing their humanity.

Human identity is a critical analytical category in studying modern Chinese history. In China as elsewhere, human identity has been thoroughly intertwined with more specific social identities. Popular paleoanthropology in twentieth-century China has worked to shore up nationalism, to define gender roles, and to celebrate the laboring classes as living embodiments of what it means to be human. However, contrary to Maoist rhetoric, discourses on humanity cannot be reduced to discourses on specific social identities. Rather, they often produce a kind of identity that is potentially inclusive, extendable across ethnic, gender, and class lines. In the early twentieth century, for people like Yán Fù, evolutionary theory was meaningful because it helped explain how China as a nation could survive in the international context of imperialism. While bolstering a nationalist cause, it also depended on an understanding of Chinese people as subject to the same laws of evolutionary change governing humans everywhere. In the Máo era, humanity became defined through manual labor, embedding human identity in class identity. But the notion that labor created humanity also supported an inclusive sense of what it meant to be human: labor was a physical experience shared by all the races of the world, and it was something that could transform even intellectuals, returning their humanity to them. Rooting human origins in a state of "primitive communism," where all shared in labor and its fruits, also helped support an even more fundamental Máo-era vision for society: the achievement of communism itself. Finally, in the post-Máo period, the ethnic nationalist attachment to Peking Man as a Chinese ancestor exists alongside more inclusive internationalist visions for science and globalist understandings of humanity. The wildly popular *yěrén* stories, while pregnant with implications for gender and ethnicity, also tell of a humanity broadly

---

2. Wáng Ruòshuǐ, *Zhìhuì de tòngkǔ* [The Pain of Wisdom] (Hong Kong: Sānlián shūdiàn, 1989), 275.

shared by all those who possess and properly express human feeling.

Human identity is further important because of the multiple ways it can define people with respect to the nonhuman world. With the introduction of Darwinism to China near the beginning of the twentieth century, many Chinese people began to see the "struggle for existence" as an inevitable consequence of humans' place in the natural order. Under Máo, they learned to revere struggle for its own sake and to celebrate the victories in "humans' struggle with nature." Engels's influential treatise "The Part Played by Labor in the Transition from Ape to Human" emphasized the "final, essential distinction" between animals and humans that labor made possible: only humans are capable of "mastering" nature, transforming it to serve our goals. Yet he also cautioned: "Let us not, however, flatter ourselves overmuch on account of our human victories over nature. For each such victory nature takes its revenge on us. . . . At every step we are reminded that we by no means rule over nature like a conqueror over a foreign people, like someone standing outside nature—but that we, with flesh, blood and brain, belong to nature, and exist in its midst, and that all our mastery of it consists in the fact that we have the advantage over all other creatures of being able to know and correctly apply its laws."[3] In the Máo era, the Chinese state glorified humans' mastery over nature, with little or no attention to Engels's caveat or concern for the destruction that often ensued.[4] Amid a growing realization in the post-Máo era of the environmental degradation such modernist policies have caused, the dominance of humans over nature is no longer so celebrated. Rather, yěrén have emerged as symbols of wilderness and as foils to expose the barbarity of human destructiveness, turning the ancient adage 惟人萬物之靈 ("humans are the leaders of the ten-thousand things") on its head.

Human identity has frequently served as a potent political resource. In the Chinese classics, the difference between humans and animals was civilization, represented by fire, hunting, and other technologies taught to the people by a series of legendary rulers. People thus owed their very humanity to the rulers who had taught them the skills and had given them the social institutions that separated them from the animals.[5] But

3. Frederick Engels, *The Part Played by Labor in the Transition from Ape to Man* (New York: International Publishers, 1950), 18–19.

4. Judith Shapiro, *Mao's War against Nature: Politics and the Environment in Revolutionary China* (Cambridge: Cambridge University Press, 2001). The effort to control nature is also a major theme of Laurence Schneider's *Biology and Revolution in Twentieth-Century China* (Lanham, Md.: Rowman and Littlefield, 2003).

5. Mark Edward Lewis, *Sanctioned Violence in Early China* (Albany: State University of New York Press, 1990), 171.

this possibility for humans to slip back into the realm of beasts some-times also served as a resource to criticize heartless rulers. In a story collected by the eighteenth-century folklorist Yuán Méi, the cruelty of the first emperor's Great Wall conscription practices forced people to flee into the wilderness, where they returned to a bestial state. Darwinist evolution and Marxist history fixed the boundary between humans and other animals in the distant past and removed the potential for humans to revert to apes. The story of human evolution was (and largely remains) a progressivist and voluntarist one that imagines humans as naturally predisposed to strive toward ever-more-developed states. Nonetheless, the enthusiasm for stories about *yěrén* in the post-Máo period, and the frequency with which *yěrén* turn out to be humans "gone wild," indicate that the gate between humans and animals continues to swing both ways in the imaginations of many Chinese people. This idea adds force to moral admonitions against certain types of behavior identified as less than human, and people have used it specifically to criticize the bar-barity of Máo-era political campaigns and post-Máo assaults on nature. *Yěrén* have tapped the deep wells of popular culture. Their voices, reso-nant with primordial authority, are worthy of consideration alongside those of post-Máo intellectuals who argue for the relevance of "human rights" and a "humanistic spirit." Just as the political dissident Wáng Ruòshuǐ found significance in the Marxist humanist implications of Engels's theory that labor created humanity, *yěrén* revive an apprecia-tion for such long-standing values as "human feeling." In such ways has human identity been of crucial significance in modern Chinese history.

Mass Science and Popular Culture

Paleoanthropology in China has owed much to the durability and influ-ence of popular culture, which has acted as a reservoir for ideas about humanity and for knowledge about natural history. Máo was right to think that "the masses" had something to offer science, but he missed the link between mass science and popular culture. His and others' pre-occupation with eradicating superstition obscured the role of popular culture and limited the acknowledgment of the masses' contributions to their experiences in production, very narrowly defined. This precluded a full appreciation of the extent to which many sciences—especially sci-ences about the nature of people—derive much from popular culture. The many ways in which popular culture influenced Chinese paleoan-thropology could have been understood in the Máo era under the rubric of mass science, if only the rubric had included cultural and intellectual contributions rather than just material and labor contributions—that is, if only popular culture had not been equated with superstition.

As one example, dragon bones have repeatedly found their way into the scientific culture of paleoanthropologists and paleontologists working in China. Scientists dubbed the Peking Man excavation site Dragon Bone Hill, and they often discuss dragon bones at length in their memoirs and other writings. I suggest that dragon bones have been important to Chinese paleoanthropologists because they invest Chinese paleoanthropology with cultural meaning, making the science more "Chinese." Yet always there is an underlying discomfort arising from the odor of superstition that clings to dragon bones, and this probably contributed to the failure of dragon bones to be explicitly identified as part of mass science.

Popular culture also offers clues about local natural history. We saw compelling evidence of this in the late Cultural Revolution, when scientists working in the Níhéwān Basin encountered a local legend attesting to the existence of a lake in the remote past. But it does not seem Chinese scientists often pursued such clues, and certainly not in any systematic way that acknowledged legends as a legitimate part of mass science. Actually, reports on the excavations in Níhéwān made no reference to the benefit of the lake story to scientists, but rather focused on how the local people "learned" about the lake from the scientists who came to excavate there.

Nothing from the rich reservoir of popular culture has influenced paleoanthropology more than the passionate concern for ancestors. The notion of national ancestors has permeated Chinese materials on human evolution produced by scientists and circulated through official channels. Yet an interest in ancestors is hardly limited to paleoanthropologists and state officials. Rather, it is broadly shared throughout the general population, especially although not exclusively in China. Picture books that valorize Peking Man as "our ancestor," students who pay their respects at the graves of famous paleoanthropologists atop Zhōukǒudiàn, and people who visit the Yuánmóu Man site to pray for blessings and protection are all putting into practice an enduring cultural commitment to honor those who have come before.

The boundary erected between science and popular culture appears to be weakening still further in the post-Máo period under the influence of market reforms. Nowhere is this more evident than in the phenomenal interest *yěrén* have sparked among scientists, peasants, poets, and other members of contemporary Chinese society. What scientists know about *yěrén* is rooted in stories passed down from imperial-era texts and still circulated orally today by rural people. *Yěrén* have thus been a very significant way in which popular culture has contributed to the exploration of meanings about humanity. They address issues of sex, wilderness, the relationship of people to nature and to one another, and the barbarity of certain past and recent political events. The questions paleoanthropologists ask about humanity (whether it is our aggressive or peaceful nature, the relationship of the sexes, or other questions) are always informed by culture. The *yěrén* literature is thus not an exception, but rather a particularly colorful way for this to happen. And because of all of this I think of the *yěrén* literature as having made popular science live up to the first half of its name. *Yěrén* are popular in every sense of the word.

How Popular Science Changes What We Think about China

Jiāng Qīng and her allies, for all their mistakes and moral failings, were right to reject "exceptionalism" in science, and China scholars would be wise to follow suit in greater numbers. Just as science is too important to be left to scientists, it has been too central in the lives of twentieth-century Chinese people to be treated as a marginal subject in history. Science dissemination and mass science in particular are intimately connected to the more recognized issues of ideology and class struggle in Chinese history.

Science dissemination, especially though again by no means exclusively in China, has been above all an attempt to modernize people's thoughts. Beginning in the early twentieth century, Chinese science dissemination has typically been intertwined with a more general struggle over cultural legacies from the past, often identified as "superstitions." In 1949, that struggle broadened to include all forms of "idealism," which in the case of human evolution particularly targeted Christianity. The issue remains critical in Chinese politics today, as scientists are enlisted and science dissemination projects funded to combat the "antiscientific" teachings of organizations like Fǎlún Gōng. Knowledge of science dissemination practices is thus necessary to understand larger questions of modernity and cultural transformation in China.

Science dissemination was a key component of the socialist state's efforts to replace old ways of thinking with revolutionary ones. This was a project in which intellectuals—and particularly scientists—had a clear and compelling role to play. It was also an area in which scientists' values overlapped with state priorities. The process of elevating scientific thought, moreover, raised the social and political status of scientists. Science dissemination thus offered scientists and some other intellectuals a kind of legitimation that must have been very attractive, given their generally precarious political position. Through cadre classes, lectures at factories, slide shows, books, magazines, films, and exhibitions, the state used science to validate the ideological lessons they required the people to learn. Materials on human evolution produced for lay audiences in the Máo era had a clear political agenda in converting people from Christian and other idealist views of human origins to the materialist perspective offered in Engels's essay "The Part Played by Labor in the Transition from Ape to Human." Other political agendas included the celebration of Peking Man and other human fossils as ancestors of the Chinese people, and the simultaneous attack on imperialism and embracing of "all the world as one human family."

Yet these materials cannot be dismissed as mere ideological indoctrination. Audiences were not simply asked to accept new ideas because Marx, Engels, Lenin, Stalin, or Máo said they were so. Rather, the new ideas gained credibility from a conspicuous display of material evidence together with clear analysis. Moreover, the empirical basis of scientific knowledge was underscored through frequent reference to questions that could not yet be resolved because of a lack of sufficient evidence. While there was much that was authoritarian in Máo-era science dissemination, these factors suggest that state officials and scientists considered it important to distinguish their propaganda from other kinds of ideological education, particularly religious teachings. They wanted to claim science on their side, and they thus emphasized material evidence and logical analysis as the means of convincing the masses to transform their thoughts. Indeed, part of the goal of this transformation was for people to think more scientifically. That this goal was undermined by authoritarianism does not entirely erase its significance.

If science dissemination was a special kind of ideological education, popular participation in science can be seen as a particularly "hard case," or strong test, of class struggle. When in 1941 Máo gave his famous "Talks at the Yán'ān Forum," he set forth a paradigm for a "mass style" in art and literature in which "the thoughts and feelings of our writers and artists should be fused with those of the workers, peasants, and soldiers" through "conscientiously learn[ing] the language of the masses."[6] The twin goals of dissemination and raising standards would be accomplished through a two-way flow of knowledge: intellectuals would disseminate

6. Mao Tse-tung, *Selected Works of Mao Tse-tung* (Peking: Foreign Languages Press, 1967–1971), 3:72.

art, but would in their own work aim "in the direction in which the workers, peasants and soldiers are themselves advancing."[7] In the early 1950s, science dissemination efforts drew on this language of dissemination and raising standards, but a mass style proved even more difficult to achieve in science than in art. Even Máo had recurring doubts about whether the natural sciences might require some immunity from the political struggles with which other fields had to contend. The pursuit in the Cultural Revolution of policies like "open-door science," "mixing sand," and having scientists "stand aside" to let the masses lead were thus in many ways the most extreme tests of how far workers, peasants, and soldiers could rise above the barriers posed by intellectual elitism to take the reins of society. That these efforts often failed not only in their outcomes but even in how seriously and systematically they were implemented indicates that ideas about the necessity of formal education and expertise remained very powerful even after decades of explicit struggle. Moreover, it was the state's own commitment to seeing the masses as superstitious that tied the hands of those who sought to promote a more radical epistemology.

While such issues are most striking during the Máo era, we can find them also in the republican and post-Máo periods. The cultural gap between intellectual elites and nonelites—so vivid in the Zhōukǒudiàn excavations—was a major issue for liberal reformers and socialists alike in the early twentieth century. That problem remains today, and there is little reason to think that it will be readily solved through science dissemination efforts that are increasingly split between elite-oriented programs and sensationalist materials designed for a market economy. The years between 1949 and 1976 thus share too much with the periods before and after to be seen as the "detour" some have suggested.[8]

Similarly, a study of popular science demonstrates that there is too much shared with other modern societies not to recognize China as part of a bigger world, even during its most isolated times. This does not mean unthinkingly applying to China categories created to describe other places, but rather, as R. Bin Wong has argued, recognizing the many categories that are shared across cultures and then, on the basis of these similarities, defining the significant differences.[9] Such an approach shows that China's struggles with ideology, elitism, and other issues

7. Mao Tse-tung, *Selected Works*, 3:80.

8. See, for example, Jiwei Ci, *Dialectic of the Chinese Revolution: From Utopianism to Hedonism* (Stanford, Calif.: Stanford University Press, 1994).

9. R. Bin Wong, *China Transformed: Historical Change and the Limits of European Experience* (Ithaca, N.Y.: Cornell University Press, 1997), 7.

from the more general commitment to such "productive" sciences. It was also due to the view of the masses as too superstitious to contribute to science intellectually or culturally, and only able to participate based on their experience in manual labor. This prevented the full extension into the natural sciences of a radical Marxist epistemology—that is, a model of scientific knowledge based on an understanding of laborers as uniquely positioned to contribute to science. Ironically, this impediment was less present among Marxist scientists in postwar, capitalist Japan.

Patterns of popular participation in paleoanthropology have been largely consistent throughout the twentieth century in China, and across national and political contexts. Paleoanthropologists and other field scientists in republican China, socialist China, postsocialist China, and nonsocialist countries alike have depended on local people to provide information and to assist in the labor of scientific research. Neither has China been alone in its engagement with the class politics of scientific knowledge. At least since Victorian times, science in the West has also been an arena in which diverse social actors have struggled over questions of authority and access, and many have successfully contested the notion that science is a privileged territory for educated elites, specifically scientists. The chief difference lies in the support the Chinese socialist state often offered to challenges from below, which significantly altered the handicaps. The Cultural Revolution's attack on the primacy of experts may well represent the greatest experiment to date in pushing the issue of participation to its apparently logical conclusion.

How, then, does knowledge of the Chinese experience help us think about popular participation in science in general? First, it is clear that it does not usually make sense for workers, farmers, and soldiers to lead scientific teams, or for scientists and workers to be placed in artificial relationships to force mutual learning. Moreover, it is counterproductive—not to mention morally contemptible—to attack scientific elitism through physical and psychological violence. Nonetheless, it does make sense to recognize the different kinds of expertise that laypeople have and to develop useful ways of incorporating this expertise into scientific research. And it also makes sense to acknowledge the degree to which meaningful scientific statements are dependent on broader forms of shared knowledge—as shown, for example, in discourse on ancestors.

When it comes to superstition, it must be admitted that not all forms of knowledge are equally beneficial to society and that many may actually be dangerous. However, the war on superstition waged by Chinese science disseminators and by people like Carl Sagan does little to help matters and is moreover based on a faulty understanding of the relationship between science and popular culture. The boundary between science and nonscience is blurry, contested, and constructed. Popular skepticism about scientific authority is fed by social and political structures to which people do not have equal access and in which people are not equally powerful. Despite their antielitism, Máo and other Chinese socialists appear never to have seen the issue in quite this way. Rejection or selective integration of scientific knowledge also results from the specific priorities laypeople hold, which often differ in important ways from those of scientists. We must allow for people's greater interest in stories that are meaningful than in those that are accurate, even as we encourage the cultivation of critical thinking. Finally, popular "misconceptions" of science are not the only dangerous conceptions. Theories of human evolution that buttress racist or nationalist policies constitute just one example of the potential hazards of expert knowledge.

This points to a broader question: Should scientific knowledge have any bearing on how we construct human identity, or for that matter any other kind of social identity? If modern humans are found to have originated in multiple regions, does this mean we have less in common today? If labor created humanity, must we labor to be human? If homosexual behaviors are found to be absent in other species, does this imply humans should likewise abstain? Conversely, if bonobos (pygmy chimpanzees) engage in homosexual behaviors, should humans follow suit? Should the homemaking responsibilities of male Garibaldi fish or the childbearing functions of male sea horses serve as inspirations for human men? Should we be encouraged to have sex changes because sheepfish do it? While some of these suggestions may sound ridiculous to some readers, to various degrees they are all active as resources in shaping how people in modern societies think about themselves and their worlds.[12]

Donna Haraway insightfully comments, "Forbidding comparative stories about people and animals would impoverish public discourse." "But," she continues, "no class of these stories can be seen as innocent, free of determinations by historically specific social relations and daily practice in producing and reproducing daily life. Surely scientific stories are not innocent in that sense."[13] We cannot stop the Chinese state from investing fossils with national significance, just as the Chinese state cannot stop people in Yuánmóu from worshipping fossils as ancestors with

12. The latter examples come from my observations of docent and visitor interpretations of exhibits at the Birch Aquarium in San Diego, 1998–2001.

13. Donna Haraway, *Simians, Cyborgs, and Women: The Reinvention of Nature* (New York: Routledge, 1991), 105–106.

the power to bless and protect. What we can do is to point out the complexity of nature, recognize the myriad interpretations it offers, and empower people to challenge interpretations that prevent them from achieving the kind of world they desire.

What Did Not Happen

Somewhat like sea horses and bonobos, Chinese history has often served as a way for Westerners to think about their own cultures. New-age philosophers have idealized Chinese harmonious relations with nature and perceptions of mind-body unity; members of the John Birch Society have demonized China as a totalitarian empire bent on destroying the American way of life. Many historians from diverse political perspectives have written modern Chinese history in the tragic mode as a story of missed opportunities—for liberalism, for capitalism, for socialism, for feminism, or for other priorities.[14] Such approaches may be unfair or even dangerous when they imagine China as the opposite of the West or when they chastise China for not living up to the authors' expectations. Nonetheless, to deprive ourselves of a view in the Chinese mirror would, in Haraway's words, "impoverish public discourse" on Chinese and Western cultures alike.

These issues are nowhere more explicit than in the historiography of premodern Chinese science, which began with a period dominated by the "Needham question" ("Why did the scientific revolution not occur in China?") and then moved into one dominated by critiques of that question. One commonly voiced concern is that potential questions about what did not happen in history are both limitless and impossible to answer conclusively.[15] At one time, I found Needham's approach so frustrating that I posed the following counterargument to my classmates in science studies: "What if we asked why acupuncture did not emerge in Europe, or why Western countries took so long (compared with China) to develop meritocratic bureaucracies?" To my surprise, while the former example

14. Jerome B. Grieder, *Hu Shih and the Chinese Renaissance: Liberalism in the Chinese Revolution, 1917–1937* (Cambridge, Mass.: Harvard University Press, 1970); James C. Thomson, *While China Faced West: American Reformers in Nationalist China, 1928–1937* (Cambridge, Mass.: Harvard University Press, 1969); Hsin-pao Chang, *Commissioner Lin and the Opium War* (Cambridge, Mass.: Harvard University Press, 1964); Maurice Meisner, *Marxism, Maoism, and Utopianism: Eight Essays* (Madison: University of Wisconsin Press, 1982); Harold R. Isaacs, *The Tragedy of the Chinese Revolution* (Stanford, Calif.: Stanford University Press, 1951 [1938]); Margery Wolf, *Revolution Postponed: Women in Contemporary China* (Stanford, Calif.: Stanford University Press, 1985).

15. Nathan Sivin, *Science in Ancient China: Researches and Reflections* (Aldershot, U.K.: Variorum, 1995), 51.

was too specific to be very useful, the latter stimulated real interest.

Since that time, I have reevaluated my perspective. Counterfactual questions are methodologically powerful because they liberate history from determinism.[16] Even counterfactual questions based, as Needham's was, on expectations derived from the experiences of other cultures are not inherently problematic. Rather, they permit investigation of possibilities otherwise obscured by the limited perspectives of historical actors. The real problem with the Needham question is that on its own it operates asymmetrically to reinforce existing patterns of cultural dominance. There has been no effort of equivalent influence to interpret Western histories based on Chinese expectations, although scholars like R. Bin Wong are moving us in that direction.

Numerous counterfactual questions have enriched this study.[17] For example, in chapter 3 I asked what can be inferred from the near absence of gender as an analytical category in science dissemination materials on human evolution. That they did not take available opportunities to highlight gender equality as a "natural" state indicates the relatively low priority assigned to feminist concerns. In chapter 5, I outlined what to me was an obvious link between the notion that labor created humanity and the suggestion that laborers had experience valuable to science. That this link was never articulated reveals the lack of a rigorous, radical reevaluation of the structure of scientific knowledge based on a Marxist standpoint epistemology. As a final example, chapter 8 concluded with a thought experiment on the possible ramifications should the recent out-of-Africa theory gain enough support in China to exclude Peking Man from the modern human lineage. I offered the hope that the growing fascination with human "cousins" like chimpanzees and especially *yěrén* might help Peking Man remain a cherished member of the human family even without ancestral status.

Questions like Needham's of "why did not" thus open the door for

---

16. Richard Evans has a point when he identifies the highly deterministic character of many examples of "what if" history. That only holds true, however, for counterfactual historians who think it is possible to derive what specifically would have happened given a set of imaginary circumstances. Richard Evans, "Telling It Like It Wasn't," *Historically Speaking* 5, no. 4 (March 2004): 11–14.

17. Counterfactual history has typically been understood to mean imagining the long-term effects of specific changes in the historical narrative: for example, how would history have been different if Napoleon had won at Waterloo? My interest in counterfactual history is very different. I aim to identify what has prevented the emergence of specific outcomes other than those that actually happened. Such an approach helps us understand the constraints under which people have acted. More important to me, however, is the room it makes for an honest exploration of the historian's own values and hopes. Contrary to Richard Evans's criticism of counterfactuals, I think there *is* room for "wishful thinking" in history if we admit that we care how history turns out. See his "Telling It Like It Wasn't."

questions many science fiction authors ask of "why not" and "what if." They allow us to imagine the worlds for whose existence we would strive. We inhabit the histories of these future worlds, and it is thus up to us to ask the questions and create the possibilities needed to engender them. While historians have a responsibility to be careful when gathering and interpreting data about the past, we can afford to be considerably more utopian when we imagine better alternatives. With all of this in mind, I will close with an unapologetic attempt to construct a future from some of the promises and some of the actual experiences of popular paleoanthropology in twentieth-century China.[18] A truly popular paleoanthropology would be one in which

- Scientists are versed in popular cultural understandings about humanity, origins, and ancestors, and they are self-reflective about how these understandings shape research questions, theories, and narratives of human origins
- Dragon bones are discussed without the quotation marks that betray discomfort about the perspective on the natural world they represent, just as Chinese pharmacists may speak of fossils without flinching
- Scientists ask local people not only about dragon bones, but also about other kinds of local knowledge about the land
- Paleoanthropology is seen as a field incorporating both mental and manual labor, and farmers' and workers' contributions—from digging to offering insights on how primitive tools were made and used—are solicited and acknowledged
- Local people who have made contributions to paleoanthropology are acknowledged in print by name and have the opportunity to continue to participate in future excavations, in dissemination activities, or in other ways
- Men and women, and people of diverse social and ethnic backgrounds, participate in all aspects of paleoanthropology, including research and dissemination
- Meanings produced about humanity through paleoanthropology do not serve as natural laws determining specific patterns of social behavior; instead, they are drawn upon to challenge or enhance ideas people cherish based on their own experiences and on other cultural resources; in addition, such scientific concepts are themselves challenged and reinterpreted through popular culture

Neither China nor any other country has resolved the social and political disagreements that prevent such a vision of science from being realized. Nonetheless, reflecting on one another's histories may yet inspire us to move in that direction.

18. This may be thought of as a vision for a "successor science," as Sandra Harding terms it in *The Science Question in Feminism* (Ithaca, N.Y.: Cornell University Press, 1986).

Bibliography

Because of the large number of primary source articles, I have not listed each article here. Instead, I have listed only the periodicals in which they appear.

ARCHIVES AND PERSONAL LIBRARIES

Běijīng Municipal Archives
Fairbank Center, Harvard University (red guard materials)
Gǔjǐzhuī dòngwù yǔ gǔrénlèi yánjiūsuǒ [Institute for Vertebrate Paleontology and Paleoanthropology—IVPP] archives
Jiǎ Lánpō personal archives
Péi Wénzhōng collection, IVPP library
Shénnóngjià archives
Yáng Zhōngjiàn collection, IVPP library
Zhōngguó kēxué jìshù xiéhuì [Chinese Association for Science and Technology] archives
Zhōu Míngzhèn collection, IVPP library

CHINESE-LANGUAGE PERIODICALS

Àomì huàbào [Mysteries Illustrated]
Bǎikē zhīshí [Encyclopedic Knowledge]
Běijīng qīngnián bào [Běijīng Youth News]
Běijīng rìbào [Běijīng Daily]
Běijīng wǎnbào [Běijīng Evening News]
Běijīng zhī chūn [Běijīng Spring]
Běipíng chénbào [Běipíng Morning News]
Bówù [Natural History, later *Zìrán yǔ rén*]
Chángjiāng [Yangtze River]
Chángyáng wénshǐ zīliào [Chángyáng Historical Materials]
Dà zìrán [Nature]

Dà zìrán tànsuŏ [Exploration of Nature]

Dàgōng bào [Impartial Gazette]

Dāngdài diànshì [Modern Television]

Dāngdài zuòjiā [Modern Writer]

Dàzhòng kēxué [Mass Science]

Diànyĭng wénxué [Film Literature]

Dōngfāng zázhì [Eastern Miscellany]

Guāngmíng rìbào [Guāngmíng Daily]

Gŭjĭzhuī dòngwù yŭ gŭrénlèi (formerly *Gŭ shēngwù xuébào*; known also as
   Vertebrata PalAsiatica and as *Paleovertebrata et Paleoanthropologia*)

Huánghé shuĭlì zhíyè jìshù xuéyuàn xuébào [Journal of the Yellow River Hydraulic
   Engineering Institute]

Huàshí [Fossils]

Jiācháng kēxué [Everyday Science]

Jiānghàn lùntán [Jiānghàn Tribune]

Jiāngxī shèhuì kēxué [Jiāngxī Social Sciences]

Jiěfàng rìbào [Liberation Daily]

Jù yĭng yuèbào [Drama and Film Monthly]

Kējì rìbào [Science and Technology Daily]

Kēpŭ chuàngzuò [Science Writings]

Kēxué àihàozhě [Science Hobbyist]

Kēxué bào [Science Times]

Kēxué dàzhòng [Popular Science]

Kēxué de Zhōngguó [Scientific China]

Kēxué huàbào [Popular Science / La Science Populaire / Popular Science Monthly]

Kēxué pŭjí gōngzuò [Science Dissemination Work]

Kēxué pŭjí tōngxùn [Science Dissemination Bulletin]

Kēxué qùwèi [Scientific Taste]

Kēxué shēnghuó [Science Life, published in Běipíng (Běijīng)]

Kēxué shēnghuó [Science Life, published in Shànghǎi]

Kēxué shíbào [Science Times]

Kēxué shìjì [Scientific Century]

Kēxué shìjiè [Science World]

Kēxué shíyàn [Scientific Experiment]

Kēxué tōngbào [Scientia]

Kēxué yŭ shēnghuó [Science and Life]

Kēxué zhī chuāng [Science Window]

Kēxué zhīshí [La Scio de Scienco / Scientific Knowledge]

Lŭxíng zázhì [China Traveler]

Mángzhòng [Grain in Ear]

Mínsú [Ethnology]

Qiǎnjiāng wǎnbào [Qiǎnjiāng (Húběi) Evening News]

Rénlèixué xuébào [Acta Anthropologica Sinica]

Rénmín rìbào [People's Daily]

Rénmín wénxué [People's Literature]

Shànghǎi shīfàn xuéyuàn xuébào [Journal of Shànghǎi Teachers' College]
Shānxī dàxué shīfàn xuéyuàn xuébào [Journal of the Teachers' College of Shānxī University]
Shēngwùxué tōngbào [Biology Bulletin]
Tàibái [Venus]
Tànsuǒ [Probe]
Wénhuì bào [Wénhuì News]
Wénhuì bào fùkān [Wénhuì News, Supplement]
Wényì bào [Literature and Art News]
Wénzhāi zhōubào [Weekly Digest]
Xiāngfén wénshǐ zīliào [Xiāngfén Historical Materials]
Xiāngnán xuéyuàn xuébào [Journal of Xiāngnán University]
Xīběi tōngxùn [Northwest Bulletin]
Xīn jiànshè [New Construction]
Xīnmín wǎnbào [New People's Evening News]
Yúnnán wénwù jiǎnbào [Yúnnán Cultural Relics Bulletin]
Yúzhòu qíguān [Wonders of the Universe]
Zhéxué yánjiū [Philosophical Researches]
Zhīshí jiùshì lìliàng [Knowledge Is Power]
Zhīshí yǔ qùwèi [Science and Taste]
Zhōngguó guójiā dìlǐ zázhì [National Geographic of China]
Zhōngguó gǔshēngwù zhì (see *Gǔjǐzhuī dòngwù yǔ gǔrénlèi*)
Zhōngguó kējì bào [China Science and Technology News]
Zhōngguó kēxué bào [China Science News]
Zhōngguó qīngnián bào [China Youth News]
Zìrán bówùguǎn qíngkuàng jiāoliú [Exhange of Ideas on the Condition of Natural History Museums]
Zìrán biànzhèng fǎ yánjiū [Research on Natural Dialectics]
Zìrán biànzhèng fǎ zázhì [Journal of Natural Dialectics]
Zìrán yǔ rén [Nature and Man, formerly *Bówù*]

ENGLISH-LANGUAGE PERIODICALS

American Anthropologist
Archaeology
Asian Theatre Journal
China Reconstructs (English-language magazine published in mainland China)
City Weekend (English-language magazine published in mainland China)
Cold Spring Harbor Symposia on Quantitative Biology
Cryptozoology
Current Anthropology
Evolution
Human Evolution
Journal of Human Evolution
Journal of the Royal Anthropological Institute of Great Britain and Ireland

Leader
Natural History
Nature
New York Times
Palaeontologica Sinica
Reporter
Science
Scientific Monthly
South China Morning Post
Women of China (English-language magazine published in mainland China)

TEXTBOOKS

Chén Dēngyuán. *Shìjiè zhōngxué jiàoběn gāozhōng běnguó shǐ* [World Secondary
    School Textbooks: Chinese History]. Vol. 1. Shànghǎi: Shìjiè shūjú, 1933.
Dàowěi wénhuà chūbǎnshè, ed. *Zhōngguó lìshǐ* [Chinese History]. Shànghǎi:
    Dàowěi wénhuà chūbǎnshè, 1949.
Fù Wěipíng. *Chūjí zhōngxué yòng fùxīng jiàokēshū běnguó shǐ* [Revived Text-
    book on Chinese History for Lower Secondary School]. Vol. 1. Shànghǎi:
    Shāngwù yìnshūguǎn, 1938.
Fù Yùnsēn, ed. *Xīn xué zhì lìshǐ jiàokēshū* [New Curriculum History Textbook].
    Shànghǎi: Shāngwù yìnshūguǎn, 1931.
Gāojí xiǎoxué kèběn: lìshǐ [Textbook for Upper Elementary School: History].
    Běijīng: Rénmín jiàoyù chūbǎnshè, 1962 (1961).
Guólì biānyì guǎn. *Chūjí zhōngxué lìshǐ* [Lower Secondary School History]. Vol. 1.
    Shànghǎi: Shāngwù yìnshūguǎn, 1948.
Jiǎng Wéiqiáo, *Jiǎnmíng Zhōngguó lìshǐ jiàokēshū* [Concise Textbook on the History
    of China]. Shànghǎi: Shāngwù yìnshū guǎn, 1927.
Jiàoyù bù biānshěn wěiyuánhuì, ed. *Guó dìng jiàokēshū chūzhōng běnguó shǐ*
    [Nationally Established Textbook for Lower Secondary School on Chinese
    History]. Shànghǎi: Huázhōng yìnshū jú, 1943.
Lǐ Shūfàn. *Gāojí xiǎoxué lìshǐ kèběn* [History Textbook for Upper Primary School].
    Běijīng: Rénmín jiàoyù chūbǎnshè, 1954 (1952).
Lǐ Zhí, ed. *Xīn Zhōnghuá lìshǐ kèběn* [New Chinese History Textbook]. Shànghǎi:
    Zhōnghuá shūjú, 1932 (1927).
Lǐ Zǐyào et al., eds. *Chūzhōng Zhōngguó lìshǐ jiǎnghuà* [Lower Secondary School
    Lectures on Chinese History]. Hángzhōu: Zhèjiāng rénmín chūbǎnshè,
    1956.
———. *Gāozhōng Zhōngguó gǔdài shǐ* [Upper Secondary School Chinese Ancient
    History]. Hángzhōu: Zhèjiāng rénmín chūbǎnshè, 1958.
Lǚ Ēnmiǎn, ed. *Gāojí zhōngxué yòng fùxīng jiàokēshū běnguó shǐ* [Upper Secondary
    School Revived Textbook on Chinese History]. Vol. 1. Shànghǎi: Shāngwù
    yìnshūguǎn, 1948 (1934).
Qián Huáběi rénmín zhèngfǔ jiàoyù bù, ed. *Gāojí xiǎoxué lìshǐ kèběn* [Upper

Elementary School History Textbook]. Vol. 1. Běijīng: Rénmín jiàoyù chūbǎnshè, 1951 (1950).

Rénmín jiàoyù chūbǎnshè, ed. *Shèhuì fāzhǎn shǐ (shàng běn)* [History of Social Development, Vol. 1]. Běijīng: Rénmín jiàoyù chūbǎnshè, 1989 (1988).

————, ed. *Chūjí zhōngxué kèběn (shìyòng běn): Shèhuì fāzhǎn shǐ (shàng běn)* [Lower Secondary School Textbook (Trial Edition): History of Social Development (Vol. 1)]. Běijīng: Rénmín jiàoyù chūbǎnshè, 1990.

Shí'èr nián xuéxiào gāojí xiǎoxué kèběn: lìshǐ (shì jiào běn) [Textbook for Upper Primary School in the Twelve-Year Curriculum: History (Trial Edition)]. Vol. 1. Běijīng: Rénmín jiàoyù chūbǎnshè, 1962.

Wéixīn zhèngfǔ jiàoyùbù, ed. *Gāojí xiǎoxué jiàokēshū* [Upper Elementary School History Textbook]. Vol. 1. N.p.: Wéixīn zhèngfǔ jiàoyù bù, 1938.

Xú Yìngyòng, ed. *Fùxīng jiàokēshū lìshǐ* [Revived Textbook in History]. Vol. 1. Shànghǎi: Shāngwù yìnshūguǎn, 1933.

Yáo Shàohuá, ed. *Xīn Zhōnghuá yǔtǐ běnguó shǐ xiángjiě* [New Chinese-Language Detailed History of China]. Vol. 1. Shànghǎi: Zhōnghuá shūjú, 1932.

————. *Xiǎoxué lìshǐ kèběn* [Elementary School History Textbook]. Shànghǎi: Zhōnghuá shūjú, 1936 (1933).

————. *Chūzhōng běnguó lìshǐ* [Lower Secondary School Chinese History]. Vol. 1. Shànghǎi: Zhōnghuá shūjú, 1941 (1933).

Zhōngguó lìshǐ yánjiū huì. *Gāojí zhōngxué Zhōngguó lìshǐ* [Upper Secondary School Chinese History]. Vol. 1. Běijīng: Xīnhuá chūbǎnshè, 1950.

Zhū ?xīn, ed. *Xīn zhǔyì gāojí xiǎoxué lìshǐ kèběn* [New Doctrine Upper Elementary School History Textbook]. Shànghǎi: Shìjiè shūjú, 1928. [First character in given name is obscured in original.]

MUSEUM PAMPHLETS AND GUEST BOOK

Dìzhì diàochásuǒ. Museum pamphlet (title and publishing information unavailable). Number Two Archives in Nánjīng, file 375:483.

The First Temporary Exhibition of the National Central Museum. Chungking: National Herald Press, 1943.

Lín Huìxiáng. *Xiàmén dàxué rénlèi bówùguǎn chénlièpǐn shuōmíng shū* [Guidebook to the Exhibited Materials at the Xiàmén University Anthropology Museum]. Xiàmén: Xiàmén dàxué rénlèi bówùguǎn, 1958.

Shànghǎi zìrán bówùguǎn. *"Cóng yuán dào rén" zhǎnlǎn jièshào* [Introduction to the "From Ape to Human" Exhibit]. N.p. [undoubtedly published between 1971 and 1978]. [Copy in author's possession.]

Tiānjīn zìrán bówùguǎn, ed. *Rénlèi de qǐyuán zhǎnlǎn jiǎnjiè* [Introduction to the Exhibit on Human Origins]. N.p. [undoubtedly published between 1971 and 1978]. [Copy in author's possession.]

Zhōngguó kēxuéyuàn gǔjǐzhuī dòngwù yánjiūshì. *Zhōngguó yuánrén zhī jiā* [Peking Man's Home]. Běijīng: Rénmín měishù chūbǎnshè, 1956.

Zhōngguó kēxuéyuàn gǔjǐzhuī dòngwù yǔ gǔrénlèi yánjiūsuǒ. *Běijīng yuánrén yízhǐ jiǎnjiè* [A Brief Introduction to the Peking Man Site]. Běijīng: Zhōngguó kēxuéyuàn yìnshūchǎng, 1972.

Zhōngguó lìshǐ bówùguǎn. *Zhōngguó lìshǐ bówùguǎn yù zhǎn shuōmíng* [China History Museum Exhibit Guide]. Běijīng: Wénwù chūbǎnshè, 1959.

———. *Zhōngguó lìshǐ bówùguǎn tōngshǐ chénliè shuōmíng* [Guide to the Comprehensive History Exhibit of the China History Museum]. Běijīng: Wénwù chūbǎnshè, 1964.

Zhōngyǎng zìrán bówùguǎn. *Gǔshēngwù, dòngwù, zhíwù chénliè jiǎnjiè* [Brief Introduction to the Paleontology, Zoology, and Botany Exhibits]. Běijīng: Zhōngyǎng zìrán bówùguǎn, 1961.

Zhōukǒudiàn Peking Man Site Exhibition Hall Guest Book, 1956.

SOVIET MATERIALS (IN TRANSLATION)

Gremiatskii, M. A. [Gélièmǐyàcíjī]. *Rénlèi shì zěnyàng qǐyuán de* [How Humans Originated]. Translated by Liú Quán and Wú Xīnzhì. Běijīng: Kēxué chūbǎnshè, 1964.

Gurev, [G. A.] [Gǔlièfú]. *Rénlèi shì zěnyàng zhǎngchéng de* [How Humans Developed]. Translated by Chén Yìngxīn. Shànghǎi: Kāimíng shūdiàn, 1950 (1946).

———. *Rénlèi shì zěnyàng zhǎngchéng de* [How Humans Developed]. Translated by Chén Yìngxīn. Běijīng: Zhōngguó qīngnián chūbǎnshè, 1953.

———. *Rénlèi shì zěnyàng qǐyuán de* [How Humans Originated]. Shànghǎi: Shànghǎi rénmín chūbǎnshè, 1957.

Ilin, M., and E. Segal. *How Man Became a Giant.* Translated by Beatrice Kinkead. Philadelphia: J. B. Lippincott Company, 1942.

Ilin, M. [Yīlín], and E. Segal [Xièjiāěr]. *Cóng yuán dào rén: Rén zěnyàng biànchéng jùrén* [From Ape to Human: How Humans Became Giants]. Translated by Shén Zhī. Shànghǎi: Dúshū chūbǎnshè, 1947.

———. *Cóng yuán dào rén: Rén zěnyàng biànchéng jùrén* [From Ape to Human: How Humans Became Giants]. Translated by Shén Zhī. Kāifēng: Zhōngyuán xīnhuá shūdiàn, 1948.

———. *Rén zěnyàng biànchéng jùrén* [How Humans Became Giants]. Běijīng: Kāimíng shūdiàn, 1951.

Nesturkh, M. F. [Nisētúěrhè]. *Rén de zǔxiān* [Human Ancestors]. Translated by Wú Xíngjiàn. Shànghǎi: Kēxué jìshù chūbǎnshè, 1956.

Nikol'skii, V. K. [Níkēěrsījī]. *Yuánshǐ shèhuì shǐ* [The History of Primitive Society]. Translated by Páng Lóng. Shànghǎi: Zuòjiā shūjú, 1953.

Plisetskii, M. S. [Pǔlièxuécíjī]. *Rénlèi qǐyuán de kēxué jiěshì yǔ zōngjiào chuánshuō* [Scientific Explanation and Religious Myths of Human Origins]. Translated by Huáng Dèngzhōng. Shànghǎi: Liányíng shūjú, 1950.

——— [Pǔlíxīcíjī]. *Rénlèi shì zěnyàng shēngchǎn hé fāzhǎn de* [How Humans Originated and Developed]. Translated by Bì Lí. Shànghǎi: Zhōnghuá shūjú, 1951.

——— [Pǔlǐxiècíjī]. *Rénlèi qǐyuán de kēxué shì zěnyàng fāzhǎn qǐlái* [How the

Science of Human Origins Developed]. Translated by Yáng Zhànlín and Sòng Jīndān. Shěnyáng: Liáoníng rénmín chūbǎnshè, 1957.

OTHER PRIMARY SOURCES IN ASIAN LANGUAGES

Ài Sīqí. *Lìshǐ wéiwù lùn: Shèhuì fāzhǎn shǐ* [Historical Materialism: The History of Social Development]. Běijīng: Shēnghuó, dúshū, xīnzhī, 1950.

Āndōng shì kēxué jìshù xiéhuì. *Kēxué jīchǔ zhīshí* [Basic Scientific Knowledge]. Shěnyáng: Liáoníng rénmín chūbǎnshè, 1959.

Běijīng kēxué jiàoyù diànyǐng zhìpiànchǎng. *Máohái* [Hairy Child]. 1978. Film.

Běijīng shūdiàn, ed., *Rén cóng nǎlǐ lái* [Where Humans Come From]. Běijīng: Běijīng shūdiàn, 1951.

Cáo Yú. *Běijīngrén* [Peking Man]. Shànghǎi: Wénhuà shēnghuó chūbǎnshè, 1950 (1941).

———. *Cáo Yú xuǎnjí* [Selected Works of *Cáo Yú*]. Běijīng: Kāimíng shūdiàn, 1951.

———. *Běijīngrén* [Peking Man]. Chéngdū: Sìchuān rénmín chūbǎnshè, 1984.

Chén Ānrén. *Rénlèi jìnhuà dàguān* [Vista on Human Evolution]. N.p., 1918.

Chén Jiānshàn. *Shǐqián rénlèi* [Prehistoric Humans]. N.p.: Zhōnghuá shūjú, 1936.

Chén Mèngléi, ed. *Gǔjīn túshū jíchéng, qínchóng diǎn* [Synthesis of Books and Illustrations Past and Present, Volume on Animals]. Vol. 51. Táiběi: Dǐngwén shūjú, 1971 (1726–1728).

Chén Yǐnghuáng. *Rénlèixué* [Anthropology]. Shànghǎi: Shāngwù yìnshūguǎn, 1923 (1918).

———. *Rénlèixué* [Anthropology]. Shànghǎi: Shāngwù yìnshūguǎn, 1934 (1918).

Chéng Míngshì and Fāng Shīmíng. *Cóng yuán dào rén tōngsú huà shǐ* [A Simple Pictorial History from Ape to Human]. Shànghǎi: Rén shì jiān chūbǎnshè, 1951.

Chéng Wànfú and Qín Juéshí, eds. *Zhōngguó yuánrén: Wǒmén wǔshíwàn nián de zǔxiān* [Peking Man: Our Ancestors 500,000 Years Ago]. Nánjīng: Mínfēng yìnshūguǎn, 1953.

Dèng Xiǎopíng. *Dèng Xiǎopíng wénxuǎn, 1975–1982* [Selected Works of Dèng Xiǎopíng, 1975–1982]. Hong Kong: Rénmín chūbǎn shè, 1983.

———. *Dèng Xiǎopíng wénxuǎn, dìsānjuàn* [Selected Works of Dèng Xiǎopíng, Vol. 3]. Běijīng: Rénmín chūbǎn shè, 1993.

Dīng Yī and Wáng Bīn. *Bái máo nǚ* [The White-Haired Girl]. Hong Kong: Hǎi-yáng shūwū, 1948.

Dǒng Shuǎngqiū. *Rén shì zěnyàng lái de* [How Humans Came to Be]. Chángshā: Húnán rénmín chūbǎnshè, 1957.

Dù Yǒnglín. *Yěrén: Lái zì Shénnóngjià de bàogào* [*Yěrén*: Report from Shénnóngjià]. Běijīng: Sānxiá chūbǎnshè, 1995.

Engels, Frederick [Ēngésī]. *Mǎkèsīzhǔyì de rénzhǒng yóulái shuō* [The Marxist Theory of Human Origins]. Translated by Lù Yīyuán. Shànghǎi: Chūncháo shūjú, 1928.

———. *Cóng yuán dào rén* [From Ape to Human]. Translated by Cáo Bǎohuá and Yú Guāngyuǎn. Shíjiāzhuāng: Jiěfàng shè, 1950 (1948).

Fāng Qiě. *Cóng yuán dào rén tòushì: Láodòng zěnyàng chuàngzào le rénlèi běnshēn hé shìjiè* [A Penetrating Look at from Ape to Human: How Labor Created Humanity Itself and the World]. Shànghǎi: Shànghǎi biānyì shè, 1950.

———. *Zhōnghuá mínzú de érnǚ* [Sons and Daughters of the Chinese Nationality]. Shànghǎi: Shànghǎi biānyì chūbǎnshè, 1951.

Fāng Shàoqīng [pseud. for Fāng Zōngxī]. *Gǔ yuán zěnyàng biànchéng rén* [How Ancient Apes Became Human]. Běijīng: Zhōngguó qīngnián chūbǎnshè, 1958.

———. *Gǔ yuán zěnyàng biànchéng rén* [How Ancient Apes Became Human]. Běijīng: Zhōngguó qīngnián chūbǎnshè, 1965.

Fāng Zōngxī. *Gǔyuán zěnyàng biànchéng rén* [How Ancient Apes Became Human]. Běijīng: Zhōngguó qīngnián chūbǎnshè, 1990.

Gān Bǎo. *Sōu shén jì quán yì* [Records of the Search for Spirits, Complete and Interpreted]. Edited by Huáng Dímíng. Guìyáng: Guìzhōu rénmín chūbǎnshè, 1991 (ca. 350).

Guō Mòruò. *Gémìng jīngshén rénlèi jīqiǎo zìrán* [With Revolutionary Spirit, Humanity Constructs Nature]. Shànghǎi: Kāimíng shūdiàn, 1928.

Guō Yīshí. *Rénlèi shì cóng nǎlǐ lái de* [Where Humans Came From]. Běijīng: Tōngsú dúwù chūbǎnshè, 1955.

Guójiā wénwù shìyè guǎnlǐ jú, ed. *Xīn Zhōngguó wénwù fǎguī xuǎnbiān* [Selected Laws on Cultural Property in New China]. Běijīng: Wénwù chūbǎnshè, 1987.

Hú Qiáomù zhuàn biānxiězǔ, ed. *Dèng Xiǎopíng de èrshísì cì tánhuà* [Twenty-Four Speeches of Dèng Xiǎopíng]. Běijīng: Rénmín chūbǎnshè, 2004.

Huá Wéidān. *Zhèndàn rén yǔ Zhōukǒudiàn wénhuà yīcè* [Peking Man and Zhōu-kǒudiàn Culture]. Shànghǎi: Shāngwù yìnshūguǎn, 1937 (1936).

Huáng Wànbō, Shěn Wénlóng, and Hú Huìqīng. *Wǒmén de zǔxiān* [Our Ancestors]. Běijīng: Běijīng kēxué pǔjí chūbǎnshè, 1958.

Huáng Wéiróng. *Zhōngguó yuánrén* [Peking Man]. Shànghǎi: Shàonián értóng chūbǎnshè, 1954.

Huáng Wèiwén. *Láodòng chuàngzào le rén: Wénmíng shìjiè de "Běijīngrén"* [Labor Created Humanity: The World-Famous "Peking Man"]. Běijīng: Shūmù wénxiàn chūbǎnshè, 1985.

Húnán rénmín chūbǎnshè, ed. *Shèhuì fāzhǎn jiǎn shǐ* [A Short History of Social Development]. [Chángshā?]: Húnán rénmín chūbǎnshè, 1974.

Jiǎ Lánpō. *Zhōngguó yuánrén* [Peking Man]. Shànghǎi: Lóngmén liánhé shūjú, 1950.

———. *Huàshí de fājué hé xiūlǐ* [Fossil Excavation and Preparation]. Shànghǎi: Shāngwù yìnshūguǎn, 1957 (1952).

———. *Běijīngrén de gùjù* [The Former Residence of "Peking Man"]. Běijīng: Běijīng chūbǎnshè, 1958.

———. *Zhōngguó yuánrén* [Peking Man]. Běijīng: Zhōnghuá shūjú, 1962.

———. *Zhōukǒudiàn: "Běijīngrén" zhī jiā* [Zhōukǒudiàn: The Home of "Peking Man"]. Běijīng: Běijīng rénmín chūbǎnshè, 1975.

———. *Zhōukǒudiàn jìshì* [Chronicle of Zhōukǒudiàn]. Shànghǎi: Shànghǎi kēxué jìshù chūbǎnshè, 1999.

———. *Rénlèi qǐyuán zhī wǒ jiàn* [My View on Human Origins]. Shànghǎi: Shànghǎi kējì jiàoyù chūbǎnshè, 2000.

Jiǎ Lánpō and Huáng Wèiwén. *Zhōukǒudiàn fājué jì* [The Excavation of Zhōukǒudiàn]. Tiānjīn: Tiānjīn kēxué jìshù chūbǎnshè, 1984.

———. *Fāxiàn Běijīngrén* [The Discovery of Peking Man]. Táiběi: Yòu shī, 1996.

Jiǎ Lánpō and Liú Xiàntíng. *Cóng yú dào rén* [From Fish to Human]. Tiānjīn: Zhīshí shūdiàn, 1951.

Jiǎ Lánpō and Zhēn Shuònán. *Qiān lǐ zhuīzōng liè huàshí* [A Thousand-Mile Trek in Search of Fossils]. Tiānjīn: Tiānjīn kēxué jìshù chūbǎnshè, 1981.

Jiǎ Zǔzhāng. *Cóng yuán dào rén* [From Ape to Human]. Shànghǎi?: Kāimíng shūdiàn, 1950.

Jiāng Tíngān and Yún Zhōnglóng. *"Yěrén" xún zōng jì* [Tracking the "Yěrén"]. Xī'ān: Shǎnxīshěng rénmín chūbǎnshè, 1983.

Jílínshěng kēxué jìshù pǔjí xiéhuì. *Zhēn yǒu shénguǐ ma?* [Do Spirits and Ghosts Really Exist?]. Chángchūn: Jílín rénmín chūbǎnshè, 1956.

Kēxué pǔjí chūbǎnshè, ed. *Kēxué pǔjí hé yánjiū gōngzuò wèi shēngchǎn dàyuèjìn fúwù* [Science Dissemination and Research Serves the Great Leap Forward in Production]. Běijīng: Kēxué pǔjí chūbǎnshè, 1958.

———. *Qúnzhòng kēxué yánjiū wénjí* [Collection of Mass Science Research]. Běijīng: Kēxué pǔjí chūbǎnshè, 1958.

Kokuritsu kagaku hakubutsukan. *Nihonjin no rūtsu o saguru: Pekin genjin ten* [Searching for Japanese People's Roots: Peking Man Exhibit]. Tokyo: Yomiuri shinbun, 1980.

———. *Nihonjin no kigen ten: Nihonjin wa doko kara kimashita ka* [Exhibit on Japanese Origins: Where Did Japanese People Come From?]. Tokyo: Yomiuri shinbun, 1988.

Lǐ Èróng, ed. *Yáng Zhōngjiàn huíyì lù* [The Memoirs of Yáng Zhōngjiàn]. Běijīng: Dìzhì chūbǎnshè, 1983.

Lǐ Hóngzhì. *Zhuàn fǎlún* [Spinning the Law Wheel]. Hong Kong: Fǎlún fó fǎ chūbǎnshè, 1997.

Lǐ Jiàn. *Yěrén zhī mí* [The Yěrén Mystery]. Wǔhàn: Zhōngguó dìzhì dàxué chūbǎnshè, 1990.

Lǐ Míngshēng and Yuè Nán. *Xúnzhǎo "Běijīngrén"* [The Search for "Peking Man"]. Běijīng: Huáxià chūbǎnshè, 2000.

Lǐ Qīngbō. *Rénlèi qǐyuán wèntí* [The Question of Human Origins]. Shànghǎi: Xīnshēng chūbǎnshè, 1951.

Lǐ Shízhēn. *Běncǎo gāngmù* [Classified Materia Medica]. Shànghǎi: Shànghǎi gǔjí chūbǎnshè, 1991 (1578).

Lín Yán. *Zhōngguó mínzú de yóulái* [The Origins of the Chinese People]. Shànghǎi: Yǒngxiáng yìnshūguǎn, 1947.

Lín Yàohuá. *Cóng yuán dào rén de yánjiū* [Research on from Ape to Human]. Běijīng: Gēngyún chūbǎnshè, 1951.

Liú Hòuyī and Liú Qiūshēng. *Fāxiàn Zhōngguó yuánrén de rén, Péi Wénzhōng* [Péi Wénzhōng, the Person Who Discovered Peking Man]. Kūnmíng: Yúnnán rénmín chūbǎnshè, 1980.

Liú Míngyù et al. *Zhōngguó máorén* [The Chinese Hairy Person]. Liáoníng kēxué jìshù chūbǎnshè, 1982.

Líu Mínzhuàng. *Jiēkāi 'yěrén' zhī mí* [Solving the "Yěrén" Mystery]. Nánchāng: Jiāngxī rénmín chūbǎnshè, 1988.

———. *Zhōngguó Shénnóngjià* [China's Shénnóngjià]. Shànghǎi: Wénhuì chūbǎnshè, 1993.

Liú Mínzhuàng, Chén Nǎiwén, Zhāng Guóyīng, et al. *Yěrén, xuěrén, húguài* [*Yěrén*, Yeti, and Lake Monsters]. Běijīng: Zhōngguó jiànshè chūbǎnshè, 1988.

Liú Xián. *Cóng yuán dào rén fāzhǎn shǐ* [A History of the Development from Ape to Human]. Shànghǎi: Zhōngguó kēxué túshū yíqì gōngsī, 1950.

"Mǎkèsīzhǔyì wényì lǐlùn yánjiū" biānjí bù, ed. *Lùn rénxìng hé réndàozhǔyì* [On Human Nature and Humanism]. Běijīng: Guāngmíng rìbǎo chūbǎnshè, 1982.

Máo Zédōng. *Jiànguó yǐlái Máo Zédōng wéngǎo* [Máo Zédōng's Manuscripts since the Founding of the PRC]. Běijīng: Zhōngyǎng wénxiàn chūbǎnshè, 1998.

Máo Zhènhuá, Zhōu Hóngyóu, Fù Wànměi, ed. *Shénnóngjià tàn qí* [Looking for Marvels in Shénnóngjià]. Běijīng: Gōngrén chūbǎnshè, 1986.

Nánjīng Zhōngyì xuéyuàn and Jiāngsū shěng Zhōngyì yánjiūsuǒ. *Zhōngyàoxué* [Chinese Pharmacology]. Běijīng: Rénmín wèishēng chūbǎnshè, 1959.

Péi Wénzhōng. *Zhōukǒudiàn dòngxué céng cǎijué jì* [Record of Excavations of the Zhōukǒudiàn Cave Strata]. Běijīng: Dìzhàn chūbǎnshè, 2001 (1934).

———. *Zìrán fāzhǎn jiǎnshǐ* [A Short History of Natural Development]. Běijīng: Liányíng shūdiǎn, 1950.

———. *Rénlèi de qǐyuán hé fāzhǎn* [Human Origins and Development]. Běijīng: Zhōngguó qīngnián chūbǎnshè, 1956.

Péi Wénzhōng and Jiǎ Lánpō. *Láodòng chuàngzào le rén* [Labor Created Humanity]. Běijīng: Zhōnghuá quánguó kēxué jìshù pǔjí xiéhuì, 1954.

Rèn Bǎoyì, ed. *Zhōngguó kēxuéyuàn gǔjǐzhuī dòngwù yǔ gǔrénlèi yánjiūsuǒ*. Běijīng: Zhōngguó kēxuéyuàn gǔjǐzhuī dòngwù yǔ gǔrénlèi yánjiūsuǒ, 1994.

Shànghǎi kēxué jiàoyù diànyǐng zhìpiànchǎng. *Zhōngguó yuánrén* [Peking Man]. 1959. Film.

———. *Zhōngguó gǔdài rénlèi* [China's Ancient Humans]. 1976. Film.

Shànghǎi kēxué pǔjí chūbǎnshè, ed. *Kēpǔ xuānchuán shǒucè* [Science Dissemination Handbook]. Vol. 1. Shànghǎi: Shànghǎi kēxué pǔjí chūbǎnshè, 1959 (1957).

Shànghǎi rénmín chūbǎnshè, ed. *Shíwàn ge wèishénme (19): Rénlèi shǐ* [100,000 Whys, Vol. 19: Human History]. Shànghǎi: Shànghǎi rénmín chūbǎnshè, 1976.

Shànghǎi zìrán bówùguǎn. *Cóng yuán dào rén* [From Ape to Human]. Shànghǎi: Shànghǎi rénmín chūbǎnshè, 1973.

———. *Láodòng chuàngzào rén* [Labor Created Humanity]. Shànghǎi: Shànghǎi rénmín chūbǎnshè, 1977.

———. *Rénlèi de qǐyuán* [Human Origins]. Shànghǎi: Shànghǎi kēxué jìshù chūbǎnshè, 1980.

Shèng Lì, Zhāng Lóng, and Shèng Lín. *Mànhuà cóng yuán dào rén* [Conversations on from Ape to Human]. Níngxià: Níngxià rénmín chūbǎnshè, 1983.

Shí Mòzhuāng. *Rénlèi de láilì* [Human Origins]. Běijīng: Běijīng rénmín chūbǎnshè, 1976.

Sòng Yōuxīng. *Yěrén de chuánshuō* [*Yěrén* Legends]. Hong Kong: Xiānggǎng hǎiwān chūbǎnshè, 1986.

Sūn Wǔ. *Tán tiān shuō dì, pò míxìn* [Talking about Heaven and Earth, Squashing Superstition]. Tiānjīn: Tiānjīn rénmín chūbǎnshè, 1964.

Tán Tán. *Yěrén*. Běijīng: Zhōngguó wénlián chūbǎnshè, 1993.

Tú Jǐngzōng and Chén Guāngyì. *Láodòng chuàngzào rén* [Labor Created Humanity]. Shànghǎi: Shànghǎi wénhuà shūdiàn, [1951?]. [Discussed in a critical review in *Wénhuì bào fùkān* (19 June 1951), but recent efforts to find this work have been unsuccessful.]

Wáng Bō. *Yěrén zhī mí xīn tàn* [New Investigations of the *Yěrén* Mystery]. Chóngqìng: Kēxué jìshù wénxiàn chūbǎnshè, 1989.

Wáng Ruòshuǐ. *Zhìhuì de tòngkǔ* [The Pain of Wisdom]. Hong Kong: Sānlián shūdiàn, 1989.

Wáng Shān. *Láodòng chuàngzào rénlèi* [Labor Created Humanity]. Shànghǎi: Shànghǎi qúnzhòng chūbǎnshè, 1951.

Wáng Xiǎoshí. *Cóng yuán dào rén: Tōngsú jiǎnghuà* [From Ape to Human: A Simple Account]. Shànghǎi: Xīnyà shūdiàn, 1950.

Wèi Jùxián. *Běijīngrén de xiàluò* [The Whereabouts of Peking Man]. Hong Kong: Shuōwénshè, 1952.

Wèi Qí [魏奇, not to be confused with 衛奇] and Wèi Shìfēng. *Zǒuchū Yīdiàn yuán: rénlèi de qǐyuán hé yǎnhuà* [Out of Eden: Human Origins and Evolution]. Xī'ān: Shǎnxī rénmín chūbǎnshè, 1994.

Wú Rǔkāng. *Rénlèi de qǐyuán hé fāzhǎn* [Human Origins and Development]. Běijīng: Kēxué pǔjí chūbǎnshè, 1965.

———. *Rénlèi de qǐyuán hé fāzhǎn* [Human Origins and Development]. Běijīng: Kēxué chūbǎnshè, 1976.

———. *Rénlèi de qǐyuán hé fāzhǎn* [Human Origins and Development]. Běijīng: Kēxué chūbǎnshè, 1980.

———. *Rénlèi de yóulái* [Human Origins]. Běijīng: Kēxué jìshù wénxiàn chūbǎnshè, 1992.

———. *Rénlèi de guòqù, xiàndài hé wèilái* [The Past, Present, and Future of Humanity]. Shànghǎi: Shànghǎi kēxué jiàoyù chūbǎnshè, 2000.

Wú Rǔkāng, Wú Xīnzhì, Qiū Zhōngláng, and Lín Shènglóng. *Rénlèi fāzhǎn shǐ* [The History of Human Development]. Běijīng: Kēxué chūbǎnshè, 1978.

Wú Xīnzhì. *Rénlèi jìnhuà zújī* [The Tracks of Human Evolution]. Běijīng: Běijīng shàonián értóng chūbǎnshè, 2002.

Wú Yùzhāng and Xǔ Lìqún. *Zhōngguó shǐ huà* [Notes on Chinese History]. Shànghǎi: Yěcǎo chūbǎnshè, 1946.

Xià Shùfāng. *Huàshí màntán* [Conversations about Fossils]. Shànghǎi: Shànghǎi kēxué jìshù chūbǎnshè, 1978.

Xī'nán jūnzhèng wěiyuánhuì wénjiào bù, ed. *Kēxué pǔjí gōngzuò shòucè* [Handbook for Science Dissemination Work]. Vol. 1. N.p.: Xī'nán jūnzhèng wěiyuánhuì wénjiào bù, 1951.

Xíng Wànlǐ, ed. *Rénlèi shénmì xiànxiàng quán jìlù* [Complete Records of Mysterious Human Phenomena]. Běijīng: Dàzhòng wényì chūbǎnshè, 1999.

Xīnhuá tōngxùn shè. *Běijīng yuánrén zhī jiā* [Peking Man's Home]. Běijīng: Běijīng rénmín chūbǎnshè, 1973.

Xióng Shílì. *Zhōngguó lìshǐ jiǎnghuà* [Talks on Chinese History]. Táiběi: Míngwén shūjú, 1984 (1939).

Xǔ Lìqún. *Zhōngguó shǐ huà* [Notes on Chinese History]. Běijīng: Xīnhuá shūdiàn, 1950 (1942).

Xú Xùshēng. *Zhōngguó gǔshǐ de chuánshuō shídài* [The Legendary Era of Chinese Ancient History]. Běijīng: Kēxué chūbǎnshè, 1960 (1943).

Xuē Hóngdá, ed. *Cóng yuán dào rén* [From Ape to Human]. Shànghǎi: Huádōng shūdiàn, 1950.

Yáng Háonìng (art) and Jiǎ Lánpō (text). *Wǒmén de zǔxiān 1:Wǒmén wǔshíwàn nián de zǔxiān* [Our Ancestors, vol. 1: Our Ancestors 500,000 Years Ago]. [Tiānjīn?]: Zhīshí shūdiàn, 1951.

———. *Wǒmén de zǔxiān 2: Wǒmén èrshíwàn nián de zǔxiān* [Our Ancestors, vol. 2: Our Ancestors 200,000 Years Ago]. [Tiānjīn?]: Zhīshí shūdiàn, 1951.

Yáng Yánliè, ed. *Fáng xiàn zhì* [Fáng County Gazetteer]. Táiběi: Chéngwén chūbǎnshè, 1976 (1865).

Yáng Yè. *Wǒmén de zǔxiān* [Our Ancestors]. Hànkǒu: Wǔhàn gōngrén chūbǎnshè, 1952.

Yè Wéidān. *Běijīngrén* [Peking Man]. Shànghǎi: Liángyǒu túshū yìnshuā gōngsī, 1933.

———. *Zhèndàn rén yǔ Zhōukǒudiàn wénhuà yīcè.* [Peking Man and Zhōukǒudiàn Culture]. Shànghǎi: Shāngwù yìnshūguǎn, 1937 (1936).

Yè Yǒngliè, ed. *Zhōngguó kēxué xiǎopǐn xuǎn, 1934–1949* [Selected Chinese Short Writings on Science, 1934–1949]. Tiānjīn: Tiānjīn kēxué jìshù chūbǎnshè, 1984.

———. *Zhōngguó kēxué xiǎopǐn xuǎn, 1949–1976* [Selected Chinese Short Writings on Science, 1949–1976]. Tiānjīn: Tiānjīn kēxué jìshù chūbǎnshè, 1985.

Yóu Jiādé. *Rénlèi qǐyuán* [Human Origins]. Shànghǎi: Shìjiè shūjú, 1929.

Yuǎn Jìn. *Yuǎngǔ yǔ wèilái* [The Past and the Future]. Chéngdū: Sìchuān kēxué jìshù chūbǎnshè, 1998.

Yuán Méi, *Zhèng xù Zǐ bù yǔ* [What Confucius Did Not Discuss, Corrected and Continued]. Táiběi: Xīnxìng shūjú, 1978.

Yúnnán kēxué jìshù pǔjí xiéhuì. *Kēxué zhīshí huìbiān* [Compilation of Scientific Knowledge]. Vol. 1. Kūnmíng: Yúnnán rénmín chūbǎnshè, 1959.

Zāng Yǒngqīng. *Yěrén mí zōng* [Tracking the *Yěrén* Mystery]. Shěnyáng: Liáoníng chūbǎnshè, 1996.

Zhāng Jīnxīng. *Zhēng zuò Gǔdàoěr, yǒng tàn 'yěrén' mí* [Striving to Be like Goodall, Bravely Exploring the "Yěrén" Mystery]. Běijīng: Zhōngguó kēxué tànxiǎn xiéhuì, 2002.

Zhāng Mín, Shèng Liángxián, and Shěn Tiězhēng. *Láodòng chuàngzào le rén* [Labor Created Humanity]. N.p.: Huádōng rénmín chūbǎnshè, 1954 (1952).

Zhāng Yìzhěn, Hán Fǔ, et al. *Jiēkāi kēxué zhī mí* [Cracking the Mysteries of Science]. Shànghǎi: Shàonián értóng chūbǎnshè, 1962.

Zhāng Zuòrén. *Rénlèi tiānyǎn shǐ* [The History of Human Evolution]. Shànghǎi: Shāngwù chūbǎnshè, 1930.

Zhào Yíng. *Zhōngguó yuánrén (dì èr cì gǎo)* [Peking Man, Second Draft]. Shànghǎi: Shànghǎi kēxué jiàoyù diànyǐng zhìpiànchǎng, 1959. Film script.

Zhào Zhènhuá. *Zhōngguó yuánrén* [Peking Man]. Hong Kong: Xiānggǎng Zhōnghuá shūjú, 1955.

Zhōng gòng Nánchāng shì xuānchuán bù. *Láodòng chuàngzào kēxué* [Labor Created Science]. Nánchāng: Jiāngxīrénmín chūbǎnshè, 1959 (1958).

Zhōngguó kēxuéyuàn gǔjǐzhuī dòngwù yánjiūsuǒ. *Zhōngguó yuánrén* [Peking Man]. Běijīng: Kēxué chūbǎnshè, 1972.

Zhōngguó rénmín dàxué. *"Běijīngrén" yízhǐ de fāxiàn yǔ yánjiū* [The Peking Man Site: Discovery and Research]. 1986. Film.

Zhōngguó shèhuì kēxuéyuàn zhéxué yánjiūsuǒ. *Rénxìng, réndàozhǔyì wèntí tǎolùn jí* [Collected Discussions on Questions of Human Nature and Humanism]. Běijīng: Rénmín chūbǎnshè, 1983.

Zhōnghuá quánguó kēxué jìshù pǔjí xiéhuì. *Zhōnghuá quánguó kēxué jìshù pǔjí xiéhuì fǎng Sū dàibiǎo tuán zīliào huìbiān* [Compilation of Materials from the All-China Scientific and Technological Knowledge Dissemination Association's Delegation to the Soviet Union]. Vol. 1. Běijīng, 1955.

Zhōnghuá rénmín gònghéguó wèishēng bù yàodiǎn wěiyuánhuì. *Zhōnghuá rénmín gònghéguó yàodiǎn 1963 nián bǎn yī bù* [Pharmacopoeia of the People's Republic of China, 1963 Edition]. Běijīng: Rénmín wèishēng chūbǎnshè, 1964.

Zhōnghuá rénmín jiěfàng jūn zǒng gāojí bùbīng xuéxiào zhèngzhì bù. *Shèhuì fāzhǎn shǐ huà jí* [An Illustrated History of Social Development]. Vol. 1. N.p.: Huádōng rénmínchūbǎnshè, 1953.

Zhōngyǎng rénmín guǎngbō diàntái kējì zǔ and Kēxué pǔjí chūbǎnshè biānjíbù, eds. *Kēxuéjiā tán kēxué* [Scientists Discuss Science]. Vol. 2. Běijīng: Kēxué pǔjí chūbǎnshè, 1982.

Zhōu Fāzēng and Gōng Qízhù. *Lìshǐ jiàoxué yǔ àiguózhǔyì jiàoyù* [History Pedagogy and Patriotic Education]. Jǐ'nán: Shāndōng jiàoyù chūbǎnshè, 1984.

Zhōu Guóxīng. *Rénlèi zěnyàng rènshí zìjǐ de qǐyuán: Rénlèi qǐyuán yánjiū shǐ huà* [How Humans Came to Know Our Origins: The History of Research on Human Origins]. Běijīng: Zhōngguó qīngnián chūbǎnshè, 1977.

———. *Lánghái, xuěrén, huó de huàshí* [Wolf Children, Yeti, and Living Fossils]. Tiānjīn: Tiānjīn rénmín chūbǎnshè, 1979.

———. *Rén zhī yóulái* [Human Origins]. Běijīng: Mínzú chūbǎnshè, 1986.

Zhōu Guóxīng and Liú Lìlì. *Rén zhī yóulái* [Human Origins]. Běijīng: Zhōngguó guójì guǎngbō chūbǎnshè, 1991.

Zhōu Jiànrén. *Kēxué zátán* [On Science]. Hángzhōu: Zhéjiāng rénmín chūbǎnshè, 1962.

Zhōu Liángpèi. *Yěrén jí* [*Yěrén* Collection]. Běijīng: Huáxià chūbǎnshè, 1992.

Zhōu Róngwéi. *Rén hé zìrán dòuzhēng* [Humans' Struggle with Nature]. Nánjīng: Mín fēng yìnshūguǎn, 1953.

Zhū Chángchāo. *Rénlèi zhī mí* [The Human Mystery]. Shànghǎi: Shànghǎi yuǎndōng chūbǎnshè, 1995.

Zhū Xǐ. *Wǒmén de zǔxiān* [Our Ancestors]. Shànghǎi: Wénhuà shēnghuó chūbǎnshè, 1950 (1940).

Zǐ Fēng, ed. *Yěrén qiú'ǒu jì* [A *Yěrén* Seeks a Mate]. Běijīng: Zhōngguó mínjiān wényì chūbǎnshè, 1988.

OTHER PRIMARY SOURCES IN ENGLISH

Akazawa, Takeru, Kenichi Aoki, and Tasuku Kimura, eds. *The Evolution and Dispersal of Modern Humans in Asia*. Tokyo: Hokusen-sha Publishing Company, 1992.

Andersson, Gunnar J. *Children of the Yellow Earth*. London: Kegan Paul, 1934.

Augusta, Josef, and Zdeněk Burian. *Prehistoric Man*. London: Paul Hamlyn, 1960.

Boaz, Noel T., and Russell L. Ciochon. *Dragon Bone Hill: An Ice-Age Saga of Homo Erectus*. New York: Oxford University Press, 2004.

Breuil, Henri. *Beyond the Bounds of History: Scenes from the Old Stone Age*. Translated by Mary E. Boyle. London: P. R. Gawthorn, 1949.

Carroll, Charles. *The Negro a Beast*. St. Louis: American Book and Bible House, 1900.

Chen, Yuan-tsung. *The Dragon's Village: An Autobiographical Novel of Revolutionary China*. New York: Penguin Books, 1980.

Clark, Geoffrey A., and Catherine M. Willermet, eds. *Conceptual Issues in Modern Human Origins Research*. New York: Andine de Gruyter, 1997.

Coon, Carleton S. *The Origin of Races*. New York: Alfred A. Knopf, 1962.

Corbey, Raymond, and Wil Roebroecks. *Studying Human Origins: Disciplinary History and Epistemology*. Amsterdam: Amsterdam University Press, 2001.

Cremo, Michael A., and Richard L. Thompson. *The Hidden History of the Human Race: Major Scientific Coverup Exposed*. Badger, Calif.: Govardhan Hill Publishing, 1994.

Darwin, Charles. *The Descent of Man, and Selection in Relation to Sex*. 2nd ed. New York: Clarke, Given, and Hooper, 1874 (1870–1871).

de Lacouperie, Terrien. *Western Origin of Early Chinese Civilization*. Osnabrück: Otto Zeller, 1966 (1894).

Engels, Frederick. *The Part Played by Labor in the Transition from Ape to Man.* New York: International Publishers, 1950.

———. *Ludwig Feuerbach and the End of Classical German Philosophy.* Beijing: Foreign Languages Press, 1976.

Fang, Lizhi. *Bringing Down the Great Wall: Writings on Science, Culture, and Democracy in China.* New York: Norton, 1990.

Feng, Jicai. *Voices from the Whirlwind: An Oral History of the Chinese Cultural Revolution.* New York: Pantheon Books, 1991.

Gao Xingjian. *Soul Mountain.* Translated by Mabel Lee. New York: Harper Collins, 2000 (1990).

Gould, Sidney, ed. *Sciences in Communist China.* Washington, D.C.: American Association for the Advancement of Science, 1961.

Haeckel, Ernst. *The History of Creation.* Translated by E. Ray Lankester. 5th ed. New York: D. Appleton Company, 1910 (1876).

Howells, W. W., and Patricia Jones Tsuchitani, eds. *Paleoanthropology in the People's Republic of China: A Trip Report of the American Paleoanthropology Delegation.* Washington, D.C.: National Academy of Sciences, 1977.

Huxley, Thomas H. *Man's Place in Nature.* New York: Random House, 2001 (1863).

Jia Lanpo and Huang Weiwen. *The Story of Peking Man: From Archaeology to Mystery.* Translated by Yin Zhiqi. Beijing: Foreign Languages Press, 1990.

Keith, Arthur. *Concerning Man's Origin.* New York: G. P. Putnam's Sons, 1928.

Li Chi. *Anyang.* Seattle: University of Washington Press, 1977.

Mao Tse-tung. *Selected Works of Mao Tse-tung.* Peking: Foreign Languages Press, 1961–1977.

———.*Quotations from Chairman Mao Tse-tung.* Peking: Foreign Languages Press, 1966.

Marx, Karl, and Frederick Engels. *Selected Correspondence.* Moscow: Progress Publishers, 1975.

———*Karl Marx, Frederick Engels: Collected Works.* Edited by E. J. Hobsbawm et al. Moscow: Progress Publishers, 1975–.

Montagu, M. F. Ashley. *Man's Most Dangerous Myth: The Fallacy of Race.* New York: Columbia University Press, 1942.

Nie Rongzhen. *Inside the Red Star: The Memoirs of Nie Rongzhen.* Translated by Zhong Renyi. Beijing: New World Press, 1988.

Oakley, Kenneth P. *Man the Toolmaker.* Chicago: University of Chicago Press, 1959 (1956).

O'Gara, Cuthbert M. *The Surrender to Secularism.* St. Louis: Cardinal Mindszenty Foundation, 1989 (1967).

Olsen, John W., and Wu Rukang, eds. *Paleoanthropology and Paleolithic Archaeology in the People's Republic of China.* Orlando: Academic Press, 1985.

Poirier, Gene, and Richard Greenwell. *The Wildman of China.* New York: Mystic Fire Video, 1990. Film.

Shackley, Myra L. *Wildmen: Yeti, Sasquatch and the Neanderthal Enigma.* London: Thames and Hudson, 1983.

Smith, Fred H., and Frank Spencer. *The Origins of Modern Humans*. New York: Alan R. Liss, 1984.

Stalin, Joseph. *Problems of Leninism*. Peking: Foreign Languages Press, 1976.

Tobias, Phillip V. et al., eds. *Humanity from African Naissance to Coming Millennia*. Florence: Firenze University Press, 2001.

Trudeau, Gary. *Doonesbury Dossier: The Reagan Years*. New York: Holt, Rinehart and Winston, 1984.

Turolla, Pino. *Beyond the Andes: My Search for the Origins of Pre-Inca Civilization*. New York: Harper and Row, 1980.

Weidenreich, Franz. *Apes, Giants, and Man*. Chicago: University of Chicago Press, 1946.

Wolpoff, Milford, and Rachel Caspari. *Race and Human Evolution*. New York: Simon and Schuster, 1997.

Wu, Xinzhi, and Frank E. Poirier. *Human Evolution in China: A Metric Description of the Fossils and a Review of the Sites*. New York: Oxford University Press, 1995.

SECONDARY SOURCES

Note: All paleoanthropology texts are included with primary sources.

Andrews, James T. *Science for the Masses: The Bolshevik State, Public Science, and the Popular Imagination in Soviet Russia, 1917–1934*. College Station: Texas A&M University Press, 2003.

Andrews, Julia. *Painters and Politics in the People's Republic of China, 1949–1979*. Berkeley: University of California Press, 1994.

Angle, Stephen. *Human Rights and Chinese Thought: A Cross-Cultural Inquiry*. Cambridge: Cambridge University Press, 2002.

Angle, Stephen, and Marina Svensson. *The Chinese Human Rights Reader: Documents and Commentary, 1900–2000*. Armonk, N.Y.: M. E. Sharpe, 2001.

Bailes, Kendall E. *Technology and Society under Lenin and Stalin: Origins of the Soviet Intellectual Intelligentsia, 1917–1941*. Princeton, N.J.: Princeton University Press, 1978.

Balibar, Etienne, and Immanuel Wallerstein. *Race, Nation, Class: Ambiguous Identities*. London: Verso, 1991 (1988).

Barmé, Geremie. *In the Red: On Contemporary Chinese Culture*. New York: Columbia University Press, 1999.

Barmé, Geremie, and John Minford, eds. *Seeds of Fire: Chinese Voices of Conscience*. New York: Hill and Wang, 1988.

Banister, Judith. *China's Changing Population*. Stanford, Calif.: Stanford University Press, 1987.

Barnett, A. Doak. *Communist China: The Early Years, 1949–55*. New York: Praeger, 1964.

Běijīng bówùguǎn xuéhuì. *Běijīng bówùguǎn niánjiàn (1912–1987)* [Yearbook of Běijīng Museums (1912–1987)]. Běijīng: Běijīng Yànshān chūbǎnshè, 1989.

Bennett, Gordon. "Mass Campaigns and Earthquakes: Hai-Ch'eng, 1975." *China Quarterly* 77 (March 1979): 94–112. Bloor, David. *Knowledge and Social Imagery.* London: Routledge and Kegan Paul, 1976.

Bodenhorn, Terry, ed. *Defining Modernity: Guomindang Rhetorics of a New China, 1920–1970.* Ann Arbor: Center for Chinese Studies, University of Michigan, 2002.

Bowler, Peter. *Theories of Human Evolution: A Century of Debate, 1844–1944.* Baltimore: Johns Hopkins University Press, 1986.

Brown, Shana. "Pastimes: Scholars, Art Dealers, and the Making of Modern Chinese Historiography, 1870–1928." Ph.D. diss., University of California, Berkeley, 2003.

Campany, Robert Ford. *Strange Writing: Anomaly Accounts in Early Medieval China.* Albany: State University of New York Press, 1996.

Chang, Hsin-pao. *Commissioner Lin and the Opium War.* Cambridge, Mass.: Harvard University Press, 1964.

Chang, Iris. *Thread of the Silkworm.* New York: Basic Books, 1995.

Chang, K. C. "Archaeology and Chinese Historiography." *World Archaeology* 13, no. 2 (1981): 156–169.

Cheater, A. P. "Death Ritual as Political Trickster in the People's Republic of China." *Australian Journal of Chinese Affairs* 26 (1991): 67–97.

Chen, Nancy. *Breathing Spaces: Qigong, Psychiatry, and Healing in China.* New York: Columbia University Press, 2003.

Chéng Yùqí and Chén Mèngxióng, eds. *Qián dìzhì diàochásuǒ de lìshǐ huígù: Lìshǐ píngshù yǔ zhǔyào gòngxiàn* [The History of the Former Geological Survey: Historical Narratives and Principle Contributions]. Běijīng: Dìzhì chūbǎnshè, 1996.

Chung, Yuehtsen Juliette. *Struggle for National Survival: Eugenics in Sino-Japanese Contexts, 1896–1945.* New York: Routledge, 2002.

Ci, Jiwei. *Dialectic of the Chinese Revolution: From Utopianism to Hedonism.* Stanford, Calif.: Stanford University Press, 1994.

Cohen, Joan. *Yunnan School: A Renaissance in Chinese Painting.* Minneapolis: Fingerhut Group Publishers, 1988.

Cohen, Paul. *China and Christianity: The Missionary Movement and the Growth of Chinese Antiforeignism, 1860–1870.* Cambridge, Mass.: Harvard University Press, 1963.

———.*Between Tradition and Modernity: Wang T'ao and Reform in Late Ch'ing China.* Cambridge, Mass.: Harvard University Press, 1974.

Conner, Clifford D. *A People's History of Science: Miners, Midwives, and "Low Mechaniks."* (New York: Nation Books, 2005).

Corbey, Raymond, and Bert Theunissen. *Ape, Man, Apeman: Changing Views since 1600.* Leiden: Leiden University, 1995.

Croizier, Ralph. *Traditional Medicine in Modern China: Science, Nationalism, and the Tensions of Cultural Change.* Cambridge, Mass.: Harvard University Press, 1968.

Dalton, Rex. "Fake Bird Fossil Highlights the Problem of Illegal Trading." *Nature* 404, no. 696 (2000).

Davis, Deborah S., Richard Kraus, Barry Naughton, and Elizabeth Perry, eds. *Urban Spaces in Contemporary China*. Cambridge: Cambridge University Press, 1995.

Davis, Deborah, and Ezra Vogel, eds. *Chinese Society on the Eve of Tiananmen: The Impact of Reform*. Cambridge, Mass.: Council on East Asian Studies, Harvard University, 1990.

Desmond, Adrian. "Artisan Resistance and Evolution in Britain, 1819–1848." *Osiris* 3 (1987): 77–110.

DeWoskin, Kenneth J., and J. I. Crump, Jr. *In Search of the Supernatural: The Written Record*. Stanford, Calif.: Stanford University Press, 1996.

Dikötter, Frank. *The Discourse of Race in Modern China*. London: Hurst and Company, 1992.

———, ed. *The Construction of Racial Identities in China and Japan: Historical and Contemporary Perspectives*. Honolulu: University of Hawai'i Press, 1997.

Dirlik, Arif, and Maurice Meisner, eds. *Marxism and the Chinese Experience: Issues in Contemporary Chinese Socialism*. Armonk, N.Y.: M. E. Sharpe, 1989.

Doolin, Dennis. "The Revival of the 'Hundred Flowers' Campaign: 1961." *China Quarterly* 8 (1961): 34–41.

Douthwaite, Julia V. *The Wild Girl, Natural Man, and the Monster: Dangerous Experiments in the Age of Enlightenment*. Chicago: University of Chicago Press, 2002.

Dreyer, June Teufel. *China's Forty Millions: Minority Nationalities and National Integration in the People's Republic of China*. Cambridge, Mass.: Harvard University Press, 1976.

Duara, Prasenjit. *Culture, Power, and the State: Rural North China, 1900–1942*. Stanford, Calif.: Stanford University Press, 1988.

Epstein, Steven. *Impure Science: AIDS, Activism, and the Politics of Knowledge*. Berkeley: University of California Press, 1996.

Esherick, Joseph, Paul Pickowicz, and Andrew Walder, eds. *The Chinese Cultural Revolution as History*. Stanford, Calif.: Stanford University Press, 2006.

Evans, Richard. "Telling It Like It Wasn't." *Historically Speaking* 5, no. 4 (March 2004): 11–14.

Fan, Fa-ti. *British Naturalists in China, 1760–1910*. Cambridge, Mass.: Harvard University Press, 2004.

———. "How Did the Chinese Become Native? Science and the Search for National Origins in the May Fourth Era." In *In Search of Modernity: Reexamining the May Fourth Movement*, ed. Kai-wing Chow and Tze-ki Hon. Lanham, Md.: Lexington Books, forthcoming.

Fausto-Sterling, Anne. *Myths of Gender: Biological Theories about Women and Men*. New York: Basic Books, 1992 (1985).

Feng, Amy Hwei-shuan. "Chinese Archaeology and Resistance to Foreign Participation in the 1920s." Unpublished paper.

Feuerwerker, Albert, ed. *History in Communist China*. Cambridge, Mass.: MIT Press, 1968.

Fogel, Joshua A. *Ai Ssu-ch'i's Contribution to the Development of Chinese Marxism.* Cambridge, Mass.: Harvard University Press, 1987.

Fokkema, D. W. "Chinese Criticism of Humanism: Campaigns against the Intellectuals, 1964–1965." *China Quarterly* 26 (1966): 68–81.

Friedman, Edward, Paul Pickowicz, and Mark Selden. *Chinese Village, Socialist State.* New Haven, Conn.: Yale University Press, 1991.

Furth, Charlotte. *Ting Wen-chiang: Science and China's New Culture.* Cambridge, Mass.: Harvard University Press, 1970.

Gān Lí. "Wú Yùzhāng hé 'Lùn qìjié'" [Wú Yùzhāng and "On Morals"]. *Hóng yán chūnqiū*, no. 3 (2004): 11–14.

Gardner, Martin. *Did Adam and Eve Have Navels? Discourses on Reflexology, Numerology, Urine Therapy, and Other Dubious Subjects.* New York: W. W. Norton, 2000.

Gladney, Dru C. "Representing Nationality in China: Refiguring Majority/Minority Identities." *Journal of Asian Studies* 53, no. 1 (1994): 92–123.

Goldman, Merle. "The Fall of Chou Yang." *China Quarterly* 27 (1966): 132–148.

———. *China's Intellectuals: Advise and Dissent.* Cambridge, Mass.: Harvard University Press, 1981.

Goldman, Merle, with Timothy Cheek and Carol Lee Hamrin, eds. *China's Intellectuals and the State: In Search of a New Relationship.* Cambridge, Mass.: Council on East Asian Studies of Harvard University, 1987.

Gould, Stephen Jay. *Ontogeny and Phylogeny.* Cambridge, Mass.: Harvard University Press, Belknap Press, 1977.

———. "Human Equality Is a Contingent Fact of History." In *The Flamingo's Smile: Reflections in Natural History*, 185–198. New York: Norton, 1985.

———. *Wonderful Life: The Burgess Shale and the Nature of History.* New York: Norton, 1989.

———. *Rocks of Ages: Science and Religion in the Fullness of Life.* New York: Ballantine, 1999.

Graham, Loren. *Science in Russia and the Soviet Union: A Short History.* Cambridge: Cambridge University Press, 1993.

Gregory, Jane, and Steve Miller. *Science in Public: Communication, Culture, and Credibility.* New York: Plenum Trade, 1998.

Grieder, Jerome B. *Hu Shih and the Chinese Renaissance: Liberalism in the Chinese Revolution, 1917–1937.* Cambridge, Mass.: Harvard University Press, 1970.

Gu, Edward. "Cultural Intellectuals and the Politics of the Cultural Public Space in Communist China (1979–1989): A Case Study of Three Intellectual Groups." *Journal of Asian Studies* 58, no. 2 (1999): 389–431.

Guldin, Gregory. *The Saga of Anthropology in China: From Malinowski to Mao.* Armonk, N.Y.: M. E. Sharpe, 1994.

Haraway, Donna. *Primate Visions: Gender, Race, and Nature in the World of Modern Science.* New York: Routledge, 1989.

———. *Simians, Cyborgs, and Women: The Reinvention of Nature.* New York: Routledge, 1991.

Harding, Sandra. *The Science Question in Feminism*. Ithaca, N.Y.: Cornell University Press, 1986.

Hayford, Charles W. *To the People: James Yen and Village China*. New York: Columbia University Press, 1990.

Hevi, Emmanuel John. *An African Student in China*. New York: Praeger, 1962.

Hevly, Bruce. "The Heroic Science of Glacier Motion." *Osiris* 11 (1996): 66–86.

Hilgartner, Stephen. "The Dominant View of Popularization: Conceptual Problems, Political Uses." *Social Studies of Science* 20 (1990): 519–539.

Hiltebeitel, Alf, and Barbara D. Miller. *Hair: Its Power and Meaning in Asian Cultures*. Albany: State University of New York Press, 1998.

Honig, Emily, and Gail Hershatter. *Personal Voices: Chinese Women in the 1980s*. Stanford, Calif.: Stanford University Press, 1988.

Hu, John Y. H. *Ts'ao Yü*. New York: Twayne Publishers, 1972.

Hua, Shiping. *Scientism and Humanism: Two Cultures in Post-Mao China, 1978–1989*. Albany, N.Y.: State University of New York Press, 1995.

Huáng Dàoxuàn, and Zhōng Jiàn'ān. "1927–1937 nián Zhōngguó de xuéshù yánjiū" [Chinese Academic Research, 1927–1937]. *Jìndài Zhōngguó yánjiū*, 30 March 2006. http://jds.cass.cn/Article/20060330092444.asp (viewed 2 August 2006).

Huot, Claire. *China's New Cultural Scene: A Handbook of Changes*. Durham, N.C.: Duke University Press, 2000.

Irwin, Alan, and Brian Wynne, eds. *Misunderstanding Science? The Public Reconstruction of Science and Technology*. New York: Cambridge University Press, 2003 [1996].

Isaacs, Harold R. *The Tragedy of the Chinese Revolution*. Stanford, Calif.: Stanford University Press, 1951 (1938).

Jen, C. K. "Science and the Open-Doors Educational Movement." *China Quarterly* 64 (1975): 741–747.

Johnson, Kay Ann. *Women, the Family, and Peasant Revolution in China*. Chicago: University of Chicago Press, 1983.

Keating, Pauline. "The Ecological Origins of the Yan'an Way." *Australian Journal of Chinese Affairs* 32 (1994): 123–153.

"Kēxuéjiā zhuànjì dà cídiǎn" biānjí zǔ, ed. *Zhōngguó xiàndài kēxuéjiā zhuànjì* [Biographies of Modern Chinese Scientists]. 6 vols. Běijīng: Kēxué chūbǎnshè, 1991–1994.

Kinkley, Jeffrey C., ed. *After Mao: Chinese Literature and Society, 1978–1981*. Cambridge, Mass.: Harvard University Press, 1985.

Kohl, Philip L., and Clare Fawcett, eds. *Nationalism, Politics, and the Practice of Archaeology*. Cambridge: Cambridge University Press, 1995.

Kraus, Richard Curt. *Class Conflict in Chinese Socialism*. New York: Columbia University Press, 1981.

Kwok, D. W. Y. *Scientism in Chinese Thought, 1900–1950*. New Haven, Conn.: Yale University Press, 1965.

Lai, Guolong. "Digging Up China: Nationalism, Politics and the Yinxu Excavation, 1928–1937." Unpublished paper.

Landau, Misia. *Narratives of Human Evolution*. New Haven, Conn.: Yale University Press, 1991.

Lee, Rensselaer W., III. "Ideology and Technical Innovation in Chinese Industry, 1949–1971." *Asian Survey* 12, no. 8 (1972): 647–661.

Legge, James, trans. *The Chinese Classics III: The Shoo King* [Shū jīng]. Hong Kong: Hong Kong University Press, 1970 (1861).

Lehr, Jane L. "Social Justice Pedagogies and Scientific Knowledge: Remaking Citizenship in the Non-science Classroom." Ph.D. diss., Virginia Tech University, 2006.

Leibold, James. "Competing Narratives of Racial Unity in Republican China: From the Yellow Emperor to Peking Man." *Modern China* 32, no. 2 (2006): 181–220.

Lewenstein, Bruce V. "The Meaning of 'Public Understanding of Science' in the United States after World War II." *Public Understanding of Science* 1 (1992): 45–68.

Lewin, Roger. *Bones of Contention: Controversies in the Search for Human Origins*. Chicago: University of Chicago Press, 1987.

Lewis, Mark Edward. *Sanctioned Violence in Early China*. Albany: State University of New York Press, 1990.

Lifton, Robert Jay. *Thought Reform and the Psychology of Totalism: A Study of "Brainwashing" in China*. New York: Norton, 1961.

Lightman, Bernard. "The Visual Theology of Victorian Popularizers of Science: From Reverent Eye to Chemical Retina." *Isis* 91, no. 4 (2000): 651–680.

Link, Perry, Richard P. Madsen, and Paul G. Pickowicz, eds. *Popular China: Unofficial Culture in a Globalizing Society*. Lanham, Md.: Rowman and Littlefield, 2002.

Litzinger, Ralph. *Other Chinas: The Yao and the Politics of National Belonging*. Durham, N.C.: Duke University Press, 2000.

Liú Wéimín. *Kēxué yǔ xiàndài Zhōngguó wénxué* [Science and Modern Chinese Literature]. Héféi: Ānhuī jiàoyù chūbǎnshè, 2000.

Lu Xiuyuan. "A Step toward Understanding Popular Violence in China's Cultural Revolution." *Pacific Affairs* 67, no. 4 (1994–1995): 533–563.

MacFarquhar, Roderick, Timothy Cheek, and Eugene Wu, eds. *The Secret Speeches of Chairman Mao: From the Hundred Flowers to the Great Leap Forward*. Cambridge, Mass.: Harvard University Press, 1989.

Madsen, Richard. *China's Catholics: Tragedy and Hope in an Emerging Civil Society*. Berkeley: University of California Press, 1998.

Martin, Emily. *The Woman in the Body: A Cultural Analysis of Reproduction*. Boston: Beacon Press, 1992 (1987).

———. *Flexible Bodies: Tracking Immunity in American Culture from the Days of Polio to the Age of AIDS*. Boston: Beacon Press, 1994.

Meisner, Maurice. *Li Ta-chao and the Origins of Chinese Marxism*. Cambridge, Mass.: Harvard University Press, 1967.

———. *Marxism, Maoism, and Utopianism: Eight Essays*. Madison: University of Wisconsin Press, 1982.

————. *The Deng Xiaoping Era: An Inquiry into the Fate of Chinese Socialism, 1978–1994.* New York: Hill and Wang, 1996.

Meserve Walter J., and Ruth I. Meserve. "'The White-Haired Girl': A Model for Continuing Revolution." *Theatre Quarterly* 24 (Winter 1976–1977): 26–34.

Mizuno, Hiromi. "Science, Ideology, and Empire: A History of the 'Scientific' in Japan from the 1920s to the 1940s." Ph.D. diss., University of California, Los Angeles, 2001.

Moser, Stephanie. *Ancestral Images: The Iconography of Human Origins.* Ithaca, N.Y.: Cornell University Press, 1998.

Munro, Donald J. *The Concept of Man in Early China.* Stanford, Calif.: Stanford University Press, 1969.

————. *The Concept of Man in Contemporary China.* Ann Arbor: University of Michigan Press, 1977.

Murphy, J. David. *Plunder and Preservation: Cultural Property Law and Practice in the People's Republic of China.* Oxford: Oxford University Press, 1995.

Nakayama, Shigeru, David L. Swain, and Yaga Eri, eds. *Science and Society in Modern Japan: Selected Historical Sources.* Cambridge, Mass.: MIT Press, 1974.

Neushul, Peter, and Zuoyue Wang. "Between the Devil and the Deep Sea: C. K. Tseng, Mariculture, and the Politics of Science in Modern China." *Isis* 91, no. 1 (2000): 59–88.

Newton, Michael. *Savage Girls and Wild Boys: A History of Feral Children.* London: Faber and Faber, 2002.

Oreskes, Naomi. "'Objectivity or Heroism?' On the Invisibility of Women in Science." *Osiris* 11 (1996): 87–113.

Péng Guānghuá. "Zhōngguó kēxuéhuà yùndòng xiéhuì de chuàngjiàn, huódòng jíqí lìshǐ dìwèi" [The Establishment, Activities, and Historical Significance of the Chinese Association for the Scientization Movement]. *Zhōngguó kējì shǐliào* 13, no. 1 (1992): 60–72.

Penley, Constance. *NASA/TREK: Popular Science and Sex in America.* London: Verso, 1997.

Perry, Elizabeth. "Rural Violence in Socialist China." *China Quarterly* 103 (1985): 414–440.

Proctor, Robert N. "Three Roots of Human Recency: Molecular Anthropology, the Refigured Acheulean, and the UNESCO Response to Auschwitz." *Current Anthropology* 44, no. 2 (2003): 213–239.

Pusey, James Reeve. *China and Charles Darwin.* Cambridge, Mass.: Harvard University Press, 1983.

————. *Lu Xun and Evolution.* Albany: State University of New York Press, 1998.

Reardon-Anderson, James. *The Study of Change: Chemistry in China, 1840–1949.* Cambridge: Cambridge University Press, 1991.

Rose, Steven, Leon J. Kamin, and R. C. Lewontin. *Not in Our Genes: Biology, Ideology and Human Nature.* New York: Pantheon Books, 1985.

Rudwick, Martin J. S. *The Great Devonian Controversy: The Shaping of Scientific*

Knowledge among Gentlemanly Specialists. Chicago: University of Chicago Press, 1985.

———. *Scenes from Deep Time: Early Pictorial Representations of the Prehistoric World*. Chicago: University of Chicago Press, 1992.

Sagan, Carl. *The Demon-Haunted World: Science as a Candle in the Dark*. New York: Ballantine Books, 1997.

Saich, Tony. "Negotiating the State: The Development of Social Organizations in China." *China Quarterly* 161 (2000): 124–141.

Sautman, Barry. "Anti-black Racism in Post-Mao China." *China Quarterly* 138 (1994): 413–437.

———. "Peking Man and the Politics of Paleoanthropological Nationalism in China." *Journal of Asian Studies* 60, no. 1 (2001): 95–124.

Schiebinger, Londa. *Nature's Body: Gender in the Making of Modern Science*. Boston: Beacon Press, 1994.

Schmidt, Anne Carlisle. "The Confuciusornis Sanctus: An Examination of Chinese Cultural Property Law and Policy in Action." *Boston College International and Comparative Law Review* 23, no. 2 (2002): 185–228.

Schneider, Laurence. *Biology and Revolution in Twentieth-Century China*. Lanham, Md.: Rowman and Littlefield, 2003.

Schram, Stuart. *Chairman Mao Talks to the People: Talks and Letters: 1956–1971*. New York: Pantheon Books, 1974.

Schurmann, Franz. *Ideology and Organization in Communist China*. Berkeley: University of California Press, 1966.

Schwartz, Benjamin. *In Search of Wealth and Power: Yen Fu and the West*. New York: Harper and Row, 1964.

Science for the People. *China: Science Walks on Two Legs*. New York: Avon Books, 1974.

Secord, Anne. "Science in the Pub: Artisan Botanists in Early Nineteenth-Century Lancashire." *History of Science* 32, no. 3 (1994): 269–315.

Secord, James A. *Victorian Sensation: The Extraordinary Publication, Reception, and Secret Authorship of "Vestiges of the Natural History of Creation."* Chicago: University of Chicago Press, 2000.

Shapin, Steven. *A Social History of Truth: Civility and Science in Seventeenth-Century England*. Chicago: University of Chicago Press, 1994.

Shapiro, Harry L. *Peking Man*. New York: Simon and Schuster, 1974.

Shapiro, Judith. *Mao's War against Nature: Politics and the Environment in Revolutionary China*. Cambridge: Cambridge University Press, 2001.

Shen, Grace. "Mining the Cave: Global Visions and Local Traditions in the Story of Peking Man." Paper presented at the annual meeting of the History of Science Society, Vancouver, 3 November 2000.

Shěn Qíyì et al. *Zhōngguó kēxué jìshù xiéhuì* [The Chinese Association for Science and Technology]. Běijīng: Dāngdài Zhōngguó chūbǎnshè, 1994.

Sherman, Daniel J. *Museums and Difference*. Bloomington: Indiana University Press, 2007.

Sidel, Victor, and Ruth Sidel. *Serve the People: Observations on Medicine in the People's Republic of China.* New York: Josiah Macy, Jr., Foundation, 1973.

Simon, Denis Fred, and Merle Goldman. *Science and Technology in Post-Mao China.* Cambridge, Mass.: Council on East Asian Studies, Harvard University, 1989.

Sivin, Nathan. *Traditional Medicine in Contemporary China.* Ann Arbor: Center for Chinese Studies, University of Michigan, 1987.

Sivin, Nathan. *Science in Ancient China.* Aldershot, U.K.: Variorum, 1995.

Smith, Steve A. "Local Cadres Confront the Supernatural: The Politics of Holy Water (*Shenshui*) in the PRC, 1949–1966." *The China Quarterly* 186 (2006): 999–1022.

———. "Talking Toads and Chinless Ghosts: The Politics of 'Superstitious' Rumors in the People's Republic of China, 1961–1965." *American Historical Review* 111, no. 2 (2006): 405–427.

Snow, Philip. *The Star Raft: China's Encounter with Africa.* Ithaca, N.Y.: Cornell University Press, 1988.

Songster, Elena. "A Natural Place for Nationalism: The Wanglang Nature Reserve and the Emergence of the Giant Panda as a National Icon." Ph.D. diss., University of California, San Diego, 2004.

Sorensen, Per K. *The Mirror Illuminating the Royal Genealogies: Tibetan Buddhist Historiography.* Wiesbaden: Harrassowitz, 1994.

Stacey, Judith. *Patriarchy and Socialist Revolution.* Berkeley: University of California Press, 1983.

Sullivan, Michael. "The 1988–89 Nanjing Anti-African Protests: Racial Nationalism or National Racism?" *China Quarterly* 138 (1994): 438–457.

Sun, Yan. *The Chinese Reassessment of Socialism, 1976–1992.* Princeton, N.J.: Princeton University Press, 1995.

Suttmeier, Richard. *Research and Revolution: Science Policy and Societal Change in China.* Lexington, Mass.: Lexington Books, 1974.

Tang, Edmond, and Jean-Paul Wiest. *The Catholic Church in Modern China: Perspectives.* Maryknoll, N.Y.: Orbis Books, 1993.

Taussig, Michael. *Shamanism, Colonialism, and the Wild Man: A Study in Terror and Healing.* Chicago: University of Chicago Press, 1987.

Taylor, Kim. *Chinese Medicine in Early Communist China, 1945–63: A Medicine of Revolution.* London: Routledge Curzon, 2005.

Thomson, James C. *While China Faced West: American Reformers in Nationalist China, 1928–1937.* Cambridge, Mass.: Harvard University Press, 1969.

Tián, Sòng. "Wéikēxué, fǎnkēxué, wěikēxué" [Scientism, Antiscience, and Pseudoscience]. *Zìrán biànzhèng fǎ yánjiū* 16, no. 9 (2000): 14–20.

Torrens, Hugh. "Mary Anning (1799–1847) of Lyme: 'The Greatest Fossilist the World Ever Knew.'" *British Journal for the History of Science* 28 (1995): 257–284.

Unschuld, Paul U. *Medicine in China: A History of Ideas.* Berkeley: University of California Press, 1985.

van den Bosch, Robert. *The Pesticide Conspiracy.* Garden City, N.Y.: Doubleday, 1978.

Wallis, Roy. *On the Margins of Science: The Social Construction of Rejected Knowledge*. Sociological Review Monograph 27. Keele, Staffordshire: University of Keele, 1979.

Wáng Hénglǐ et al., eds. *Zhōngguó dìzhì rénmíng lù* [Biographical Dictionary of Chinese Geology]. Wǔhàn: Zhōngguó dìzhì dàxué chūbǎnshè, 1989.

Wang Hui. *China's New Order: Society, Politics and Economy in Transition*. Edited by Theodore Huters. Cambridge, Mass.: Harvard University Press, 2003.

Wáng Shùntóng et al. *Zhōngguó kēxué jìshù xiéhuì* [Chinese Association for Science and Technology]. Běijīng: Dāngdài Zhōngguó chūbǎnshè, 1994.

Weidman, Nadine. "Popularizing the Ancestry of Man: Ardrey, Dart, and the Killer Instinct." Paper presented at the annual meeting of the History of Science Society, Vancouver, 4 November 2006.

White, Gordon. "Prospects for Civil Society in China: A Case Study of Xiaoshan City." *Australian Journal of Chinese Affairs* 29 (1993): 63–87.

Whyte, Martin King. *Small Groups and Political Rituals in China*. Berkeley: University of California Press, 1974.

Williams, James H. "Fang Lizhi's Expanding Universe." *China Quarterly* 123 (1990): 459–484.

———. "Fang Lizhi's Big Bang: Science and Politics in Mao's China." Ph.D. diss., University of California, Berkeley, 1994.

Winter, Alison. "Mesmerism and Popular Culture in Early Victorian England." *History of Science* 32, no. 3 (1994): 317–343.

Wolf, Margery. *Revolution Postponed: Women in Contemporary China*. Stanford, Calif.: Stanford University Press, 1985.

Wong, R. Bin. *China Transformed: Historical Change and the Limits of European Experience*. Ithaca, N.Y.: Cornell University Press, 1997.

Wú Zéyán, Huáng Qiūyún, and Liú Yèqiū, eds. *Cí yuán, dàlù bǎn* [Origin of Words, Mainland Edition]. Táiběi: Táiwān shāngwù yìnshūguǎn, 1993 (1989).

Wynne, Brian. "Knowledges in Context." *Science, Technology, and Human Values* 16, no. 1 (1991): 111–121.

Xī'ān Bànpō bówùguǎn "niánjiàn" biānjí wěiyuánhuì, ed. *Xī'ān Bànpō bówùguǎn niánjiàn, 1958–1998* [Xī'ān Bànpō Museum Yearbook, 1958–1998]. N.p. [Copy in author's possession.]

Xu, Jian. "Body, Discourse, and the Cultural Politics of Contemporary Chinese Qigong." *Journal of Asian Studies* 58, no. 4 (1999): 965–966.

Yan Jiaqi and Gao Gao. *Turbulent Decade: A History of the Cultural Revolution*. Translated and edited by D. W. Y. Kwok. Honolulu: University of Hawai'i Press, 1996.

Yáng Jiǎo. *Zhōngguó kēpǔ zuòjiā cídiǎn 1* [Dictionary of Chinese Popular Science Writers, Vol. 1]. Hā'ěrbīn [Harbin]: Hēilóngjiāng kēxué jìshù chūbǎnshè, 1989.

Yao, Shuping. "Chinese Intellectuals and Science: A History of the Chinese Academy of Sciences (CAS)." *Science in Context* 3, no. 2 (1989): 447–473.

Yao, Yusheng. "The Making of a National Hero: Tao Xingzhi's Legacies in the People's Republic of China." *Review of Education, Pedagogy, and Cultural Studies* 24 (2002): 251–281.

Zhāng Zǐzhāng. *Rénxìng yǔ "kàngyì wénxué"* [Human Nature and "Protest Literature"]. Táibéi: Yòushī wénhuà shìyè gōngsī, 1984.

Zhōngguó kēpǔ yánjiūsuǒ, ed. *Zhōng wài kēpǔ chuàngzuò bǐjiào yánjiū jiēduànxìng yánjiū bàogào, guónèi kēpǔ chuàngzào bùfēn* [Preliminary Research Report on Research Comparing Chinese and Foreign Science Dissemination Materials, Section on Chinese Science Dissemination Materials]. Běijīng: Zhōngguó kēpǔ yánjiūsuǒ, 2002.

Zilsel, Edgar. "The Sociological Roots of Modern Science." *American Journal of Sociology* 47 (1942): 544–562.

Index

African fossils, 5, 273

African origins. *See under* human origins

African people, Chinese perceptions of, 108–10, 277, 279, 285

agriculture: and Chinese identity, 284; and compulsory labor, 139; and discovery of dragon bones or fossils, 36, 203; as labor, 198; and Lysenkoism, 73; and mass science, 118, 127, 139, 173, 295; origins of, 100; and science dissemination, 32, 147, 198

Ài Sīqí 艾思奇, 61, 62, 190

alienation, 87, 287–88

All-China Association for the Dissemination of Scientific and Technological Knowledge 中華全國科學技術普及協會, 39, 68–70, 114, 117, 127, 184

All-China Federation of Scientific Societies 中華全國自然科學專門學會聯合會, 68

All-China Federation of Trade Unions 中華全國總工會, 117, 127

all-round Communist, 156

amateurs: in China, 161, 175–76, 204–5; in Japan, 172; in North America, 161. *See also* fans; hobbyists; laypeople; one-person institutes

American Museum of Natural History, 46, 99n47, 102

American Paleoanthropological Delegation (to the PRC), 64n21, 131, 153n45

ancestors: animals as, 22, 50, 53; common human, 11, 257, 261, 278; contrasted with offshoots or cousins, 50, 97–98, 258, 286; life of, 100–101; of minority nationalities, 206, 279; and nationalism, 278, 288, 291; personal, 278, 282, 284–85; and popular culture, 291, 296, 300; and race, 104, 247, 256, 277; worship of, 10, 190, 268, 281–82, 284, 286, 291, 297. *See also under* Peking Man

Andersson, J. Gunnar, 35–37, 42

Andrews, Peter, 261

Andrews, Roy Chapman, 26n36, 35n64

animal husbandry, 48, 100

animals and humans: evolutionary relationship between, 20–25, 71, 74, 290; labor as difference between, 74–75, 92–93, 140, 152, 195, 287, 289; Máo on, 88; other differences between, 22, 24, 89–93, 175, 195, 289–290; slippery boundary between, 14, 16, 20–21, 23–24, 60, 83, 211–12, 215, 238, 243, 290

Anning, Mary, 128

anthropology. *See* imperialism: and anthropology; Marx: and anthropology; paleoanthropology

anthropomorphism 擬人論, 80n94, 81

antiquarianism 金石學, 49